MW00837648

GETTING TO KNOW
ArcGIS® Desktop 10.8
SIXTH EDITION

GETTING TO KNOW

ArcGIS®

DESKTOP 10.8

SIXTH EDITION

MICHAEL LAW │ AMY COLLINS

Esri Press
REDLANDS │ CALIFORNIA

Esri Press, 380 New York Street, Redlands, California 92373-8100

Copyright © 2022 Esri

All rights reserved. First edition 2004. Fifth edition 2018.

26 25 24 23 22 1 2 3 4 5 6 7 8 9 10

ISBN: 9781589485778

The information contained in this document is the exclusive property of Esri or its licensors. This work is protected under United States copyright law and other international copyright treaties and conventions. No part of this work may be reproduced or transmitted in any form or by any means, electronic or mechanical, including photocopying and recording, or by any information storage or retrieval system, except as expressly permitted in writing by Esri. All requests should be sent to Attention: Director, Contracts and Legal Department, Esri, 380 New York Street, Redlands, California 92373-8100, USA.

The information contained in this document is subject to change without notice.

US Government Restricted/Limited Rights: Any software, documentation, and/or data delivered hereunder is subject to the terms of the License Agreement. The commercial license rights in the License Agreement strictly govern Licensee's use, reproduction, or disclosure of the software, data, and documentation. In no event shall the US Government acquire greater than RESTRICTED/LIMITED RIGHTS. At a minimum, use, duplication, or disclosure by the US Government is subject to restrictions as set forth in FAR §52.227-14 Alternates I, II, and III (DEC 2007); FAR §52.227-19(b) (DEC 2007) and/or FAR §12.211/12.212 (Commercial Technical Data/Computer Software); and DFARS §252.227-7015 (DEC 2011) (Technical Data – Commercial Items) and/or DFARS §227.7202 (Commercial Computer Software and Commercial Computer Software Documentation), as applicable. Contractor/Manufacturer is Esri, 380 New York Street, Redlands, California 92373-8100, USA.

Esri products or services referenced in this publication are trademarks, service marks, or registered marks of Esri in the United States, the European Community, or certain other jurisdictions. To learn more about Esri marks, go to: links.esri.com/EsriProductNamingGuide. Other companies and products or services mentioned herein may be trademarks, service marks, or registered marks of their respective mark owners.

For purchasing and distribution options (both domestic and international), please visit esripress.esri.com.

Library of Congress Control Number: 2022930374

Contents

Preface

Getting to Know ArcGIS® Desktop 10.8 is a workbook for students and professionals who are getting started in the field of geographic information systems (GIS). Its detailed, step-by-step exercises teach you the core functionality of ArcGIS Desktop software: how to make maps, carry out spatial analysis, and build and edit spatial databases in the context of realistic projects. The exercises are supported by conceptual discussions at the start of each chapter and as needed throughout the book. Abundant color graphics confirm your progress along the way.

Getting to Know ArcGIS Desktop 10.8 is a hands-on workbook that's meant to be a practical manual for classroom lab work or on-the-job training. If you have no GIS background, chapter 1 will give you a quick overview. If you have no ArcGIS software experience, chapter 2 will provide a background on these Esri® GIS software products. The remaining 18 chapters require you to work through software exercises as you read. Each of these chapters contains two to four exercises that focus on a particular GIS task or problem. Many common tasks are covered, including navigating digital maps, finding GIS data online, setting map projections, symbolizing and labeling maps, classifying data, making map layouts, building spatial databases, editing data, geocoding addresses, querying maps, processing data, and analyzing spatial relationships.

About the software and data

Exercises are designed to be performed using ArcGIS Desktop 10.8. If you are using an older version of ArcGIS, you will be able to do most of the exercises, but your results may be somewhat different. To complete chapter 20, you will need to install and enable the ArcGIS Spatial Analyst™ extension. If you are using older software, you should download the latest trial software to use with this book. To make sure your Windows system meets ArcGIS system requirements, visit go.esri.com/ArcGISProSysReqs.

You can download data for the book exercises at go.esri.com/GTKDesktopData. In addition, you will find information about trial software and additional resources at go.esri.com/GTKDesktopResources.

How to use this book

Each exercise in *Getting to Know ArcGIS Desktop 10.8* is a fresh starting point, with the maps and data you need already prepared for you; however, we advise you to complete the chapters in order, because tools and functions introduced and fully described in early chapters may be

referred to without description in later chapters. The exercises will work, however, no matter which chapter you start with. Each chapter takes one to two hours to complete.

What's new in this edition

This edition of *Getting to Know ArcGIS Desktop 10.8* retains all the new elements introduced in the previous editions, including learning objectives, chapter summaries, questions for critical thinking, software tips, independent-challenge tasks ("On your own"), and a glossary of key terms. Glossary terms appear in a different color in the text.

This edition has been tested and updated to reflect the most recent software release.

We hope you enjoy *Getting to Know ArcGIS Desktop 10.8*.

Michael Law
Toronto, Canada

Amy Collins
Napa, California

Acknowledgments

The authors thank the following individuals and organizations for contributing data and images to this book:

- The World Bank provided the air pollution data used in chapters 3 and 4. http://data.worldbank.org
- World Wildlife Fund provided the terrestrial ecosystem data featured in chapter 7. www.worldwildlife.org
- The Florida datasets (chapters 7 and 9) were obtained from the Florida Geographic Data Library (FGDL). www.fgdl.org/metadataexplorer
- The shipwreck points used in chapters 7 and 9 are based on data obtained from the Automated Wreck and Obstruction Information System (AWOIS) used by the Office of Coast Survey (OCS), a division of the National Oceanic and Atmospheric Administration (NOAA). www.nauticalcharts.noaa.gov/hsd/awois.html
- The National Archive of Criminal Justice Data, the Inter-university Consortium for Political and Social Research, and the US Department of Justice, Federal Bureau of Investigation provided crime report data featured in chapter 8. www.icpsr.umich.edu/icpsrweb/NACJD
- Natural Earth provided the raster imagery featured in chapters 7, 9, and 10. www.naturalearthdata.com
- State of Queensland, Department of Science, Information Technology, Innovation and the Arts, Queensland, Australia, provided bioregion and bio-subregion data featured in chapter 10. www.qld.gov.au/dsitia/
- New York City health areas, streets, neighborhoods, and buildings, used in chapters 11 through 13, provided courtesy of NYC Department of City Planning. www.nyc.gov/html/dcp
- New York City subway routes and subway stops (chapters 11 through 13) courtesy of Steven Romalewski, CUNY Mapping Service at the Center for Urban Research.www.gc.cuny.edu
- The Philadelphia data used in chapter 14 was obtained from the City of Philadelphia's Office of Innovation and Technology and The School District of Philadelphia's Office of Capital Programs. www.phila.gov/it http://webgui.phila.k12.pa.us
- Oswald Eichberger, senior director of Consumer Intelligence and Client Services of Generation5 (a Motricity company), provided consumer market data featured in chapter 15, with consultation from Aaron Heppleston, data management specialist of Geomedia Inc. http://www.generation5.ca, www.geomedia.ca
- City of Toronto and Open Data provided street data featured in chapter 15. www.toronto.ca/open

- Tim Pudoff, GIS manager of Sonoma County, provided data and input for the Russian River flood zone exercises (chapter 16) and the clinic suitability study (chapter 20). Thanks also to JoAnn Borri of the Sonoma County Department of Health Services.
 https://gis.sonoma-county.org
- The food desert spreadsheet used in chapter 17 was obtained from the US Department of Agriculture, Economic Research Service.
 www.ers.usda.gov/data-products/food-access-research-atlas
- The Oregon Department of Forestry provided fire risk data featured in chapters 18 and 19.
 www.oregon.gov/ODF

All other datasets (cities, countries, states, rivers, and census data) were obtained from Data and Maps for ArcGIS 2010.

Thanks to everyone at Esri Press.

We are deeply indebted to the original author team who made *Getting to Know ArcGIS Desktop* a best seller for many years—Tim Ormsby, Eileen Napoleon, Rob Burke, Carolyn Groessl, and Laura Bowden. No small amount of the original text remains in this edition, and we truly hope their straightforward teaching style and intent—making sophisticated software accessible to all levels of learners—remain intact.

CHAPTER 1
Introducing GIS

Learning objectives

- Learn the uses of geographic information systems (GIS)
- Learn how GIS layers work
- Differentiate between GIS features and surfaces
- Obtain preliminary understanding of feature locations, scale, and attributes
- Obtain preliminary understanding of spatial relationships between features

GIS allows you to study and explore everything about a place. With a GIS, you can identify and analyze the land features, climate, boundaries, population, resources, and many other things about places that interest you. In the past, you needed to use paper maps, globes, and a variety of hard-copy data sources to perform these analyses. Now, a GIS makes the maps, globes, data, and analysis tools available on your computer, allowing you to perform sophisticated analysis, map your results, and store and share your information digitally.

An example of the type of map you could study or create in a GIS is a population density map, which shows the distribution of population in relation to specific areas (e.g., cities, census tracts, counties, states, countries). Population density information is important to consider when you are mapping areas that vary greatly in size; for example, 200,000 people living in an 8-square-mile county and 200,000 people living in an 80-square-mile county are very different conditions. Population density, in turn, provides context for other maps in your study, which may show demographics, economic statistics, or housing prices.

Another analysis you can perform with a GIS is to find locations that meet specific criteria. For example, if you work for a real estate company, you might have access to a GIS database of properties with associated property values that you could use to generate sales data. If you work for a utility company, you might use a GIS to coordinate, map, and manage the thousands of repairs you have to complete each year.

You can also use a GIS to analyze what is happening around a specific area. For example, if you work for an environmental agency, you might map wildlife species' habitats and take inventory of

1
2
3
4
5
6
7
8
9
10
11
12
13
14
15
16
17
18
19
20

vegetation across a region to better understand the relationship between them. If you work for a water conservation agency, you might analyze areas adjacent to stream watersheds to gauge how natural features, seasonal water flow, and existing habitats change.

You can also use a GIS to map how and where things evolve over a period of time to learn about subject movement and behavior. If you work for a public health agency, for example, you might track an influenza outbreak geographically to try to determine its origin, find disease clusters, and identify areas of high risk. You can use a GIS to evaluate current situations and anticipate future conditions. For example, you could track and model hurricane data to provide early warning for evacuation schedules that can help save people's lives.

Basic GIS principles

The following sections will introduce you to some basic principles of GIS. These concepts will become clearer and more understandable as you work with the software in chapters 3 through 20.

A GIS map contains layers

On a paper map, you cannot peel cities away from countries or countries away from the ocean, but on a GIS map, you can. A GIS map is made up of layers, or collections of geographic objects that are alike and defined by a visual symbol. To make a map, you can add as many layers as you want. On the map shown, the transmission lines are layered on top of the electricity generation circles, which are then layered on top of the coal basins, and finally are layered on top of the terrain and ocean layers. This creates an important layer hierarchy on the map.

This map of North America shows state boundaries, lakes, rivers, terrain, and the ocean. It also shows the amount of electricity generation, transmission lines, and locations of coal basins. The map is accompanied by a legend that shows some of its layers. Map courtesy of Kansas Geological Survey at the University of Kansas. Data from Regional Carbon Sequestration Partnerships.

Layers may contain features or surfaces

On the map shown, the coal basins layer includes many basins, and the transmission lines layer contains many lines. The same is true of the states and lakes layers—each layer contains many of the same types of things. Each geographic object in a layer—each state, lake, transmission line, or coal basin—is called a feature.

Not all layers contain features. The ocean layer is not a collection of geographic objects as are the other layers. It is considered a single, continuous expanse that changes from one location to another according to the depth of the water. Similarly, the terrain layer shows the elevation of the landscape. A geographic expanse such as this is called a surface.

This topographic map shows the terrain of a landmass and the seafloor. Map made with Natural Earth.

Features have shape and size

Geographic objects have an endless variety of shapes; however, they can all be represented as one of three geometrical forms: a polygon, a line, or a point.

Polygons represent objects that are large enough to have boundaries, such as countries, lakes, and tracts of land. Lines represent objects that are relatively too narrow to be polygons, such as rivers, roads, and pipelines. Points are used for objects that are relatively too small to be polygons, such as cities, schools, and fire hydrants. The same object may be represented by a polygon in one layer and a line or a point in a different layer, depending on the scale of the layer.

Polygons, lines, and points, collectively, are called vector data.

Earthquake epicenters and seismic hazards in Indonesia. Polygons are used to represent landmasses, points to represent earthquakes, and lines to represent subduction zones. Map and data courtesy of US Geological Survey.

Surfaces have numeric values rather than shapes

Unlike countries or rivers, natural phenomena, such as elevation, slope, temperature, rainfall, and wind speed, have no distinct shape. What they have instead are measurable values for any location on the earth's surface. For instance, wherever you go, you are either at sea level or a number of meters above or below it. Geographic phenomena such as this are easier to represent as surfaces than as features.

The most common kind of surface is a raster, a matrix of identically sized square cells. Each cell represents a unit of surface area—for example, 10 square meters—and contains a measured or estimated value for that location.

A close look at this raster of ocean depth shows that it is composed of square cells. Each cell holds a numeric value indicating ocean depth. Map made with Natural Earth.

The world is not divided neatly into features and surfaces. Many things can be looked at either way—as a feature or as a surface. For example, polygons are often used to mark the boundaries of different vegetation types in a region, but this implies that the change from one type to another is more abrupt than it probably is because one type of vegetation often blends into another. Vegetation can be represented not only as a polygon, but also as a raster surface, with each cell value standing for a type of vegetation.

Features have locations

If you were asked to find Berlin, Germany, on a map of the world, it probably wouldn't take you very long. But suppose Berlin wasn't shown on the map. Could you make a pencil mark exactly where it should be?

Now suppose you could lay a grid or graticule over the world map and you knew that Berlin was a certain number of marks up from and to the right of a given starting point. In that case, it would be easy to put your pencil on the right spot. A grid of this kind is called a coordinate system, and it is what a GIS uses to put features in their proper place on a map.

On this map, a coordinate system has an x-axis and a y-axis. The intersection of the axes is called the origin. Feature locations are specified by their distance from the origin in meters, feet, or a similar unit of measure—in this case, lines of latitude and longitude. Data from Data and Maps for ArcGIS 2010, courtesy of ArcWorld, DeLorme, Esri®, and ArcWorld Supplement.

The location of a point feature on a map is defined by a pair of coordinates. Berlin has coordinates of 52°30′ N (latitude) and 13°25′ E (longitude). A straight line needs two pairs of coordinates—one at the beginning of the line and one at the end. If the line bends, like a river, there must be a pair of coordinates at every location where the line changes direction. The same holds true of a polygon, which is simply a line that returns to its starting point.

Features can be displayed at different sizes

On a GIS map, you can zoom in to see features at a closer range. As you do, the scale of the map changes.

Scale, commonly expressed as a ratio, is the relationship between the size of the features on a map and the size of the corresponding places in the world. If the scale of a map is 1:1,000,000, it means that features on the map are 1/1,000,000th of, or 1 million times smaller than, their true size.

1
2
3
4
5
6
7
8
9
10
11
12
13
14
15
16
17
18
19
20

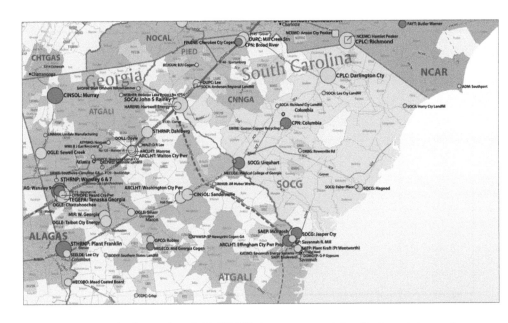

A utilities map at a scale of approximately 1:1,000,000. County, state, and shoreline boundaries are shown.

Copyright Platts, a division of the McGraw-Hill Companies Inc., 2008. All rights reserved. Data from Energy Advantage and POWERmap from Platts.

Zooming in gives you a closer view of features within a smaller area of the map. The amount of detail in the features does not change, but what details you can see changes. A river has the same bends, and a coastline has the same crenulations, whether you are zoomed in and can discern them or are zoomed out and cannot.

A city transit map at a scale of approximately 1:12,000. Building footprints, roads, and transit lines are visible.

Copyright RMSI and Universal Map Group LLC, 2009. Reproduced by permission.

How much detail the features have depends on the layer you are using. Just as a paper map of the world generalizes the boundaries of Brazil more than a map of South America does, so do different GIS layers contain more feature detail or less.

Internet basemaps, such as those in ArcGIS® Online, cover a wide range of map scales—and a wide level of detail. A map can have several levels of detail that correlate to the scale range. As you zoom in, layers at small scale with less detail will be turned off, and new layers at large scale with more detail will be turned on. Each zoom level can be perceived as being made up of seamless data—each level, however, is made up of layers composed of several datasets, all with varying amounts of detail, and all of which are contained within a spatial database.

Features are linked to information

A feature is more than its physical shape and location. A feature also includes everything else that might be known about that feature. For a country, this might be its population, capital, system of government, leading imports and exports, average rainfall, and mineral resources. For a road, it might be the speed limit, the number of lanes, whether it is paved or unpaved, and whether it has one-way or two-way traffic. A great deal of information exists about any feature, whether it be a humble length of sewer pipe or the vastness of an ocean.

Information about the features in a layer is stored in a table. The table has a record (row) of each feature in the layer and a field (column) for each category of information. These categories are called attributes, and the table is called an attribute table.

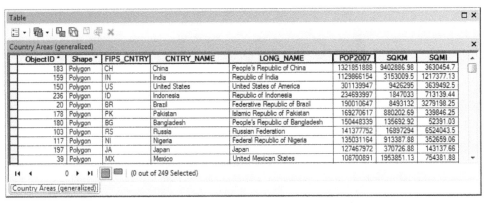

The attribute table of a layer of countries includes each feature's shape, ID number, name, population, and area, among other things. Data from Data and Maps for ArcGIS 2010, courtesy of ArcWorld Supplement and Esri.

Features on a GIS map are linked to the information in the feature's attribute table. If you select Brazil on a map, you can bring up all the information stored about Brazil in the attribute table of countries. If you select a record in the table, you will see the corresponding feature highlighted on the map.

Object ID *	Shape *	FIPS_CNTRY	CNTRY_NAME	LONG_NAME	POP2007	SQKM	SQMI
183	Polygon	CH	China	People's Republic of China	1321851888	9402886.98	3630454.7
159	Polygon	IN	India	Republic of India	1129866154	3153009.5	1217377.13
150	Polygon	US	United States	United States of America	301139947	9426295	3639492.5
236	Polygon	ID	Indonesia	Republic of Indonesia	234693997	1847033	713139.44
20	Polygon	BR	Brazil	Federative Republic of Brazil	190010647	8493132	3279198.25
178	Polygon	PK	Pakistan	Islamic Republic of Pakistan	169270617	880202.69	339846.25
180	Polygon	BG	Bangladesh	People's Republic of Bangladesh	150448339	135593.83	52391.03
103	Polygon	RS	Russia				
117	Polygon	NI	Nigeria				
197	Polygon	JA	Japan				
39	Polygon	MX	Mexico				

Brazil is highlighted in the attribute table of countries, and it is similarly highlighted on the map. Data from Data and Maps for ArcGIS 2010, courtesy of ArcWorld, DeLorme, Esri, and ArcWorld Supplement.

The link between features and their attributes makes it possible to ask questions about the information in an attribute table and display the answers on the map. You can use this ability to locate and identify specific features.

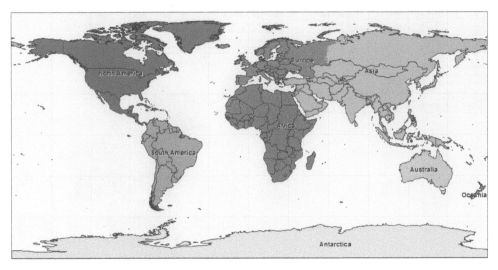

Which countries belong to the continent of Asia? Data from Data and Maps for ArcGIS 2010, courtesy of ArcWorld, DeLorme, Esri, and ArcWorld Supplement.

Which countries have a population of 1 billion or greater? Data from Data and Maps for ArcGIS 2010, courtesy of ArcWorld, DeLorme, Esri, and ArcWorld Supplement.

Which cities in Asia are national capitals, and which are not? Data from Data and Maps for ArcGIS 2010, courtesy of ArcWorld, DeLorme, Esri, and ArcWorld Supplement.

Which cities in Asia are capitals and have populations of 5 million or greater? Data from Data and Maps for ArcGIS 2010, courtesy of ArcWorld, DeLorme, Esri, and ArcWorld Supplement.

Similarly, you can use attributes to create thematic maps–maps that use colors or other symbols to indicate the nature of feature attributes.

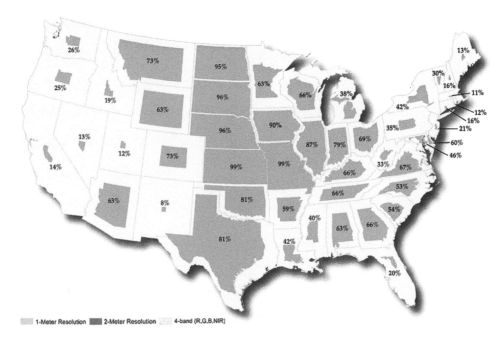

This map shows the percentage of a state covered by the common land unit, the smallest unit of agricultural land that can be enrolled in a farm program administered by the US Department of Agriculture. Map and data courtesy of US Department of Agriculture Farm Service Agency Aerial Photography Field Office (USDA-FSA-APFO), 2009.

Features have spatial relationships

In addition to asking questions about the feature information stored in attribute tables, you can ask questions about the spatial relationships between features–for example, which features are nearest others, which cross others, and which are contained by others. The coordinates of features are used in a GIS to compare their locations. On the map shown, the relationships between parcel locations, waterlines, and valves are established in the GIS.

This utility map shows parcel locations, waterlines, and valves for Cobb County Water System in Marietta, Georgia. Map courtesy of Cobb County Water System. Data from Cobb County GIS.

It is critical to the utility's distribution operation to ensure the isolation of the high-pressure system from the normal-pressure system. The map provides detail for each of the essential valves located in the high-pressure water system. Which water valve belongs to which parcel? Which waterline crosses a main road? Which valve affects the water pressure of which waterline? By taking a close look at the map, engineers can quickly make informed decisions on pressure problems reported in the area and determine whether extending the high-pressure zone is a potential solution.

New features can be created from areas of overlap

Queries about attributes and spatial relationships identify existing features based on whether they have or do not have certain qualities. To solve some geographic problems, however, it is necessary to use a GIS to create features from existing ones.

On the map shown, a GIS database containing layers such as soils, slopes, rivers and streams, wetlands, floodplains, parks and natural areas, significant habitat inventory, greenways, and natural hazard data was used to identify regionally significant areas.

Regionally significant landscape features are derived from other features in the area surrounding Portland, Oregon. Map courtesy of Matthew Hampton, Oregon Metro. Data from Metro Regional Land Information System, Esri, National Park Service Natural Earth, Nature Conservancy, and Oregon Natural Heritage Program.

To find places that meet the specified conditions, the cooperative expertise of various federal, state, local, and private organizations was used in a GIS to find areas of overlap among features in the different layers. The map shows significant natural systems and land patterns that define the quality and character of the region and capture its sense of place. This allows for resource protection at a larger landscape and ecosystem scale and helps define where future growth should and should not occur. The new feature's boundary is the area of overlap that is different from the boundaries of the features from which it was created. The result of the analysis is a new layer that shows regionally significant landscape features.

Data requires management

GIS data can quickly grow complicated. Early in the planning stages of a GIS project, you must decide how to organize your spatial data, for example, by type, geography, or granularity. Regardless of which approach you use, recording metadata (information about data) for your spatial data is a key tool in your data management strategy.

Where to store data is often another problem. Small projects can be stored on a local drive, but some datasets are extremely large and need to be accessed by many people. In this case, storing data on a central server is a good solution. Large-scale shared data strategies, such as those employed by government agencies, private companies, and individuals who collect large datasets, use server GIS or cloud networks to ensure secure data governance, reduced redundancy, and compatibility with a wide range of systems.

You now have some idea of what a GIS is and what it can do. In chapter 2, you will learn more about ArcGIS® Desktop.

Terms

GIS
vector
raster
graticule

1
2
3
4
5
6
7
8
9
10
11
12
13
14
15
16
17
18
19
20

CHAPTER 2
Introducing ArcGIS

Learning objectives

- Name the ArcGIS Desktop applications used for general mapping and data management
- Briefly describe some ArcGIS extensions
- Know the components of the ArcGIS system
- Browse through Esri data sources

In ArcGIS Desktop, you can query data; analyze spatial relationships, such as distance, intersection, and containment among map features; and overlay layers to discover how different types of data are interrelated at a particular location.

You can also create and edit data with a complete set of drawing and editing tools. Multiple users can take advantage of versioned data capabilities, allowing more than one person to edit data simultaneously. You can also check data out of a database, edit it in the field, and then check it back in.

Additionally, ArcGIS Desktop offers a full set of spatial analysis tools to perform advanced GIS data analysis, modeling, and data conversion. It also provides high-end cartographic tools and advanced capabilities for data translation, creation, and conversion of a wide range of spatial file formats.

Applications and tools

ArcGIS Desktop comprises several applications in which you will do your work. The ArcMap™ and ArcCatalog™ applications are the most frequently used. The ArcScene and ArcGlobe applications are used for three-dimensional (3D) visualization. Core ArcGIS Desktop geoprocessing tools are available through the ArcToolbox window and the ModelBuilder application, which are accessible from both ArcMap and ArcCatalog.

ArcGIS Desktop applications

GIS tasks can be broadly divided into two categories: (1) mapmaking, editing, and spatial analysis; and (2) database design and data management. This division of tasks is reflected in the functionalities of the ArcMap and ArcCatalog applications, the core components of ArcGIS Desktop.

ArcMap is the ArcGIS application for making maps and analyzing data.

ArcMap has a data view for creating, symbolizing, and analyzing maps. Map from the Florida Geographic Data Library, courtesy of the University of Florida GeoPlan Center.

ArcMap has a layout view for composing maps to be printed. You can add a title, scale bar, legend, and other elements. Map from the Florida Geographic Data Library, courtesy of the University of Florida GeoPlan Center.

ArcCatalog is the ArcGIS application for data management.

In ArcCatalog, *behind*, you can organize, browse through, document, and search for spatial data. ArcCatalog also has tools for building and managing GIS databases. Although there is a separate ArcCatalog application, the Catalog window in ArcMap, *front*, can meet most of your data organization needs. Data from the Florida Geographic Data Library, courtesy of the University of Florida GeoPlan Center.

ArcGIS Desktop also offers two 3D visualization environments: ArcScene and ArcGlobe™. ArcScene allows you to view your GIS data in three dimensions. ArcGlobe, part of the ArcGIS 3D Analyst extension (discussed later in this section), creates seamless visualizations of raster and vector features and can display them at varying levels of detail. The data is organized into image tiles and can be saved to system memory (the cache) for quick display and navigation.

THE ARCGIS PRO APPLICATION

ArcGIS Desktop users will notice an additional application included with the Desktop suite: **ArcGIS®
Pro**. ArcGIS Pro is an all-in-one product for GIS professionals who want to display, create, analyze,
and share both 2D and 3D geospatial data. It is built using new system architecture that improves
performance, and it offers a new interface and modern user experience.

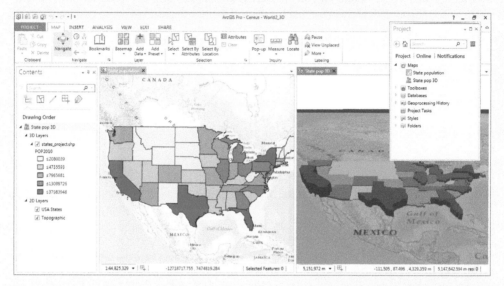

The ArcGIS Pro mapping application has an intuitive interface and the ability to display data in two and three
dimensions, simultaneously. Data from Data and Maps for ArcGIS 2010, courtesy of ArcUSA, US Census, Esri (Pop2010). Basemap
sources: Esri, HERE, DeLorme, TomTom, Intermap, increment P Corp., GEBCO, USGS, FAO, NPS, NRCAN, GeoBase, IGN, Kadaster NL, Ordnance
Survey, Esri Japan, METI, Esri China (Hong Kong), swisstopo, MapmyIndia, © OpenStreetMap contributors, and the GIS user community.

In ArcGIS Pro, all your work and data are organized into projects (.aprx). A project contains maps,
layouts, layers, tables, tasks, tools, and connections to servers, databases, folders, and styles.
Existing Esri file formats such as map documents (.mxd), scenes (.sxd), and globes (.3dd) can be
imported into ArcGIS Pro. Projects are not backward compatible with ArcMap; however, the data
used by the application can be accessed by ArcMap through the geodatabase created by ArcGIS
Pro. Services published using ArcGIS Pro can be used and shared with ArcMap.

This book focuses on the ArcMap and ArcCatalog applications of ArcGIS Desktop. To see how
ArcGIS Pro compares with ArcMap, see *Getting to Know ArcGIS Pro 2.8* by Michael Law and Amy
Collins (Esri Press, 2021).

1
2
3
4
5
6
7
8
9
10
11
12
13
14
15
16
17
18
19
20

ArcGIS Desktop tools

Although ArcMap has many geoprocessing tools that are accessible through menus and various toolbars, the ArcToolbox window contains a collection of geoprocessing tools, models, and scripts organized in one interface. When you run a geoprocessing tool from ArcToolbox, a dialog box appears that contains parameters for you to specify. Unlike ArcMap and ArcCatalog, which are stand-alone applications, ArcToolbox must be started from within one of these two applications.

ArcToolbox includes a collection of geoprocessing tools, models, and scripts. Here, the Extract toolset is highlighted.

ModelBuilder helps you create models to solve spatial analysis problems. You can use the ModelBuilder interface to organize the inputs, tools, and outputs needed to perform geospatial analysis. You can use ModelBuilder to string together geoprocessing tools from ArcToolbox to create a model to perform a complex workflow with a single click.

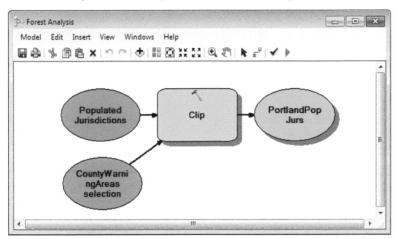

ModelBuilder can be used to create models to automate geoprocessing tasks.

Extending ArcGIS Desktop functionality

The extension products that are fully integrated with ArcGIS Desktop add increased functionality and capabilities. These specialized tools allow you to perform tasks such as raster geoprocessing, 3D analysis, and map publishing. Several extensions are available with the trial software that comes with this book (available at go.esri.com/GTKDesktopEVATrial). Three commonly used extensions are described below. The ArcGIS Spatial Analyst extension is used in chapter 20 of this book. You can explore the other extensions on your own by using the ArcGIS Desktop Help and tutorials.

ArcGIS Spatial Analyst is primarily used for raster analysis, or for deriving from vector input a raster output in which each cell stores a relevant value. For instance, you can create slope and aspect layers from an elevation raster, or perform density analysis on input points or lines. You can also estimate values at unmeasured locations by mathematical interpolation of known sample values, and analyze groundwater and hydrology data, among other things.

ArcGIS 3D Analyst allows you to see spatial data in three dimensions. You can "fly through" terrain and examine it from any height or angle. You can model cities and neighborhoods by drawing buildings at their correct heights. Analysis tools allow you to solve visibility, volume, and downhill path problems: Which areas can be seen from the mountaintop? Can tower 1 be seen from tower 2? How much earth is in the hill? Which course will water follow down a surface? Some geoprocessing tools are found in both the 3D Analyst and the Spatial Analyst toolboxes, but if you are using 3D Analyst, you can model solutions in a 3D space.

ArcGIS Network Analyst provides network-based transportation spatial analysis tools to map fleet routing, travel directions, closest facility, service areas, and location allocation (determining optimal location sites). You can dynamically model realistic network conditions, including one-way streets, turn and height restrictions, speed limits, and variable travel speeds based on traffic data. You can easily build transportation networks from your existing GIS data and use it to define service areas based on travel time and distance.

The ArcGIS system

ArcGIS Desktop is just one part of ArcGIS, which also includes ArcGIS Enterprise, ArcGIS Online, ArcGIS for Developers, ArcGIS apps, and more.

ArcGIS Enterprise

ArcGIS Enterprise is used to build web services that deliver maps, apps, and information—such as driving directions or retail locations—over the internet. ArcGIS Enterprise brings GIS to people who may have no GIS software.

1 2 3 4 5 6 7 8 9 10 11 12 13 14 15 16 17 18 19 20

1

2

3

4

5

6

7

8

9

10

11

12

13

14

15

16

17

18

19

20

ArcGIS Enterprise is also designed for cloud computing, a rapidly emerging technology that allows you to use a cloud provider's computing infrastructure instead of your own. With cloud computing, you get the same capabilities for managing and delivering GIS services and applications as you do when you run ArcGIS Enterprise at an actual physical location.

ArcGIS Online

ArcGIS Online is a family of cloud-based web apps and resources that enhance and extend the ArcGIS user experience. Using ArcGIS Online (www.arcgis.com), you can create maps; access libraries of ready-to-use maps; store and manage maps, data, and other geospatial content; perform spatial analytics; purchase maps and data from vendors; and share your content. ArcGIS Online content and sharing capabilities are fully integrated into ArcGIS Desktop. You can search for and add data from the cloud, as well as upload map and layer packages that you create using either ArcGIS Desktop or ArcGIS Enterprise.

ArcGIS for Developers and ArcGIS apps

ArcGIS for Developers is used when building custom solutions. A suite of developer tools can be used to extend and configure existing ArcGIS applications or build and configure custom apps.

You can use ArcGIS apps on desktops, smartphones, and tablets to navigate maps, collect and report data, and perform GIS analysis. The Collector for ArcGIS app uses Global Positioning System (GPS) capabilities to capture, edit, and display spatial data. It helps field staff who may have no GIS experience easily collect data that can then be shared through ArcGIS Online and ArcGIS Enterprise. You can browse through map galleries from ArcGIS Online, use hundreds of ready-to-use apps specific to your industry, and get apps through ArcGIS Marketplace, developed by providers.

Esri data

What is software without high-quality data? In addition to extensive location data (country and state boundaries, major cities, lakes, rivers, streets, and so on), Esri provides you with comprehensive demographic, lifestyle segmentation, consumer spending, and business data and delivers this data in a variety of ways.

Esri demographic data includes census data and demographic projections you can use to gain valuable insights about population changes and trends that can affect planning and business decisions.

Tapestry™, Esri's segmentation data, classifies US neighborhoods into 65 segments based on their socioeconomic and demographic composition, which can help you analyze markets, profile customers, evaluate competitors, and more.

Esri consumer data provides insight on consumer spending, market potential, and supply and demand of the retail marketplace. It allows you to accurately assess what products and services consumers want versus what they actually buy.

Esri business data provides business locations by industry, bank locations, major retail shopping centers, traffic counts, and crime statistics.

Esri data is also offered in other ways. ArcGIS Online Map Service and StreetMap Premium for ArcGIS deliver basemaps, reference and specialty maps, and enhanced street datasets that facilitate geocoding (finding street addresses based on geographic location), routing, and high-quality cartographic display for the entire United States, Canada, and Europe. ArcGIS Online World Imagery map service provides one meter or better satellite and aerial imagery coverage for many areas of the world.

Data and Maps for ArcGIS is a set of annual map data that is included with ArcGIS software and is preconfigured to work with ArcGIS products.

In this book, you will learn the basics of ArcGIS Desktop and ArcGIS Online. To find out more about ArcGIS Desktop extensions, Esri data, ArcGIS Online, and the rest of the ArcGIS system, visit the Esri website at www.esri.com.

Terms

ArcMap
ArcCatalog
tiles
cache
ArcGIS Pro
projects
cloud computing
GPS

1
2
3
4
5
6
7
8
9
10
11
12
13
14
15
16
17
18
19
20

CHAPTER 3
Interacting with maps

Display map data
Navigate a map
Use basic tools
Look at feature attributes

Learning objectives
- Navigate the ArcMap interface
- Locate and open ArcMap windows, menus, and tools
- Turn map layers on and off
- Reorder layers in a map
- Change a layer name
- Examine layer properties
- View the source of map layers
- Explore a map document using Zoom and Pan tools
- View attribute information for individual features
- Turn on MapTips
- Turn on feature labels and set a scale dependency
- Examine alternative views of the map
- Search for features on a map
- Create a spatial bookmark
- Measure distances on a map
- Open and explore feature layer attribute tables
- Examine attribute statistics

ArcMap is the most frequently used component of ArcGIS software. It is an application for displaying maps and investigating them, for analyzing maps to answer geographic questions, and for producing maps that make persuasive arguments. The ArcMap application consists of a map display for viewing spatial data, a table of contents for listing the layers shown in the display, and a variety of toolbars for working with the data.

Main menu bar

Standard toolbar

Tools toolbar

Table of contents

Map display

The ArcMap application of ArcGIS Desktop. Data from Data and Maps for ArcGIS 2010, courtesy of ArcWorld, ArcWorld Supplement, DeLorme, and Esri.

You can change the way ArcMap looks to suit your preferences and the kind of work you do. Toolbars can be hidden or shown. New commands can be added. Toolbars can be docked at different places in the application window or float independently of it.

TIP

To dock a floating toolbar, drag it to any outer edge of the application. To undock it, use the dashed vertical gray bar at the toolbar's left edge to drag the toolbar away from its docked position. To hide or show a toolbar, click the Customize menu, point to Toolbars, and click to select or clear the desired toolbar option.

Most embedded windows (Catalog, Search, ArcToolbox, Table of Contents, and Python) can be floating, docked, or collapsed into a tab for easy access. The exception is the ModelBuilder window, which is designed to be used only as a separate, floating window.

TIP

To dock a floating window, click and drag it over the application. The window turns blue, and four arrows point to locations where the window can be docked. Drag the window to an arrow, and then release it to choose the docking location. For windows with the Auto Hide option, when the window is docked, click the Auto Hide button (it looks like a pushpin) to activate the tabbed behavior. The window will appear when you hover over the tab.

ArcMap toolbars and windows can be easily docked and undocked. Data from Data and Maps for ArcGIS 2010, courtesy of ArcWorld, ArcWorld Supplement, DeLorme, and Esri.

Most of the graphics in this book reflect ArcMap's default interface. It is assumed that the Standard and Tools toolbars, which are used a great deal, are always turned on.

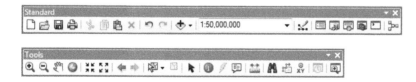

Top to bottom. The Standard and Tools toolbars are generally turned on.

Changes you make to the interface are applied to subsequent ArcMap sessions. If you dock a toolbar or window in one session, it will be docked the next time you start ArcMap. If you resize the application window, it will keep the new size in the next session. Changes such as these will not significantly affect the exercises but may give you slightly different results for operations that are influenced by the size and scale of the map display.

You can further customize the appearance of the ArcGIS Desktop user interface by removing unused tools, renaming tools, moving specialty extension tools to the Tools toolbar where they can be quickly accessed, creating custom toolbars that pertain to your organization's workflow, and more. For more information about tailoring the interface to your preferences, go to ArcGIS Desktop Help. On the Contents tab, navigate to Get started > Customize > About configuring the user interface.

TIP

To perform the exercises as written in this book, you must have ArcGIS Desktop 10.8 and the exercise data. Keep in mind that the interface may look different, and certain tools, functions, windows, dialog boxes, or menu options may or may not be present. If you are using a different software version, you may be able to do many of the exercises, but you may encounter steps where a work-around will be necessary.

For the print book, you can authorize a 180-day trial version of ArcGIS at go.esri.com /GTKDesktopEVATrial. You will need the authorization code found on the inside back cover of this book to download the software trial.

The exercise data for this book is available at go.esri.com/GTKDesktopData. It is shared with an ArcGIS Online group in the Learn ArcGIS organization. Download and install this data to your computer. For additional resources, visit go.esri.com /GTKDesktopResources.

Data

The exercises in this chapter use the following datasets, stored in a geodatabase called World.gdb:

- Background—a grid of polygons used to display a single-color background behind land features, such as the ocean *(Source: Esri)*
- Cities—point features representing major cities of the world with populations greater than 1 million *(Source: ArcWorld)*

- Countries–polygon features representing world countries, including demographic data and air pollution estimates *(Sources: ArcWorld; US Census International Division; CIA Factbook; The World Bank)*
- Latlong–line features representing latitude and longitude *(Source: Esri)*

Exercise 3a
Display map data

In this exercise, you will learn how to display data in ArcMap. You will explore a map document that contains world data and learn the basics of how map documents work. The exercise instructions assume you have installed the data for *Getting to Know ArcGIS Desktop* to the default directory (C:\EsriPress\GTKArcGIS). If you have installed the data elsewhere, you will need to substitute the correct paths.

1. Start ArcMap (located in the ArcGIS suite).

When ArcMap is open, the ArcMap - Getting Started dialog box is in the main application window. This dialog box allows you to quickly start a new map or open an existing map or template.

2. On the Getting Started dialog box, under the Existing Maps section, click Browse For More.

3. **On the Open ArcMap Document dialog box, browse to your C:\EsriPress\GTKArcGIS\ Chapter03 folder (or to the folder where you installed the GTKArcGIS data). Click ex03a.mxd, as shown in the graphic, and click Open.**

When the map document is open, observe the display area (where you have the map—a map of the world, in this case), the toolbars, and the table of contents on the left.

TIP

Your application may look somewhat different from the graphics.

You have a world map with grid lines and dark-blue circles. Taking a quick look at the table of contents and with some modest deciphering, you can surmise that the dark-blue circles represent cities with populations greater than 1 million (with increasingly larger circles representing increasingly larger populations), and the grid lines represent lines of latitude and longitude. Countries are yellow, and oceans are light blue. Recall from chapter 1 that a **layer** is a spatial dataset represented by symbols on a GIS map. The visible layers on this map are named Cities, Latlong, Countries, and Ocean. You also see additional layers in the table of contents (World Population and Air Pollution by Country), although they are not visible on the map at the moment. Next, you will take a closer look at each layer.

4. **In the table of contents, select the check box next to Cities to turn the layer off.**

The cities disappear from the map.

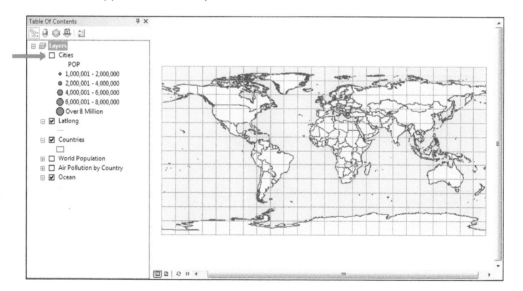

5. **Turn Cities back on. In the same way, turn the following layers off and then back on: Latlong, Countries, and Ocean.**

All map layers in ArcMap can be turned off and on with a click. Next, you will observe the effects of changing the layer drawing order. Layers are drawn, one on top of the other, like a stack of translucent papers. Imagine that the layers you are looking at are actually opaque symbols drawn on translucent paper. If you placed cities, drawn as filled-in dots,

below the countries, which are filled-in blocks of color, you wouldn't see them. You can experiment with this concept in ArcMap.

6. **In the table of contents, drag Cities to the bottom of the list of layers, below Ocean. As you drag the layer, a horizontal black bar indicates its position in the list.**

 When you let go of the mouse button, Cities is at the bottom of the table of contents and is no longer visible on the map. The polygon layers using solid fill colors obscure the point layer of cities.

Always layer lines and points on top of polygons to avoid this problem. When you add data to ArcMap, the software automatically places points and lines above polygons.

7. **Move Cities back to its original location at the top of the table of contents.**

ON YOUR OWN

Experiment by moving the Latlong layer to different locations in the table of contents. What happens when you drag it below the Countries layer? What happens when you drag it below the Ocean layer? When you are done, move the layer back to its original position, below Cities.

The layer name Cities isn't ideal, because it implies that all cities of the world are included in the dataset. So next, you will give the layer a more descriptive name.

8. In the table of contents, right-click the Cities layer name. This opens a shortcut (context) menu. Many ArcMap operations are started from these shortcut menus. On the shortcut menu, click Properties to open the Layer Properties dialog box.

9. On the Layer Properties dialog box, click the General tab if it is not already active. (ArcMap remembers the last tab used and makes this the active tab; the very first time you open the properties, the General tab should be active.)

The Layer Properties dialog box has several tabs for setting layer properties, most of which you will use in this book.

10. In the Layer Name text box, the name Cities is highlighted. Type **Cities > 1 Million** in its place. Click OK.

The layer is renamed in the table of contents.

TIP

Here is another way to rename a layer. Just click the layer name in the table of contents to highlight it, click again to activate the cursor, and type the new name.

You should know that when you change a layer name in a map document, you are not changing the name of the underlying source dataset. A layer name does not have to match the name of the source dataset. You will learn more about the connection between a layer and its source data shortly.

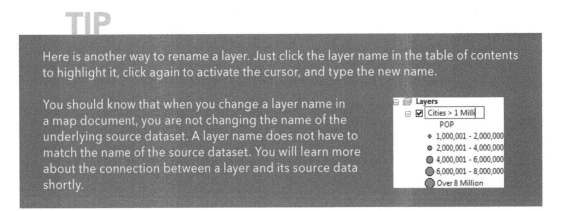

11. Using the method taught in the preceding tip, in the table of contents, below Cities > 1 Million, change the legend name POP to **Population**.

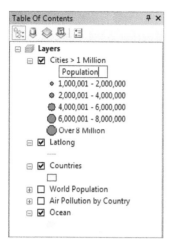

This makes it clear that the size of a blue circle corresponds to the size of a city's population.

ON YOUR OWN

Rename the Latlong layer to Graticule, the technical name for lines of latitude and longitude on a map.

12. In the table of contents, select the check box next to World Population to turn the layer on.

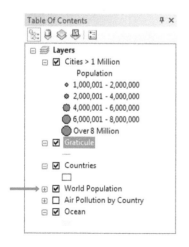

1
2
3
4
5
6
7
8
9
10
11
12
13
14
15
16
17
18
19
20

Nothing is changed in the display. Can you figure out why? If you think it may have something to do with the drawing order, which was discussed earlier in the exercise, you are correct. The World Population layer is composed of polygons that are the same shape as those in the Countries layer. Since Countries is the first polygon layer in the list of layers, it is obscuring the World Population layer below it.

13. Turn off the Countries layer. The World Population layer is now visible.

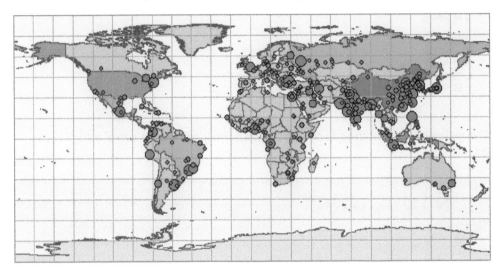

Now you can see something new on the map. In the World Population layer, countries appear in different colors. To understand what the colors mean, you will examine the layer's legend. But first, you will turn off the points and lines to get a clearer picture.

14. Turn off the layers Cities > 1 Million and Graticule.

15. Click the plus sign (+) next to World Population to expand the legend for this layer. Examine the colors on the map and compare them with the legend.

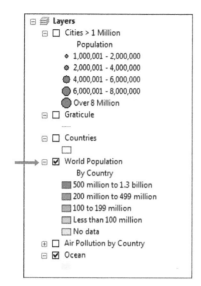

TIP

If you cannot see the entire layer or legend text, you can widen the table of contents. Hover over the right edge of the pane and drag to resize.

Two countries display populations greater than 500 million. Can you name them?

To make sure you answered the preceding question correctly, you will next use the Identify tool.

16. **On the Tools toolbar, pause the mouse pointer over the Identify tool (a blue circle with the letter i). Notice that the tool name and a short description appear.**

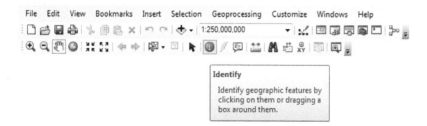

1
2
3
4
5
6
7
8
9
10
11
12
13
14
15
16
17
18
19
20

When you use the Identify tool, the pointer is accompanied by an *i* symbol as you move it around the map display. When you click a feature, it opens the Identify window.

17. Click the Identify button, move the pointer over one of the two areas filled with the darkest shade of orange, and click in that area.

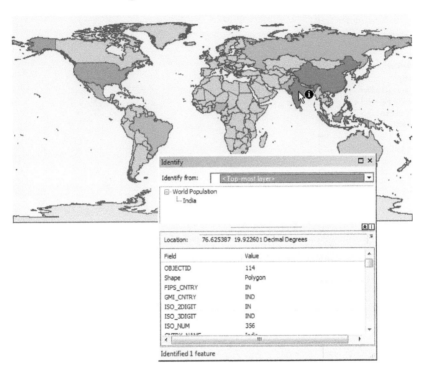

The Identify window shows the name of the country you clicked, as well as other information called **attributes**, which are stored in a table that is part of this dataset. (Depending on which country you clicked, you may have different results.)

18. Close the Identify window and turn off the World Population layer. Turn on the Air Pollution by Country layer, and expand its legend.

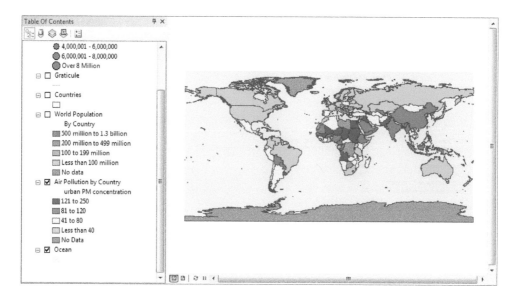

This layer shows concentrations of particulate matter (PM) in the air. Data was gathered for urban areas, from which countrywide estimates were calculated.

19. Turn the World Population layer on and off to compare it with Air Pollution.

> *Does high pollution always correlate with high population?*
> *What might be some reasons for this?*

Now that you have seen examples of various types of data being displayed on a map, you will next investigate the connection between *source data* and *map layers*.

20. In the table of contents, right-click Countries and click Properties. On the Layer Properties dialog box, click the Source tab and examine the Data Source information.

Notice that a layer does not have to be visible on the map for you to see its properties.

TIP

Later in the book, you may be asked to simply "open the layer properties." You will need to remember how to do this. You can also double-click a layer in the table of contents to open its properties.

What is the source data type?

A feature class is a collection of geographic features that have the same geometry type (such as point, line, or polygon), the same attribute fields, and the same spatial reference. Feature classes can be stored in a geodatabase or as shapefiles, coverages, or other data formats. You will learn more about data formats in chapter 4.

What is the name of the source feature class,
and in which geodatabase is it located?

21. **Using the same approach, look at the source dataset for World Population and Air Pollution by Country.**

What are the names of the source feature classes for these two layers?

It is important to understand that a map may include multiple layers that point to the same underlying data. In the case of the current map document, the Countries, World Population, and Air Pollution layers are based on the same data source—a geodatabase

feature class called Countries. (You will learn more about data types later in this book.) Layers and source datasets sometimes have the same name, but they don't have to, as demonstrated in step 10.

Despite sharing the same source dataset, the layers look different from each other. Why? Because they have different symbology schemes. Symbology refers to the appearance of features on a map–color, line thickness, fill motif, and so on. Unlike source data, layers store symbology information in a map document. For instance, the Countries layer is symbolized using a single color for all country polygons. The World Population layer has graduated symbols based on population data, and the Air Pollution layer has graduated symbols based on particulate matter calculations. Population and particulate matter are all stored as attributes of the Countries feature class. (There are many other attributes stored with the feature class.) You will learn more about attribute data later in this chapter, and you will learn how to create varied symbology in chapters 7 and 8.

22. Close the layer properties. At the top of the table of contents, click the List By Source button . Pause the mouse pointer over the data path to see the path in its entirety.

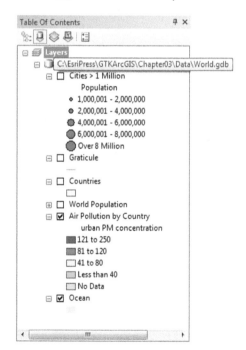

TIP

The List By Source option is especially helpful when you have data from different locations in the same map document.

1 2 3 4 5 6 7 8 9 10 11 12 13 14 15 16 17 18 19 20

23. Click the List By Drawing Order button [image] to restore the table of contents to its default appearance.

24. To save your work, on the menu bar, click File, and click Save As. Browse to C:\EsriPress\GTKArcGIS\Chapter03\MyData. Give the file a new name, and click Save. Saving your work is optional. When you begin a new exercise, however, it is important that you open the new map document specified in the instructions. The exercises are not always continuous, and if you work in the same map document throughout a chapter, your results may not match the book.

25. If you are continuing to the next exercise, leave ArcMap open. Otherwise, click the File menu and click Exit. Click No if prompted to save your changes.

Exercise 3b
Navigate a map

Now that you understand some basic principles of how map documents work, it's time to do some exploring using ArcMap navigation tools. In this exercise, you will learn how to zoom and pan around a map, use MapTips, and identify features and examine their attributes.

1. Start ArcMap, if necessary.

2. If you get the Getting Started dialog box, click Cancel. On the menu bar, click File > Open. Browse to your C:\EsriPress\GTKArcGIS\Chapter03 folder. Click ex03b.mxd, and click Open.

TIP

If you would rather not see the Getting Started dialog box when you start ArcMap, select the check box at the bottom of the dialog box that says, "Do not show this dialog in the future."

This map document looks similar to the one you worked with in exercise 3a but with fewer layers.

3. On the Tools toolbar, click the Zoom In tool ⊕.

4. Move the mouse pointer over the map. The pointer becomes a magnifying glass. Press and drag the pointer to draw a box around the country of Australia, approximately as shown in the graphic. If you make a mistake, click either the Full Extent button ⊙ or the Go Back To Previous Extent button ◄, both on the Tools toolbar, and try again.

1

2

3

4

5

6

7

8

9

10

11

12

13

14

15

16

17

18

19

20

The display is zoomed in on Australia.

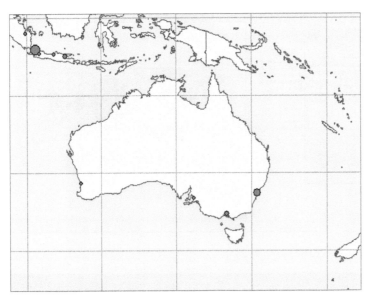

Zooming in or out changes the display scale, which is shown on the Standard toolbar. When the map showed the entire world, the scale was approximately 1:250,000,000. This means that map features are displayed at 1/250,000,000th of their actual size. The scale should now be somewhere in the vicinity of 1:30,000,000. Scale is also affected by the size of the ArcMap application window.

TIP

You can use "continuous zoom" by rotating the wheel of the mouse, if you have one. Try it to see how it works.

5. To avoid accidental zooming, it is best to use the Select Elements tool. On the Tools toolbar, click the Select Elements tool ⌘.

TIP

Remember, you can pause the mouse pointer over a tool to see its name.

6. Move the mouse pointer over the Australian continent. The country name is displayed as a MapTip.

MapTips are a layer property. They can be turned on or off on the Display tab on the Layer Properties dialog box. You can see MapTips no matter which tool is active.

7. On the Tools toolbar, click the Pan tool ![hand icon]. When you move the mouse pointer over the display area, it turns into a hand symbol.

8. Pressing and holding the mouse button, drag the display to the southeast (so your travel direction is northwest) until you can see Iran (use the MapTips for help identifying the country). You will have to press and then release the mouse button a couple of times until you arrive at your intended destination.

1 2 3 4 5 6 7 8 9 10 11 12 13 14 15 16 17 18 19 20

TIP

Another way to pan is to drag the map with the middle mouse button or mouse wheel pressed down.

9. On the Tools toolbar, click the Identify tool , and click Iran on the map.

Notice that the feature flashes green when you click it with the Identify tool.

In the Identify window, you will find feature-specific information that is stored with the Countries dataset, including International Organization for Standardization (ISO) numbers and codes, population, area, and many other statistics.

What is the ISO number for Iran?

10. With the Identify window still open, click several other countries in the display area. Notice that the information in the Identify window is updated automatically as you click different features—you do not have to close the window and reopen it each time you want to identify a new feature.

11. Close the Identify window.

12. Click the Fixed Zoom In button ⬚ once. The Fixed Zoom buttons maintain the center point of the display area as they increase or decrease the display scale.

13. Click the Fixed Zoom In button three more times and observe the scale bar. Now click the Fixed Zoom Out button ⬛ several times.

14. Click the Go Back To Previous Extent button several times and then the Go To Next Extent button ➡ several times.

ON YOUR OWN

Use the following tools to move around the map until you are comfortable with ArcMap navigation: Zoom In, Zoom Out, Pan, Full Extent, Fixed Zoom In, Fixed Zoom Out, Previous Extent, and Next Extent.

15. When you are done navigating, zoom to the full extent of the map, and click the Select Elements tool.

TIP

It's a good habit to activate the Select Elements tool whenever you finish using a tool. This will prevent unwanted zooming, measuring, or other tool activity once you are finished using the tool. Another option is to access tools by right-clicking the map. A context menu appears with navigation and other commonly used commands that you can select without changing your current active tool.

You didn't make any changes to this map, so there is no need to save your work.

16. If you are continuing to the next exercise, leave ArcMap open. Otherwise, on the menu bar, click File, and click Exit. Click No when prompted to save your changes.

Exercise 3c
Use basic tools

In this exercise, you will learn about many common ArcMap tools and functions that facilitate data exploration, beyond navigation. You will learn how to turn on MapTips, create feature labels that turn on only at a specified map scale, create spatial bookmarks to save a specific view of a map, measure distances between features, and use the viewer windows to see alternative views of the data simultaneously.

1. **In ArcMap on the Standard toolbar, click the Open button 🖼️. (Dismiss the Getting Started dialog box if necessary.) Browse to your C:\EsriPress\GTKArcGIS\Chapter03 folder, click ex03c.mxd, and click Open.**

TIP

If you don't see the.mxd file extension, check your folder option settings in File Explorer.

In Windows 8 and 10: Right-click the Start button and click File Explorer. Click the View tab, and select the File Name Extensions check box to turn it on.

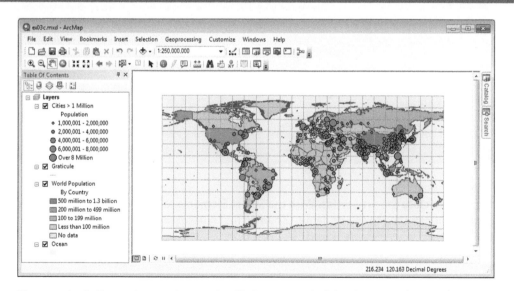

The map is similar to the one in exercise 3b, but instead of the Countries layer, it has a World Population layer, which you saw in exercise 3a. You will use MapTips to quickly obtain city names.

2. Open the properties for Cities > 1 Million. Click the Display tab and select the Show MapTips Using The Display Expression check box.

TIP

Notice that the Display Expression field is set to CITY_NAME. This is the attribute field ArcMap will use to pull the information when you use MapTips. You can change this field so that ArcMap will show different types of information when you pause the mouse pointer over features. By default, ArcMap selects the first string-type field in the layer's attribute table that includes the text "name" (case insensitive). If no field with that text in its name exists, ArcMap defaults to the first field of string type, then the first field of integer type, and then the first field of any type. (You will learn more about field types in chapter 11.) This is the case for both MapTips and labels.

3. Click OK to close the properties. With the Select Elements tool active, pause the mouse pointer over any city to see its name. (Zoom in first if necessary, and then zoom back to the full extent when you are done.)

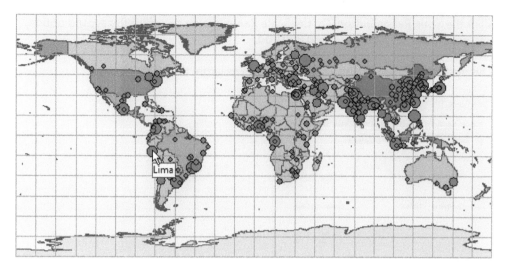

MapTips are helpful, but many times you will want to see permanent labels on your map. You can turn labels on so they are always visible. However, at the full extent of the map, seeing all the city names would result in a cluttered map that is difficult to read. To avoid this problem, you will next set a visibility scale so the labels will appear only when the map is zoomed to a certain scale.

4. **Open the properties of the Cities > 1 Million layer and click the Labels tab. Near the top of the dialog box, select the Label Features In This Layer check box. Notice that the label field is set to CITY_NAME.**

5. Click the Scale Range button. On the Scale Range dialog box, click the second option, and next to Out beyond, type **1:80,000,000**.

This means that labels will be visible only at scales larger than 1:80,000,000.

6. Click OK to return to the Labels tab and OK again to close the Layer Properties dialog box.

7. To see your labels, click the Fixed Zoom In button several times, until the city labels appear. Then look at the scale bar and notice that it is set to 1:80,000,000 or less (as the denominator gets smaller, the scale gets larger).

TIP

To quickly turn on labels for any layer without setting a scale range, right-click the layer in the table of contents and click Label Features.

8. Return to the map's full extent. On the Tools toolbar, click the Create Viewer Window button. The mouse pointer turns into a crosshair symbol so that you can draw a box around an area you want to observe in a separate window.

9. Drag the mouse pointer to draw a box around Europe, creating a Viewer window. Depending on the size of your screen and the dimensions of the box you draw, your Viewer window may or may not display city labels.

The Viewer window allows you to examine different parts of the map without changing the main display. You can move the Viewer window so it doesn't obscure the main map, or you can enlarge it–this is helpful when you have a large computer monitor (or two monitors) and you want to see two views of the same map simultaneously.

10. Use the basic navigation tools in the Viewer window–not the tools on the ArcMap Tools toolbar–to manipulate the display of the window.

11. When you are comfortable navigating in the Viewer window, click the right arrow in the upper-right corner of the window, and click Magnifier.

1 2 3 4 5 6 7 8 9 10 11 12 13 14 15 16 17 18 19 20

Now the window acts like a magnifying glass.

12. Drag the Magnifier window title bar to move the window around the main map display. The area seen through the window is magnified four times (or 400 percent–a number you can modify).

ON YOUR OWN

On the main menu bar, click Help > ArcGIS Desktop Help. Over the left column, click the Search tab. In the text box, type viewer window and click Ask. Choose the first returned topic, "Working with magnifier, viewer, and overview windows." Read about these windows to find out more about how they work.

Suppose, for this exercise, you are a pilot and intend to fly a small plane from Shanghai to New York. You want to measure the distance between the cities and determine the shortest route, so you will do that next.

13. Close any open windows. On the Tools toolbar, click the Find tool **M**.

14. On the Find dialog box, for Find, type **Shanghai**, and for In, select the Cities > 1 Million layer.

15. Click Find. The bottom of the dialog box expands to show the results. Right-click Shanghai and click Zoom To. If you get a message that asks you to sign in to ArcGIS Online to use the ArcGIS Online World Routing Service, click Cancel. You'll learn more about ArcGIS Online in chapter 5.

16. Close the Find dialog box.

1 2 3 4 5 6 7 8 9 10 11 12 13 14 15 16 17 18 19 20

The map is centered on Shanghai.

17. On the menu bar, click Bookmarks > Create Bookmark.

18. Replace the existing Bookmark name with **Shanghai**, and click OK.

19. Now use the Pan tool to go westward over Asia and the Atlantic Ocean until you reach the point representing New York City.

20. On the Tools toolbar, click the Measure tool 📐. This opens the Measure window. Click the Choose Units arrow, point to Distance, and click Kilometers if it is not already selected. Keep the Measure window open.

21. Move the mouse pointer over the display. The pointer becomes a ruler with small crosshairs. Place the pointer at the center of the point that represents New York City, and click to begin a line. Without clicking anywhere else on the map, go to the Bookmarks menu and click Shanghai. (The Measure tool will remain active when you click the Bookmarks menu.)

TIP

If you accidentally click the wrong spot, double-click to end the measurement, and start again.

22. When you get the point representing Shanghai, double-click to end the measure line.

Shanghai is over 11,000 kilometers from New York. Due to differences in exact click locations, your results may vary slightly.

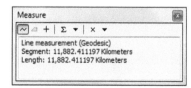

The Measure tool calculates the shortest distance between two points. Is the shortest distance between the cities found by traveling over the Atlantic Ocean or the Pacific Ocean?

ON YOUR OWN

Zoom to the map's full extent and again measure the distance between New York and Shanghai. Notice the behavior of the measurement line. Why do you think this is happening? You will have more opportunity to think about the mechanics of placing a spherical earth on a flat map surface in chapter 6.

23. To save your work, on the menu bar, click File, and click Save As. Browse to C:\ EsriPress\GTKArcGIS\Chapter03\MyData. Give the file a new name and click Save.

24. If you are continuing to the next exercise, leave ArcMap open. Otherwise, click the File menu and click Exit. Click No if prompted to save your changes.

Exercise 3d
Look at feature attributes

In a GIS, a feature on a map may be associated with a great deal of information—more than can be displayed at any given time. This information is stored in an attribute table. A layer's attribute table contains a row (or record) for each feature in the layer and a column (or field) for each attribute or category of information. When you used the Identify tool in exercises 3a and 3b, you saw a feature's attributes in the Identify window.

In this exercise, you will examine the attribute table of the world's largest cities.

1. In ArcMap, open the ex03d map document from your C:\EsriPress\GTKArcGIS\ Chapter03 folder. This is a replica of the map you worked with in exercise 3c.

2. In the table of contents, right-click Cities > 1 Million, and click Open Attribute Table.

TIP

Later in this book, you may simply be asked to "open the attribute table." You will need to remember how to do this.

The attribute table shows you all the data for the cities included with this map. It can be resized to your preference.

Shape*	CNTRY_NAME	CITY_NAME	GMI_ADMIN	FIPS_CNTRY	STATUS
Point	Brazil	Brasilia	BRA-DFD	BR	National and provincial capit
Point	Brazil	Goiania	BRA-GOI	BR	Provincial capital
Point	Chile	Santiago	CHL-RMT	CI	National and provincial capit
Point	Ecuador	Guayaquil	ECU-GUA	EC	Provincial capital
Point	Ecuador	Quito	ECU-PCH	EC	National and provincial capit
Point	Peru	Lima	PER-LIM	PE	National and provincial capit
Point	Brazil	Belem	BRA-PAR	BR	Provincial capital
Point	Brazil	Manaus	BRA-AMZ	BR	Provincial capital
Point	Bolivia	Santa Cruz de La Sierra	BOL-SCR	BL	Other
Point	Paraguay	Asuncion	PRY-CNT	PA	National capital

Table

Cities > 1 Million

0 ► ►I (0 out of 256 Selected)

Cities > 1 Million

3. Scroll down through the table and examine the records. (Each row in the table is called a record.) There are 256 records, one for each city with a population greater than 1 million. Now scroll across the table to look at the attributes, or fields. (Each column is called a field.)

There are 10 attribute fields. The first is the ObjectID field—it contains a unique identification number for each record. The Shape field describes the feature geometry. Both of these fields are created and maintained by ArcGIS Desktop. The rest of the fields are added and maintained by the user. Other fields include the name of each city, its country, status, population, population rank, population class, and administrative abbreviations.

The intersection of a record and a field is a cell. A cell contains an attribute value. For example, the attribute value of the CITY_NAME field for the first record is "Brasilia." You can change the width of any field by dragging the divider between fields.

Next, you will narrow the display width of the CITY_NAME field so you can see more attributes without having to scroll across the table.

4. Scroll back all the way to the left. Place the mouse pointer on the vertical black bar between the CITY_NAME and GMI_ADMIN fields. The pointer changes to a double-headed arrow.

5. Press and hold the mouse button and drag the pointer to the left. As you drag, the original field width is marked by a vertical red line and the new width by a vertical black line. Release the mouse button at some point before you start cutting off city names.

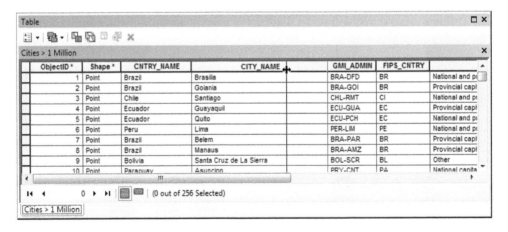

Some of the numbers of the POP_CLASS field are cut off, so next, you will widen this field.

6. Scroll to the right side of the table. Place the mouse pointer on the vertical black bar to the right of the POP_CLASS field. Drag the pointer slightly to the right to see the entire name of the field.

The FIPS_CNTRY field contains country abbreviations designated by Federal Information Processing Standards. If you don't want to see a field in an attribute table, but you don't want to delete it permanently either, you can simply hide it, which you will do next.

7. Scroll to the left and place the mouse pointer on the vertical black bar between the FIPS_CNTRY and STATUS fields. Drag the pointer to the right edge of the GMI_ADMIN field, and release the mouse button. The FIPS_CNTRY field is hidden. (Actually, you have simply made it so narrow that it can no longer be seen.)

If you wanted to restore the hidden field, you would double-click the border between the GMI_ADMIN and STATUS field names to see the field again.

TIP

You can also hide fields by right-clicking the field heading and clicking Turn Field Off. With this option, to restore hidden fields, you must click the Table Options arrow, and click Turn All Fields On.

ON YOUR OWN

Turn off the GMI_ADMIN field.

You can rearrange the order of fields as well. It would be more natural to have the city name appear before the country name, so next, you will rearrange them.

8. Click the column heading of the CNTRY_NAME field. The field is highlighted in cyan.

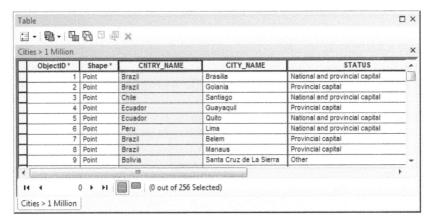

9. Drag the column heading to the right. The pointer becomes an arrow with a small rectangle, showing that a field is being moved. When the vertical red line is between the CITY_NAME and STATUS fields, as shown in the graphic, release the mouse button.

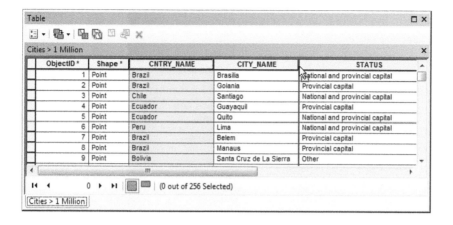

The field names are rearranged.

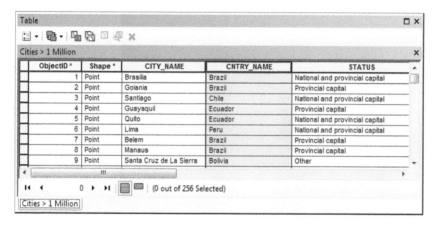

TIP

You can also freeze fields to make it easier to explore and understand feature attributes. To do so, right-click a field header and click Freeze/Unfreeze Column. This allows you to scroll across fields and maintain a view of the frozen column. It is especially useful for tables with many fields. Repeat the action to return the table to its default behavior.

Records, as well as fields, can be highlighted. When a record is highlighted in the table, the corresponding feature is highlighted on the map. A highlighted record or feature is said to be selected.

1
2
3
4
5
6
7
8
9
10
11
12
13
14
15
16
17
18
19
20

10. Click the gray tab to the left of the first record in the table. The record is selected, shown by the cyan highlighting.

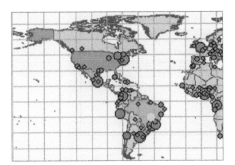

11. Move the attribute table so it doesn't cover the map display. Brasilia is highlighted on the map.

Next, you will clear the current selection, and then select the cities with a population of 10 million or greater.

12. At the upper left of the attribute table, click the Table Options arrow, and click Clear Selection.

13. In the attribute table, scroll to the POP field. Right-click POP and click Sort Descending.

The records are ordered from the largest population to the smallest. Resize the attribute table if necessary to see more fields.

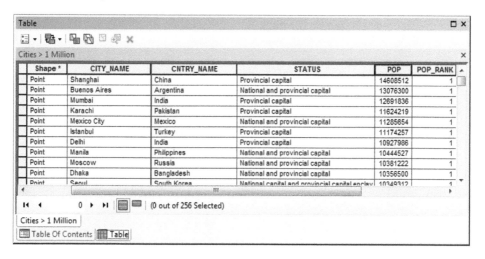

Next, you will select multiple records.

14. Press Shift, and click the tabs to the left of the first and twelfth records in the table to select them all at once (all cities with populations of 10 million or greater). Alternatively, press Ctrl and click the tabs to the left of each record up to the twelfth row in the table.

Table				□ ✕

Cities > 1 Million ✕

CITY_NAME	CNTRY_NAME	STATUS	POP	⌃
Shanghai	China	Provincial capital	14608512	
Buenos Aires	Argentina	National and provincial capital	13076300	
Mumbai	India	Provincial capital	12691836	
Karachi	Pakistan	Provincial capital	11624219	
Mexico City	Mexico	National and provincial capital	11285654	
Istanbul	Turkey	Provincial capital	11174257	
Delhi	India	Provincial capital	10927986	
Manila	Philippines	National and provincial capital	10444527	
Moscow	Russia	National and provincial capital	10381222	
Dhaka	Bangladesh	National and provincial capital	10356500	
Seoul	South Korea	National capital and provincial capital enclave	10349312	
Sao Paulo	Brazil	Provincial capital	10021295	
Lagos	Nigeria	Other	9000000	⌄

I◀ ◀ 15 ▶ ▶I ▦ ▭ | (12 out of 256 Selected)

Cities > 1 Million

Look at the bottom of the table. Notice that 12 of 256 records are selected.

15. Move the table out of the way to see the map. The 12 cities with populations of 10 million or greater are selected on the map.

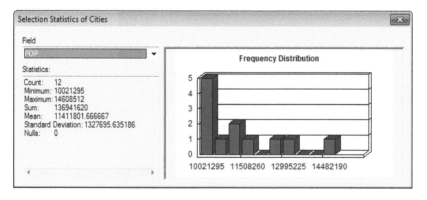

141	Point	Karachi	Pakistan
33	Point	Mexico City	Mexico
82	Point	Istanbul	Turkey
144	Point	Delhi	India
223	Point	Manila	Philippines
123	Point	Moscow	Russia
181	Point	Dhaka	Bangladesh
217	Point	Seoul	South Korea
14	Point	Sao Paulo	Brazil

ON YOUR OWN

At the bottom of the table, click the Show Sele—
table to show only the records of immediate in

To restore the complete table, click the Show
All Records button on the left.

I◄ ◄ 1 ► ►I ▦ ▣ (12 out of 256 Selected)

Cities > 1 Million

Sorting a field is useful for seeing high and low values, but you can get more detailed information from running statistics on a field. If records are selected in a table, the statistics will be calculated for those selected records only. Next, you will run some statistics.

16. With the 12 records still selected in the table, right-click the POP field and click Statistics.

Selection Statistics of Cities

Field
POP ▼

Statistics:
Count: 12
Minimum: 10021295
Maximum: 14608512
Sum: 136941620
Mean: 11411801.666667
Standard Deviation: 1327695.635186
Nulls: 0

Frequency Distribution

(chart x-axis values: 10021295 11508260 12995225 14482190)

The statistics window displays the number of records for which statistics are being calculated (12) and the minimum, maximum, sum, mean, and standard deviation values.

What is the total number (sum) of people living in the 12 largest cities?

The Frequency Distribution chart represents the distribution of values graphically.

How many selected cities have values under 11,508,260? (Compare your answer to the table to be sure you are reading the chart correctly.)

17. Close the statistics window and the attribute table. On the menu bar, click Selection > Clear Selected Features.

1
2
3
4
5
6
7
8
9
10
11
12
13
14
15
16
17
18
19
20

18. To save your work, on the menu bar, click File > Save As. Browse to your C:\EsriPress\ GTKArcGIS\Chapter03\MyData folder. Give the file a new name, and click Save.

19. Click the File menu and click Exit. Click No if prompted to save your changes.

Summary

This chapter gave you a broad introduction to map exploration using ArcMap, the mapping application for ArcGIS Desktop. By now, you should feel comfortable navigating a map, opening shortcut menus to choose commands, and using basic tools that are commonly used to explore GIS data. You should now understand that GIS data includes feature geometry, as well as attribute data. You can obtain valuable information not only from the map display, but also from the feature attribute tables.

Review the learning objectives at the beginning of the chapter. Do you feel that you are able to perform the tasks listed? If not, review some sections of this chapter as a reminder. You received a lot of introductory information in this chapter. If you feel a bit overwhelmed, don't worry—in the chapters ahead, you will get more practice performing these basic tasks and learning new ones.

Terms

layer
shortcut (context) menu
attributes
feature class
symbology
display scale

CHAPTER 4
Interacting with data

Browse through map data
Add data to a map
Work with map layers

Learning objectives

- Examine the contents of your GIS workspaces in ArcCatalog
- Differentiate data types according to their icons
- Preview geographic data
- Preview tabular (attribute) data
- Navigate and identify data in preview mode
- Create thumbnail graphics
- Read metadata
- Browse through the data catalog from within ArcMap
- Add data using various methods
- Search for data
- Fix broken data paths in a map document
- Work with data frames in layout view
- Copy, paste, and remove layers

ArcCatalog is an ArcGIS Desktop application for managing geographic data. In ArcCatalog, you can copy, move, and delete data; search for data; look at data before deciding whether to add it to a map; create data; and read detailed information about data. The ArcCatalog application window includes the catalog display for looking at spatial data, the Catalog tree for browsing through data, and several toolbars.

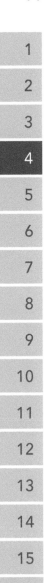

Main menu bar

Standard toolbar

Geography toolbar

Location bar

Metadata toolbar

Catalog display (preview mode)

Catalog tree

ArcCatalog interface. Data from Data and Maps for ArcGIS 2010, courtesy of ArcWorld, ArcWorld Supplement, DeLorme, and Esri.

You can also manage your data from within ArcMap by opening the embedded Catalog window. This is the easiest way to manage data. From here, you can access the core functionality of ArcCatalog without having to open a separate application.

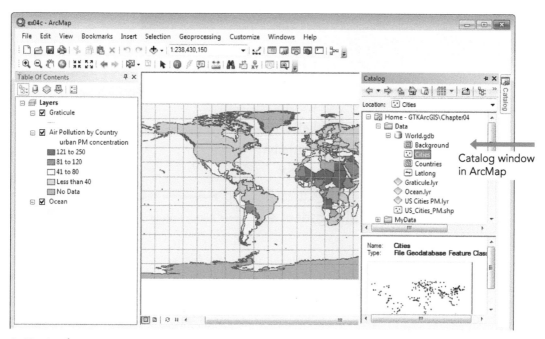

ArcMap interface. Data from Data and Maps for ArcGIS 2010, courtesy of ArcWorld, ArcWorld Supplement, DeLorme, and Esri.

TIP

Although the Catalog window in ArcMap can meet most of your data-browsing needs, the ArcCatalog application includes 3D navigation tools that aren't available in the ArcMap Catalog window. So if you need to preview 3D data, you will use the ArcCatalog application. You will use both methods in this chapter, but in subsequent chapters where data browsing is required, the Catalog window method will be used.

Spatial data comes in many formats, including geodatabase feature classes, coverages, shapefiles, CAD (computer-aided design) files, rasters, and TINs (triangulated irregular networks). Each format is identified by its own icon in ArcCatalog. The icon for a geodatabase, for example, is a cylinder. The feature classes within the geodatabase are represented by rectangle icons, with different patterns distinguishing point, line, and polygon features. (Geodatabases can also contain other special datasets such as annotation features, topologies, and relationship classes.)

In this book, you will use only the most common vector data types: feature classes stored in a data container called a geodatabase, as well as some shapefiles. You will occasionally work with raster surfaces, and you will also use layer files, which are not spatial datasets, but rather instructions for displaying spatial datasets by using certain colors, symbol markers, line widths, and so on. You will learn more about layer files and spatial data types throughout the book.

Geodatabase icons.

Data

Chapter 4 uses the same datasets as those used in chapter 3, as well as one additional shapefile:

- US_Cities_PM.shp–point features representing US cities with a population of 100,000 or greater, including demographics and pollution data *(Sources: ArcWorld; The World Bank)*

Exercise 4a
Browse through map data

ArcCatalog gives you more information about spatial data than you can get from Windows Explorer or other file browsers. It can show you which folders contain spatial data and what kinds of spatial data they contain. It allows you to preview features and attributes of data before you add the data to a map. It also allows you to examine and edit metadata, which is information about your data, such as when and how it was created.

TIP

It is never a good idea to use the system file browser (such as File Explorer) to view or move GIS data. Spatial datasets appear differently in Explorer, with many files, whereas in ArcCatalog, you get only one. If you don't copy all the files, the dataset will be corrupted. ArcCatalog (or the Catalog window in ArcMap) is the best option for viewing, moving, and managing your GIS data.

In this exercise, you will use ArcCatalog to look at GIS data and get information about it.

The exercise instructions assume you have installed the data for *Getting to Know ArcGIS Desktop* to the default directory (C:\EsriPress\GTKArcGIS). If you have installed the data elsewhere, you will need to substitute the correct paths.

1. **Start ArcCatalog (located in the ArcGIS suite).**

 This opens the ArcCatalog application window.

 If this is the first time you have opened the ArcCatalog application, you may need to display the Catalog tree.

2. **If necessary, click the Catalog Tree button 🖳 on the Standard toolbar.**

 The Catalog tree lists the data and services to which ArcCatalog is connected. Your application may look different from the preceding graphic, depending on the connections you may have already created.

You can connect to any local drive on your computer, as well as database servers, internet servers, and other services, organized by folders in the Catalog tree. ArcMap also maintains a connection to ArcToolbox, which you will learn about later in this book.

Once you connect to a folder, you can access the GIS data it contains. In this exercise, it will be helpful to see the file extensions of different spatial datasets.

3. **On the menu bar, click Customize > ArcCatalog Options. This opens the ArcCatalog Options dialog box.**

The ArcCatalog Options dialog box allows you to specify the types of data ArcCatalog displays and the information it shows about this data (file name, file size, date modified, and so on). You can distinguish folders that contain spatial data from those that don't and make many other customizations to the way data is displayed.

4. **Make sure the General tab is selected. Uncheck the Hide File Extensions check box if selected, and click OK.**

Next, you will create a connection to the GTKArcGIS exercise data so you can work with it in ArcCatalog.

5. **On the Standard toolbar, click the Connect To Folder button.**

6. **On the Connect To Folder dialog box, browse to the local C drive to view its contents. (Systems are different, but you may find the C drive in a directory called My Computer or Computer.) Expand the EsriPress folder, click the GTKArcGIS folder, and click OK.**

You have made a connection to C:\EsriPress\GTKArcGIS.

7. In the Catalog tree, expand C:\EsriPress\GTKArcGIS. Then expand the Chapter04 folder. It contains two subfolders, Data and MyData, and a map document called ex04c.mxd. Expand the Data folder.

The Data folder includes a geodatabase called World.gdb and a shapefile called US_Cities_PM.shp.

8. Expand World.gdb.

The geodatabase contains four feature classes. A feature class is a group of points, lines, or polygons that represent geographic objects of the same kind. The Cities feature class contains point features, the

Latlong feature class contains line features, and the Background and Countries feature classes contain polygon features. Each type of feature class is symbolized by a specific icon.

The US_Cities_PM shapefile is a different data type. It contains points that represent cities and has more attributes than the Cities feature class stored in the geodatabase. It is not unusual to work with several types of data, although at some point in your workflow, you may choose to convert these data types to the same format.

9. On the Standard toolbar, make sure the Details button ▦ is active.

10. In the Catalog tree, click World.gdb. In the catalog display, make sure the Contents tab is active.

Four adjacent buttons on the Standard toolbar change how files look on the Contents tab. The Large Icons button displays large icons horizontally. The List button displays small icons vertically. The Details button is like the List button, except that it also shows the file type–in this case, file geodatabase feature classes. The Thumbnails button allows you to view small graphics of spatial datasets. Whenever a folder or a geodatabase is highlighted in the Catalog tree, these four buttons are activated.

11. On the Standard toolbar, click each of the buttons that change the look of the Contents tab: Large Icons ▦, List ▦, Details ▦, and Thumbnails ▦.

TIP

At any time, pause the mouse pointer over a button to see its ToolTip. This also works in ArcMap and other ArcGIS Desktop applications.

The display changes to show thumbnail graphics of the content of each file. A thumbnail graphic of a dataset can help you decide quickly whether you want to use the data. Notice that there is no thumbnail graphic for countries. When a thumbnail graphic hasn't been created, you will see only a larger version of the feature class icon. Next, you will create a thumbnail graphic for the Countries feature class.

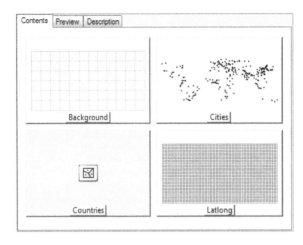

12. In the Catalog tree, click the Countries feature class in the World geodatabase. In the catalog display, click the Preview tab.

The Countries polygons are displayed in pale yellow.

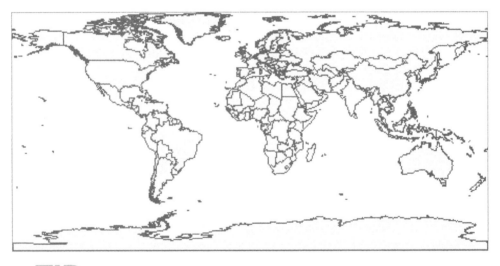

TIP

All polygons are displayed in pale yellow when previewed in ArcCatalog. Points are displayed in black and lines in navy blue.

13. On the Geography toolbar, click the Create Thumbnail button.

An image of the data shown on the Preview tab is saved as a thumbnail graphic.

14. In the catalog display, click the Contents tab to see the new thumbnail graphic.

A thumbnail graphic gives you a quick visual of the data, but you may want to view the data in more detail. The Preview tab and the tools on the Geography toolbar allow you to investigate a dataset further before you decide whether you want to use it on a map.

15. Click the Preview tab to see the Countries dataset at full size.

16. On the Geography toolbar, click the Zoom In tool ⊕. Drag the tool to draw a box around the area north of Australia, as shown in the graphic.

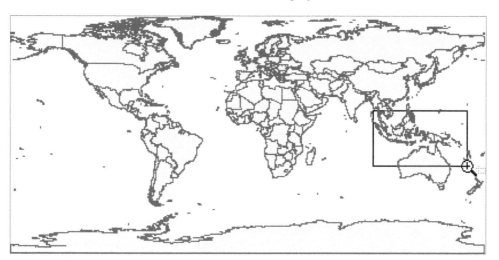

17. On the Geography toolbar, click the Identify tool ⓘ. Click Papua New Guinea (the feature indicated in the graphic) to identify it. If you don't see the feature flash green on the map, move the Identify Results window away from the display, and click the feature again.

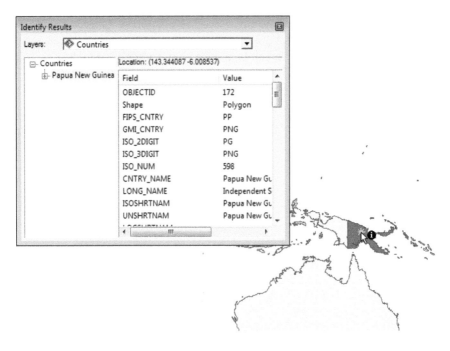

Is Papua New Guinea a UN member state? (You may need to scroll down in the Identify Results window to view the Status attribute.)

18. Close the Identify Results window. Use the Pan tool from the Geography toolbar to move around the map. The Pan tool works here just as it does in ArcMap.

19. Click the Full Extent button to zoom to the full extent of the data.

You can preview the attributes of a dataset as well as its geography.

20. At the bottom of the catalog display, click the Preview menu, and click Table.

The display shows the attribute table of the Countries feature class.

21. Scroll through the attribute columns.

Sometimes attributes have cryptic names. Many organizations maintain a data dictionary, which decodes and describes attributes.

22. When you are finished looking, click the Preview menu again, and click Geography.

The third tab on the catalog display is the Description tab, which displays metadata—essentially, data about data. ArcCatalog doesn't require you to maintain metadata, but it is considered best practice to do so, and many organizations require metadata to ensure data standardization and quality.

23. In the catalog display, click the Description tab. If needed, scroll down to see the thumbnail graphic you created, as well as keyword tags, summary, description, credits, and use limitations.

TIP

Notice the metadata shows a different name (World Countries instead of Countries). In geodatabases, any changes to feature class names are not automatically reflected in the metadata. One technique for documenting data sources is to maintain the original name (World Countries) in the metadata, even if you rename the actual feature class (Countries). Or, you can change the original name in the metadata using the Edit function described in step 24.

24. Click the Edit button on the uppermost bar inside the description to access metadata about the Countries feature class.

Metadata documentation can be created and edited on the Description tab.

25. Scroll down to see what type of metadata can be entered.

TIP

> Many organizations require adherence to strict metadata standards. You can choose different styles of metadata to be in accordance with various national or international standards. For more information about metadata and the styles supported by ArcGIS, consult ArcGIS Desktop Help. On the main menu, click Help, and click ArcGIS Desktop Help. On the Contents tab, browse to Manage Data > Data Types > Metadata.

26. In the catalog display, click the Contents tab. In the Catalog tree, click World.gdb. On the Standard toolbar, click the Details button.

This restores the display to its initial state.

27. If you are continuing to the next exercise, leave ArcCatalog open. Otherwise, exit the application.

Exercise 4b
Add data to a map

You can easily add data to ArcMap from the ArcCatalog application or, easier still, from the Catalog window within ArcMap. Once feature classes are added to ArcMap, they become map layers.

ArcGIS applications include a Search window that allows you to search for geographic data on disk or across a network. You can search by name, file type, geographic location, creation date, or other properties stored in the metadata.

Suppose you want to create a map that presents worldwide air pollution data. In this exercise, you will add layers to a new map document and search for more data that may be a useful addition to the world population map you studied in chapter 3.

1. **Start ArcCatalog if not already open. If necessary, in the Catalog tree, browse to your C:\EsriPress\GTKArcGIS folder connection and expand Chapter04\Data\World.gdb.**

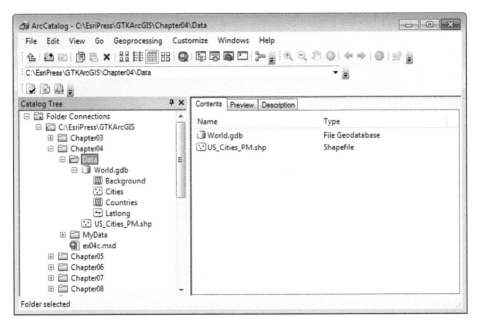

2. **On the Standard toolbar, click the ArcMap button to start ArcMap.**

3. **If you have a Getting Started dialog box, under New Maps, click My Templates. In the My Templates panel, click the Blank Map symbol, and click OK. If ArcMap is already open, click the New button to open a blank map.**

Next, you will add layers to your new, empty map document.

4. Position the ArcCatalog and ArcMap application windows so that you can see both the Catalog tree and the ArcMap display window.

5. From the Catalog tree, drag the Latlong feature class file anywhere over the ArcMap display window and release the mouse button. Bring ArcMap forward by clicking its title bar.

The layer is added to ArcMap. ArcMap chooses random symbol colors when new datasets are added to a map, so your results probably won't match the graphics in the book.

6. Close ArcCatalog. You will add the rest of the map layers from within ArcMap.

7. In ArcMap, look for a Catalog tab on the right side of the display window. Click the tab to see the Catalog window. If you do not see a small Catalog tab, on the Standard toolbar, click the Catalog button to open the window. Dock the window to the right of the map display.

1
2
3
4
5
6
7
8
9
10
11
12
13
14
15
16
17
18
19
20

TIP

Here is another way to open the Catalog window: On the menu bar, click Windows > Catalog.

Most of the functionality of ArcCatalog is available in the ArcMap Catalog window. The window can be docked or undocked. If docked, you can use the Auto Hide function to collapse the window when you're not using it. The Catalog window shows your folder connections and also includes a Home folder, which points to the current map document's workspace, if you are working in a saved map document.

ON YOUR OWN

Turn the Auto Hide function on and off by clicking the Auto Hide button (pushpin icon) and notice the resulting behavior of the window when you click the map display. When you've tried it both ways, set the option to the one you prefer.

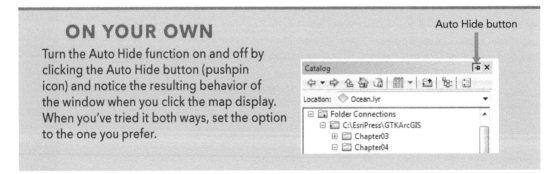

Auto Hide button

8. In the Catalog window, go to your C:\EsriPress\GTKArcGIS folder connection, and expand Chapter04\Data\World.gdb. Click the Countries feature class to highlight it.

9. On the Catalog window toolbar, click the Toggle Contents Panel button 🗗 to see both the file tree and the preview. (You may need to drag the divider between the panes to see the preview in its entirety.)

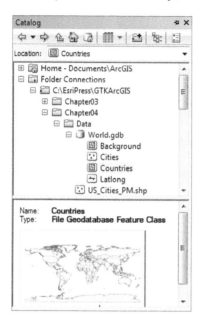

10. Drag the Countries feature class to the ArcMap display.

You can also add data using the Add Data button.

11. On the ArcMap Standard toolbar, click the Add Data button ⊕. (Click just the button, not the arrow next to it.)

12. On the Add Data dialog box, browse to \GTKArcGIS\Chapter04\Data. Double-click World.gdb. Click the Background feature class, and click Add.

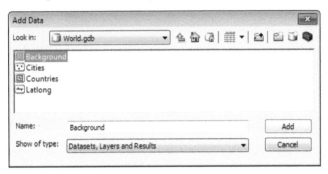

13. In the ArcMap table of contents, drag Background to the bottom of the list so it does not obscure Countries.

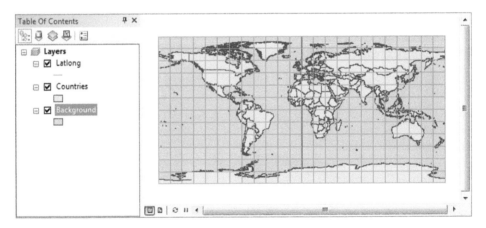

Your map clearly needs work—the default symbols need to be modified so the map looks nice and shows the attributes you are interested in. In particular, you want the Countries layer to show air pollution levels. However, you won't concern yourself with creating symbology in this chapter. That is presented in chapter 7.

ON YOUR OWN

Open the attribute table of the Countries layer and verify that the dataset includes the air pollution field called pm10.

Perhaps you want to add more data that pertains to the worldwide pollution data you already have, but you don't know if there's anything available. Next, you will use the Search window, which is a tool to find maps, data, and tools that match your search criteria.

14. On the Standard toolbar, click the Search button ▣.

This opens the Search window. By default, it is placed to the right of the display area, as a tabbed window residing in the same space as the Catalog window. It can remain there, fixed or with auto-hide turned on, or you can make it float in its own space, just like the Catalog window and the table of contents.

15. Arrange the Search window as you prefer.

You will concentrate your search for pollution data using keywords, or tags, stored in the metadata. If any matches are found, the dataset and its description will be returned in the search results. First you will create a search index to facilitate the Search tool.

16. Click the Search Options button ▣▾. Click the Add button, browse to the Chapter04 folder, and click Select. Then click Index New Items at the bottom of the dialog box. The index may take a few minutes to complete.

1
2
3
4
5
6
7
8
9
10
11
12
13
14
15
16
17
18
19
20

Search Options

Index | General | Advanced

Register Folders and Server Connections

C:\EsriPress\GTKArcGIS\Chapter04 Add...

 Remove

Indexing Options

Index new items (fast) every: 60 minute(s)

Re-Index from scratch (slow) every: 1 day(s)

 Starting at: 12:00:00 AM

☐ Create thumbnails

☑ Skip server connections in registered folders

☑ Skip mosaic dataset items

Number of parallel indexers: 1

Indexing Status

Items Indexed: 7
Indexing Status: Active
Last Indexing Start Time: 06/02/2015 11:21 PM
Last Indexing Duration: 52 seconds
Next Indexing Start Time:

Pause Indexing | Delete Index | Index New Items

 Re-Index From Scratch

 OK | Cancel | Apply

17. When indexing is complete, click OK to close the dialog box. In the Search window, make sure the menu is set to Local Search. In the text box, type pollution and press Enter. When the results are returned, you can expand the window for easier viewing if necessary.

The Search function returns several matches: a map document, two feature classes, and several tools.

18. In the search results listing, click the description (navy-blue text) of the US_Cities_PM shapefile.

This opens an Item Description window, which includes the metadata.

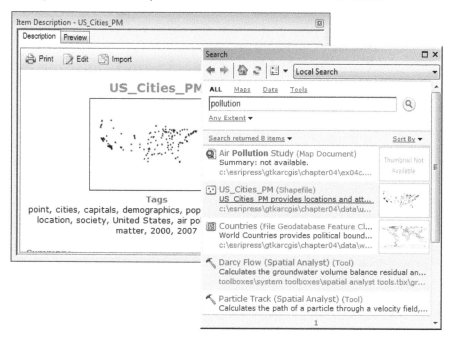

19. In the Item Description window, read the Summary and Description statements.

The metadata informs you this feature class shows air pollution measurements for most major US cities (population of 100,000 or greater). If you hadn't used the Search function, you might not have realized this shapefile includes pollution information.

TIP

The Search function highlights the importance of good metadata. Without good metadata, including relevant tags, an item would be returned only if the keywords are found in the item's name.

20. In the Item Description window, click the Preview tab.

Notice the tab has geography navigation tools so that you can explore your data more closely, just the same as in the ArcCatalog application.

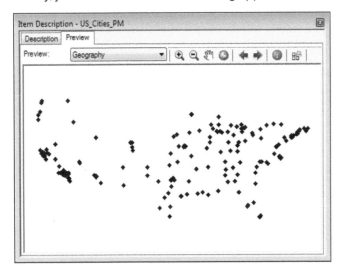

21. In the Preview menu, click Table. Scroll to the right to see the last field, pm10.

This field stores particulate matter measurements in urban areas, estimated by a research team at The World Bank.

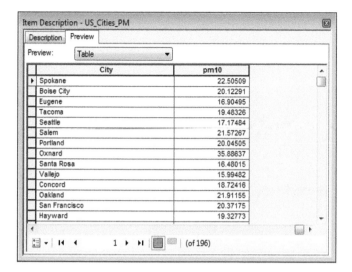

22. Close the Item Description window. From the Search window, drag US_Cities_PM (the light-blue link) to the map display.

Notice that the points go automatically to the top of the layers list in the table of contents.

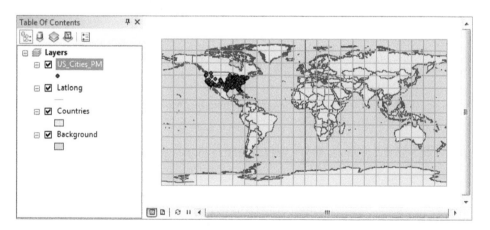

23. In the table of contents, right-click US_Cities_PM and click Zoom To Layer.

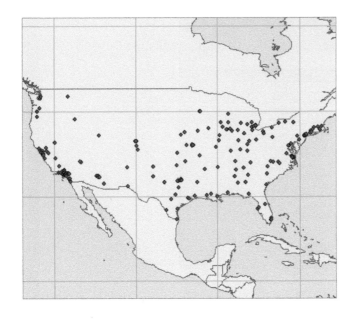

ON YOUR OWN

Identify ⓘ a few point features, and find the pm10 attribute value.

24. To save your work, on the menu bar, click File > Save As. Give your file a new name and save to your MyData folder for chapter 4.

25. If you are continuing to the next exercise, close the Search window and leave ArcMap open. Otherwise, exit the application.

In exercise 4c, you will organize the layers to create a meaningful map layout.

Exercise 4c
Work with map layers

When you add data to a map document, you can examine it as a navigable map display, as you did in chapter 3. This is called the data view of the map. You can also look at it as if it were on a page that you send to a printer. This is called the layout view. When you are in layout view, you can still interact with the map layers, but this view is best used for creating static presentation maps.

In layout view, you have map layers organized in one or more rectangles on a larger background rectangle. The smaller rectangles are called data frames, and the background is called the *virtual page*.

A map document can have one data frame or many. Data frames can contain different sets of data (usually related to a common subject), or they may show different views of the same data. If you wanted a layout to show more than one map view, you would use multiple data frames.

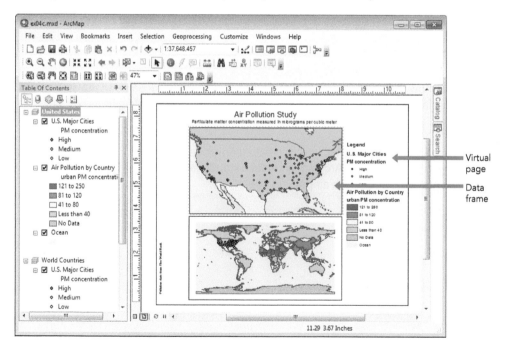

In data view, you have just one data frame at a time.

1 2 3 4 5 6 7 8 9 10 11 12 13 14 15 16 17 18 19 20

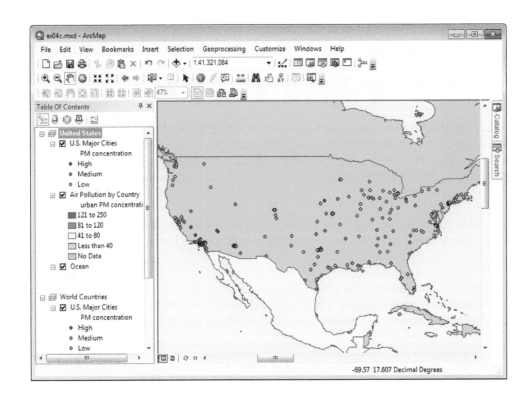

The display window shows the contents of the active data frame. The name of the active data frame is in boldface in the table of contents. In the graphic, the active data frame is called United States. The default name given to data frames is Layers.

In chapter 10, you will learn in detail how to create maps for presentation. For now, you will learn how to work with data frames in ArcMap to create a simple layout. Remember, your objective is to create a presentation that displays air pollution data.

1. **In ArcMap, open the ex04c map document from your C:\EsriPress\GTKArcGIS\ Chapter04 folder.**

There are several things to note about this map document, but of most pressing importance are the red exclamation points in the table of contents. This is the sign of a broken data path, and it often occurs if a map document is not saved with relative paths. If a user creates a map document from data stored on the computer's desktop location, and saves the map with absolute paths, ArcMap will look for that user's desktop location in search of the source data, but to no avail.

2. To fix the problem, in the table of contents, right-click U.S. Major Cities, point to Data, and click Repair Data Source.

TIP

A quick way to repair the data source is to uncheck the U.S. Major Cities check box to disable it, and then check it again to trigger the Set Data Source dialog box.

3. **On the Data Source dialog box, browse to C:\EsriPress\GTKArcGIS\Chapter04\Data. Click US_Cities_PM.shp, and click Add.**

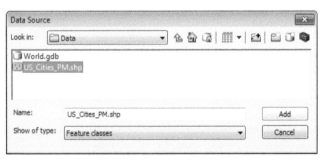

The broken path is repaired. In fact, ArcMap repaired the other broken links, too, by finding the correct source datasets in the folder you just specified.

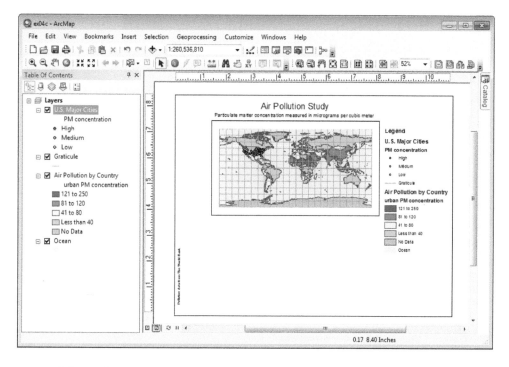

To avoid this problem in the future, you will save the map document with relative paths.

4. On the menu bar, click **File > Map Document Properties**. On the lower portion of the dialog box, check the **Store Relative Pathnames To Data Sources** check box, and click **OK**.

Now that you have fixed the data paths, you can take a look at the map display. The map is currently open in layout view. It includes the same data you worked with in exercise 4b, but the layers now have meaningful symbology, and they have been given more descriptive names. The map appears as if it was on a piece of paper. A new toolbar, the Layout toolbar, is open. It can be docked or free floating.

TIP

If you don't see the Layout toolbar, on the menu bar, click Customize > Toolbars, and check the Layout check box to turn on the toolbar.

TIP

If you resize your ArcMap window in layout view, you can restore the appearance of the layout by clicking the Zoom Whole Page button ⊠ on the Layout toolbar.

The map document has a single data frame occupying the upper half of the virtual page. Some other elements have been added to the layout for you. There is a title above the data frame, a subtitle, and a legend. The U.S. Major Cities points cannot be discerned on the map. You will zoom in to fix this; however, you also would like to show the world data. To do this, you need two data frames.

5. In the table of contents, right-click U.S. Major Cities and click Zoom To Layer.

6. On the menu bar, click Insert > Data Frame.

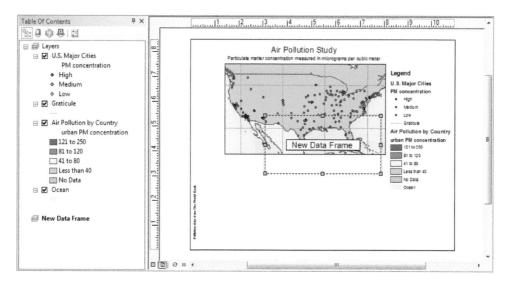

A data frame is added to the layout, and its name, New Data Frame, appears in the table of contents. The name in the table of contents is in boldface, indicating that it is the active data frame—the one to which new layers will be added.

7. In the layout window, move the mouse pointer over the new data frame. The pointer changes to a four-headed arrow. Drag the new data frame below the first data frame.

1 2 3 4 5 6 7 8 9 10 11 12 13 14 15 16 17 18 19 20

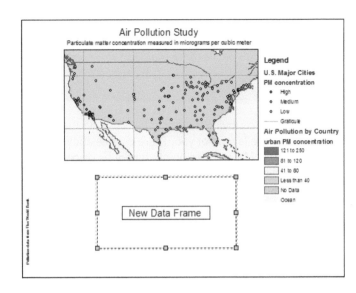

8. Move the mouse pointer over the lower-right anchor (the blue square). The pointer changes to a double-headed arrow. Drag the anchor to the lower-right corner of the layout page, enlarging the data frame window as shown in the graphic.

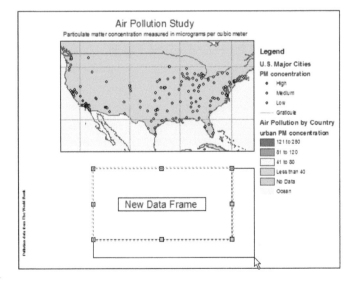

9. Now drag the upper-left anchor so the new data frame is approximately the same size as the one above it.

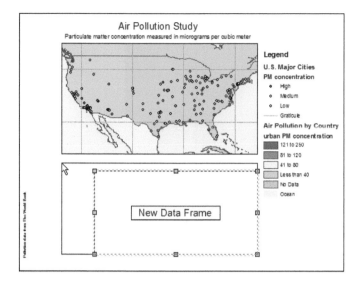

The name of a data frame is independent of any title you may choose to add to the layout. By default, the first data frame in a map document is called "Layers," the second is called "New Data Frame," the third is called "New Data Frame 2," and so on. It is often helpful to rename a data frame to something more descriptive, as described next.

10. In the table of contents, right-click the first data frame (Layers), and click Properties. On the Data Frame Properties dialog box, click the General tab. For name, enter **United States**, and click OK.

11. Click the second data frame in the table of contents, and click again to make it editable. Rename the second data frame **World Countries**.

12. Click the Data View button in the lower-left corner of the layout window. You use this button and the Layout View button next to it to switch views.

Data Layout
view view

TIP

You can also change views from the View menu. On the menu bar, click View > Data View or View > Layout View.

The display window is empty because the active data frame (World Countries) contains no layers.

TIP

To activate a data frame, right-click the frame name in the table of contents, and click Activate. Or use the keyboard shortcut Alt+click. In the layout display, a light-gray dashed line surrounds the active data frame. The navigation tools on the Layout toolbar work with whatever frame is activated.

13. In the table of contents, right-click Air Pollution by Country, and click Copy. Right-click the World Countries data frame, and click Paste Layer(s).

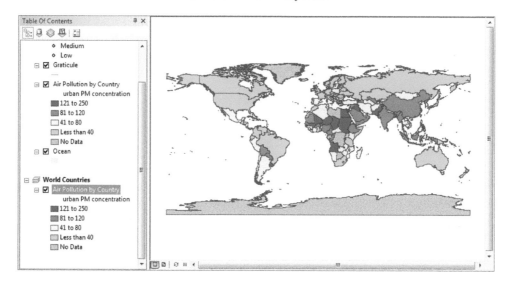

Now you have the same layer in two data frames–one layer is zoomed to the United States (the one in the United States data frame), and the other is zoomed to the full extent (the one in the World Countries data frame).

You also want copies of the Ocean and Graticule layers in the second data frame.

ON YOUR OWN

Copy the Ocean and Graticule layers to the World Countries data frame. Use the Ctrl key to select both layers at once. Reorder the layers as needed for proper drawing in the World Countries data frame.

14. Switch again to the layout view.

15. Finally, click anywhere in the white area of the virtual page to clear the selection of the lower frame. (Depending on which frame you last selected, it will have a dashed outline to show it is the active data frame.)

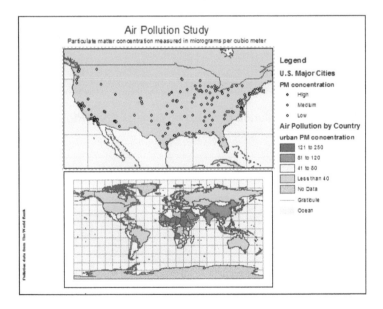

Your layout presents the air pollution data nicely. However, you decide the Graticule layer is unnecessary for the purpose of this map. You could just turn off the layer for both data frames in the table of contents. Another option is to remove it from the map document instead, to declutter the table of contents.

16. In the table of contents, under the United States data frame, right-click Graticule, and click Remove. Remove the Graticule layer from the World Countries data frame as well.

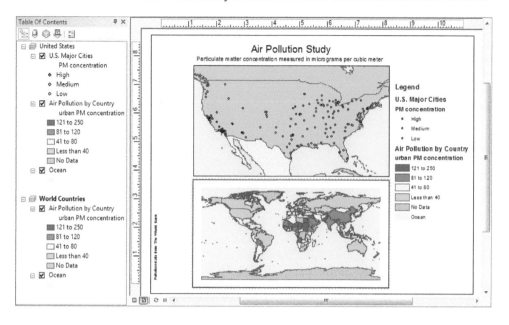

TIP

To close the Layout toolbar, right-click in an empty space near the ArcMap toolbars. From the list that appears, click Layout to uncheck the check box and turn off the layer.

17. To save your work, on the menu bar, click File > Save As. Give your file a new name and save to your MyData folder for chapter 4.

18. Exit ArcMap.

Summary

You now know how to bring data easily from ArcCatalog into ArcMap or—even easier—from the ArcMap Catalog window. Using the Search window, you learned how to find relevant data stored locally (or on the web or available servers). Finally, you practiced working with multiple data frames, which is necessary to create a printed product with more than one map. The difference between data view and layout view should now be clear.

1

2

3

4

5

6

7

8

9

10

11

12

13

14

15

16

17

18

19

20

Terms

geodatabase
shapefiles
metadata
data dictionary
data view
layout view
data frames
layout

CHAPTER 5
Exploring online resources

Create a web map
Merge online and local layers
Share a map package

Learning objectives

- Register for an ArcGIS account
- Browse the ArcGIS website
- Open and create an online map
- Explore ArcGIS basemaps and thematic layers
- Examine online map descriptions and properties
- Search for shared online data
- Add web map layers to ArcMap
- Upload a map package to ArcGIS Online

Esri's solution for GIS in the cloud, ArcGIS Online, is a web portal for working with maps and other types of geographic information. It functions as an online repository of maps, data, apps, and tools and allows anyone to use a web browser to create, analyze, and share GIS information—no software download is required. It also hosts developer resources for building web mapping applications. ArcGIS Online is an essential part of ArcGIS—it is integrated into the ArcGIS Desktop applications, which makes it easy to search for and access geospatial information when you need it.

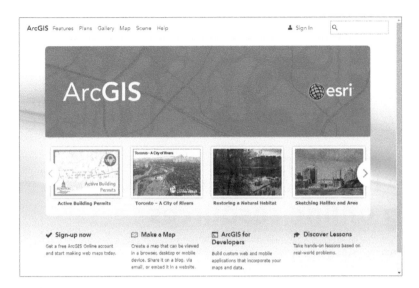

ArcGIS Online home page. Visit www.arcgis.com/home.

The ArcGIS Online Gallery features live, up-to-date basemaps, imagery, and maps created and shared by other users like you.

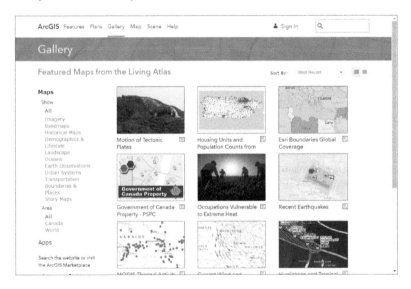

ArcGIS Online Gallery.

The web map viewer can be used to explore different parts of the world using the **basemaps** that are provided. (A basemap refers to a noneditable background map depicting reference information, such as landforms, roads, landmarks, political boundaries, or satellite imagery, onto which other GIS layers are placed.) You can add other layers, and save and share your map.

The Amazon Basin, as seen in the ArcGIS Online map viewer. Map and data courtesy of Esri, DeLorme, FAO, USGS, NOAA.

ArcGIS Online offers a variety of configurable apps. ArcGIS Online also offers ready-to-use apps for smartphones, tablets, and desktops. These downloadable apps are available to all ArcGIS Online users.

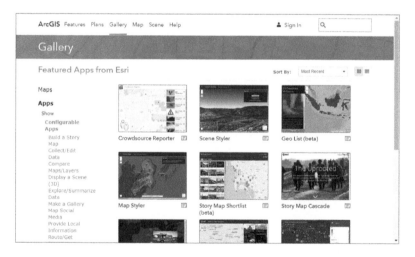

Share information and maps with ArcGIS apps that run on desktops, tablets, and smartphones.

With an ArcGIS account, you can view and join different ArcGIS Online groups. Through your group, you can collaborate, discuss, and share your content with others in your organization, the ArcGIS Online community, or the public through social media sites such as Twitter and Facebook.

A group can share maps, layers, apps, tools, and files in ArcGIS Online.

You can manage your own storage space and content—such as web maps, feature layers, service definitions, ArcMap documents, and map packages—from the My Content tab.

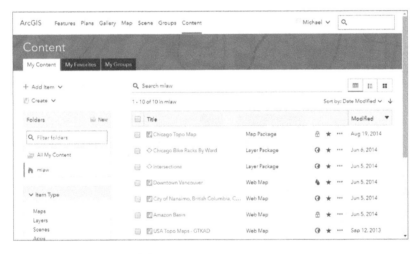

My Content space in ArcGIS Online.

TIP

To access the features of ArcGIS Online, you must have a working internet connection and an ArcGIS account. You can either have a subscription account (the common choice for organizations) or a free public account. Subscriptions come with varying levels of usage credits that you need to use the online geoprocessing and analysis tools. You can also use your Esri account login information to sign up for a free public account with limited usage. For this book, the free public account will suffice.

ArcGIS Online truly brings GIS to the world, offering community-hosted web maps highlighting endless sectors, problems, and areas of interest. Maps are interactive, allowing you to create and store information, explore a vast collection of data and maps, and perform some GIS functions, such as measuring distance and area, adding notes and bookmarks, editing features, and geoprocessing data.

ArcGIS Online is fully integrated with ArcGIS Desktop software. In ArcMap, you can add basemaps served from ArcGIS Online, and you can publish ArcMap documents to ArcGIS Online, sharing your work with members of your organization or with the user community.

In this chapter, you will explore ArcGIS Online to understand how it is used in conjunction with ArcGIS Desktop.

Note: Due to the dynamic nature of websites, the graphics used in the book may not reflect the current appearance and features of ArcGIS Online.

Data

This chapter uses the following datasets:

- World Topo Map—basemap includes administrative boundaries, cities, water features, physiographic features, parks, landmarks, highways, roads, railways, and airports overlaid on land cover and shaded-relief imagery for added context. The map provides coverage of the world. *(Sources: US Geological Survey [USGS], US Environmental Protection Agency [EPA], US National Park Service [NPS], Food and Agriculture Organization of the United Nations [FAO], Department of Natural Resources Canada [NRCAN], GeoBase, Agriculture and Agri-Food Canada, NAVTEQ, TomTom, Automotive Navigation Data [AND], DeLorme, and Esri)*
- World Imagery—basemap of low-resolution imagery of the world and high-resolution imagery of the United States and other areas around the world. *(Sources: NASA Blue Marble, i-cubed, Getmapping, AeroGRID, GeoEye IKONOS, US Department of Agriculture [USDA], National Agriculture Imagery Program [NAIP], US Geological Survey [USGS], and Digital Ortho Quarter Quad [DOQQ])*
- Bike Racks—point layer bike rack locations in Chicago, Illinois.
 (Source: City of Chicago)
- Bike Routes—line layer showing bike routes in Chicago, Illinois.
 (Source: City of Chicago)

1
2
3
4
5
6
7
8
9
10
11
12
13
14
15
16
17
18
19
20

Exercise 5a
Create a web map

ArcGIS Online provides a robust web map viewer you can use to quickly create web maps containing a basemap and layers found throughout the site. You can create a web map in just three steps: (1) choose an area to display, (2) add content, and (3) save your map. You can change the geographic area of interest (for example, you can zoom out from a country view to a continental view), explore the metadata behind the map, and add and remove layers.

You always start a web map with a basemap layer. Then you can add operational layers, which are the layers you can manipulate. This process is sometimes called a mashup, because you are combining layers of information, often from different sources, to create a single map. These content layers can be any map, image, or feature service hosted by ArcGIS Online—these might be layers provided by Esri or by any member of the GIS community who has made data accessible to ArcGIS users. Map services make maps, features, and attribute data created in ArcMap available to the web using ArcGIS. Image services serve raster datasets, such as surfaces or imagery. Feature services serve vector features and provide the symbology for displaying the features.

What is unique about these web maps is that you can view, create, and share them from within ArcGIS Online. In addition, other clients such as smartphones, tablets, and ArcGIS Desktop can access these maps without any additional programming or third-party applications or plug-ins.

Note: This exercise requires you to be connected to the internet.

1. In a web browser, go to www.arcgis.com. In the upper-right corner, click Sign In. If you already have an Esri account (the general account used for accessing personalized Esri content), sign in and skip to step 3.

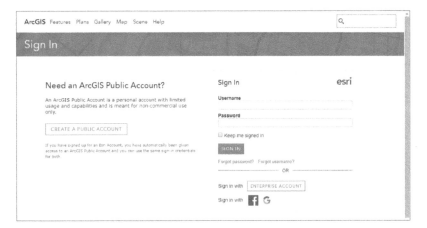

2. If you don't have an Esri account, click the **Create A Public Account** button. Fill out the form, accept the terms of use, and create the account. Optionally, you can edit your profile.

3. On the main navigation bar at the top, click **Content**.

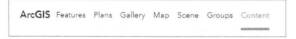

The Content page contains all the tools necessary to add, edit, share, and access all your saved content. Your map viewer may not show any content yet, but in the following steps, you will learn how to create a map and save it to your Content page.

TIP

The Edit option is not visible from the Content page. It is available on the detailed description page of existing content.

4. On the main navigation bar, click **Map**. If you have been viewing or working with any maps in the current session, a dialog box may appear to ask if you want to open a new map. Click **Yes**.

The web map viewer displays a basemap showing a topographic map of the world. To navigate the map, use standard mouse pointer operations. (This is probably already intuitive.)

5. Drag the pointer to pan around the map.

6. Double-click a location on the map to center and zoom.

7. Rotate the scroll wheel of the mouse up and down to zoom in and out, respectively.

TIP

If you are using a tablet or a touch-screen monitor, use standard touch-screen navigation methods: swipe one finger to pan, and use finger and thumb in conjunction to zoom (pinch finger and thumb together to zoom out; slide finger and thumb away from each other to zoom in).

8. Pan and zoom to the Himalayas, a mountain range in Asia separating India and China.

The Himalayas run west-northwest to east-southeast, in an arc 2,400 kilometers (1,500 miles) long. It is home to Mount Everest, the planet's highest peak at 8,848 meters (29,029 feet). To get a better sense of the land cover, you will change the basemap.

9. On the map viewer toolbar, click the Basemap button ⊞⊞. The menu that appears shows previews of available basemaps. Click Imagery Hybrid.

The basemap changes to show satellite imagery, country borders, and labels for countries and major cities. To find out more about this basemap, you will look at its description and metadata.

10. On the Details panel (if it is not visible, click the Details button), click the Content button ▤. (You may have to widen the Details pane to see the options.)

11. Hover over the Imagery Hybrid text. Notice that an arrow appears to the left of the label. Click the arrow to expand it. Hover over World Imagery and click the three dots that appear to open the properties menu. Click Description.

This opens a new window or tab. The service description lists information about the image sources, resolution, tiling detail, spatial extent, and ownership information. Notice the row of View In hyperlinks. You can click these links to download the basemap as readable files for various applications, including ArcMap. Or you can get more metadata about individual layers.

12. Scroll down the page to view more metadata. Stop and review the Tile Info section.

How many levels of detail are there in this basemap?

Compare the resolution and scale values for two different levels. What does this tell you?

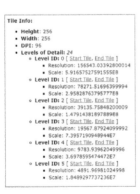

The basemaps in ArcGIS Online are composed of individual image tiles. Multiple levels of detail provide the appropriate amount of detail at each map scale.

13. If your description was opened on a new browser tab or in a new window, close it after you finish viewing the metadata. If you did not return to the exact same location, set the basemap to Imagery Hybrid, and zoom to the Himalayas again.

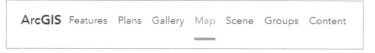

The map is now ready to be saved.

14. On the map viewer toolbar, click the Save button 🖫, and click Save. On the Save Map dialog box, enter the title: **Himalayas**. Enter the tags: **himalayas, mountain range, mount everest**. Enter the summary: **A basemap of the Himalayan range**, and click the Save Map button. The Save in folder list shows your own sign-in name.

TIP

Tags, or keywords, are used to categorize content and make it searchable. Appropriately defined tags can help users find the content in ArcGIS Online or elsewhere on the web.

The map is saved, and the title of it is changed at the top of the map viewer.

When you click the About button ⓘ, a short summary appears along with the type of document (Web Map) and the author (you). The rating system indicates which maps are popular.

15. To the left of the title Himalayas, click Home, and in the list, click Content.

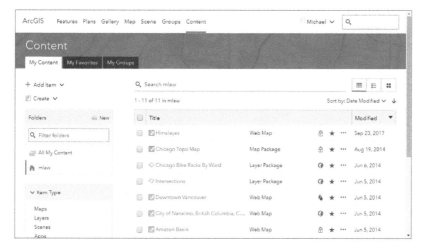

Notice that your Himalayas map is now listed, showing its document type, modified date, and shared status. If you previously saved maps, they will be listed here as well. Next, you will view more map details and options.

16. Click the Himalayas map title.

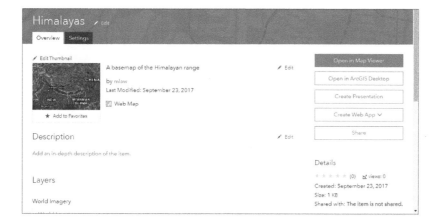

This page is used to manage your saved map. It displays a thumbnail graphic of the map, its description, hyperlinks to the content used in the map (in this case, web services), shared status, tags, credits, file size, spatial extent, and other map settings. In addition, this page allows you, or anyone you share your map with, to comment about the map. To further extend the reach of your map, you can share it with your social network through Facebook and Twitter.

TIP

If you want other ArcGIS users to see your map, you must first share it with the public or a group. To avoid a high volume of duplicate maps, it's best not to share with the ArcGIS community right now. But feel free to share your map with your personal network.

Next, you will view the edit page of this map to see its options.

17. Below the Description heading, click the Add An In-Depth Description Of The Item link. Alternatively, click the Edit link to the right of the heading.

The edit page provides form fields where the thumbnail graphic, title, summary, description, access and use constraints, tags, credits, and spatial extent can be entered or modified. Next, you will add some information to the Description box and save it.

18. In the Description box, type This map shows satellite imagery and labels of the Himalayan mountain range.

19. Click Save. On the main navigation bar, click Content to return to the content list. If you are continuing to exercise 5b, leave the browser window open. Otherwise, you may close it now.

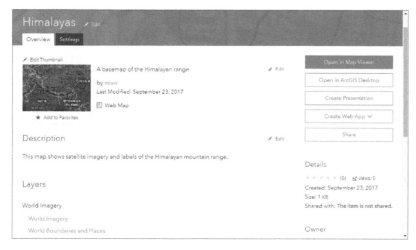

ON YOUR OWN

On the main navigation bar, click Groups. Browse through some of the content hosted by featured groups.

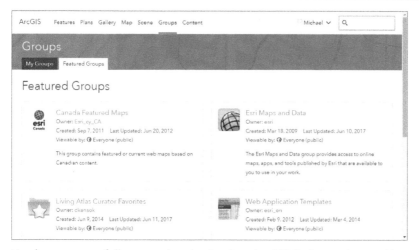

You have successfully created and edited an ArcGIS Online web map. You can edit the information in your map at any time, and the information will be updated in real time. This means that when it is shared with the public or a group, everybody will be able to see your edits immediately. This is very useful when working on a project with others in your community.

1

2

3

4

5

6

7

8

9

10

11

12

13

14

15

16

17

18

19

20

Exercise 5b
Merge online and local layers

ArcMap has the ability to load basemaps directly from ArcGIS Online. This feature complements the online services available from Esri—the ArcGIS system now fully integrates all aspects of the web, allowing you to share information and easily collaborate with others in your organization or industry. GIS professionals can compile and publish rich, interactive maps and geographic information straight from ArcMap.

Note: This exercise requires you to be connected to the internet.

1. Start ArcMap. If you get a Getting Started dialog box, click My Templates and double-click Blank Map. Or if ArcMap is already open, click the New button to open a blank map.

2. On the menu bar, click File > Sign In.

3. On the ArcGIS Sign In dialog box, enter your public account user name and password, and click Sign In.

TIP

You can still access maps and data if you don't sign in. You won't, however, be able to load your own maps and data, access groups, or share layers. For this exercise, it is recommended that you sign in to your account, but this is not necessary.

4. On the menu bar, click File > ArcGIS Online.

This opens the ArcGIS Online window. Featured maps are listed in a tile view, each showing a thumbnail graphic, a short description, the publisher, and the date it was published. (Your screen will be different from the graphic shown.)

TIP

In ArcMap, you can also click the Add Data arrow, and click Add Data from ArcGIS Online.

5. In the Search box, type **Chicago Topo Map**, and press Enter. Your search results may be limited to your organization. To search all of ArcGIS Online, uncheck the Only Search In [Your Organization] check box.

6. In the returned maps, locate the one titled Chicago Topo Map (a basemap authored by Esri Atlas), and click the Details link.

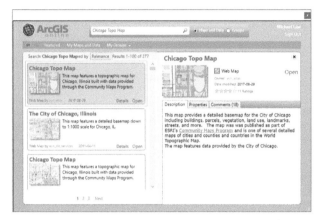

To the right is a description of the map that the publisher has provided. Notice that this map, and others like it, is interactive, allowing the community to rate it and leave comments. Through these shared user experiences, the community can see which maps are the most useful. This process also leads to improved maps and encourages publishers to update or publish new ones.

7. In the Description panel, click Open. If ArcMap asks you to save changes to your untitled map, click No. If a message about improving drawing performance with hardware acceleration appears, click Yes.

Only one basemap can be loaded at a time. You can, however, load as many data layers as you want.

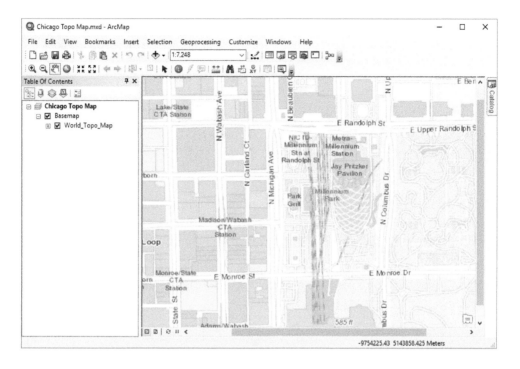

This loads the Chicago Topo Map from ArcGIS Online into ArcMap. Depending on the size of your ArcMap window, your map may look different from the graphic shown. Also, depending on your internet connection speed, your map may take a few moments to load.

8. Access the World_Topo_Map layer properties.

1

2

3

4

5

6

7

8

9

10

11

12

13

14

15

16

17

18

19

20

Notice the layer description and source credits, authored in ArcGIS Online, are imported with the data. Map document properties are also imported. To see these, go to File > Map Document Properties.

ON YOUR OWN

Use the Scale Range to limit the visible range of the layer. This can be useful in cases where the basemap needs to be visible only at certain zoom ranges.

Adding data to overlay an ArcGIS Online basemap is the same as adding data to any ArcMap map document. You will next add several layers to the map to see how each one is displayed and then explore the different zoom levels of the map.

9. Close the Layer Properties dialog box. On the Standard toolbar, click the Add Data button. Browse to C:\EsriPress\GTKArcGIS\Chapter05\Data, and add Bike Routes.lyr.

TIP

If your data is not saved to the default location, you will need to correct the data source for the layer file. On the layer properties, click the Source tab and click Set Data Source. Browse to the bike_routes feature class in \Chapter05\Data\Chicago.gdb, click Add, and click OK.

A line layer called Bike Routes is added to the map and displayed with preset symbols for each route. These routes indicate what kind of road the bike route shares, including recommended bike routes, existing bike lanes, cycle tracks, and off-street trails.

10. On the Tools toolbar, click the Fixed Zoom Out button until the map scale is approximately 1:450,000. If necessary, pan the map so the bike routes line layer is centered in the view.

As you zoom out, notice how the different zoom levels affect how the map details look. This map and other maps in ArcGIS Online are loaded as tiles with multiple zoom levels at varying scales, also known as levels of detail. These tiles are cached by the system to load quickly when the level of detail is changed. The topographic layers that are drawn depend on the scale and effectively cover views from the city level to the world level. The basemap creators defined the map scales at which certain details are visible—for example, small streets are visible at scales greater than 1:125,000, and building footprints are visible at scales greater than 1:15,000.

The bike routes are not affected by the zoom, because no rules for display based on map scale are set. Next, you will add another layer to the map from ArcGIS Online.

11. On the menu bar, click File > ArcGIS Online. In the Search box, type **Chicago bike racks by ward**, and press Enter. In the search results, several layer packages called Chicago Bike Racks By Ward may be available for download. At the bottom of the tile, note the text indicating the user account that uploaded the file. Select the item uploaded by the account associated with esripress_GettingToKnowArcGISDesktop10.8 Click the Details link to see a short description of it.

12. In the ArcGIS Online window, click the Add button.

You added a layer package. Layer packages encapsulate the data, symbology, and other properties of the layer as authored in ArcMap in one easy-to-share package. From the layer package, a point layer called Bike Racks is added to the map. It displays all the bike racks in the city, symbolized by ward. This information could be useful for cyclists who want to know where bike racks are located along their planned route. Next, you will zoom in closer to see the details.

13. Click the Zoom In tool and draw a rectangle around the downtown portion of Chicago, as indicated in the graphic.

Depending on how close you zoomed in, your map scale may be approximately 1:36,000. At this scale, it's easier to see how the basemap, bike racks layer, and bike routes layer come together.

14. Zoom in until the map scale is at least 1:5,000.

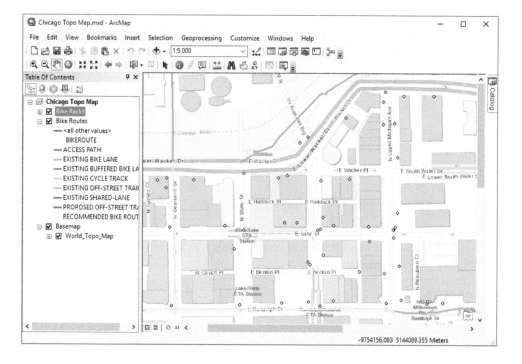

This zoom level has the greatest amount of detail included in the Chicago ArcGIS Online map. The building footprints show detail, and some are individually labeled. You can even tell on which side of the street bike racks are located.

15. **Click File > Save As. Save the map document to your MyData folder for chapter 5.**

As long as you have an internet connection, the Chicago basemap and the bike racks by ward will be recalled the next time you open the map because the information is stored in ArcGIS Online.

16. **If you are continuing to the next exercise, leave ArcMap open. Otherwise, exit the application.**

ON YOUR OWN

Explore ArcGIS Online and find out how to create your own group. You can decide whether you want a public group, a nonmember public group, or a private group. Your choice depends on the task you want to achieve and the level of access to your information you want to provide. A public group can accept member requests and is open to all who want to participate. A nonmember public group shares content but doesn't allow other users to join or share. A private group is inaccessible to the public; only the owner can invite people to join.

Once you have created your group, add some content by learning how to create, manage, store, and share information in ArcGIS Online. Then, invite your peers to the group, or accept membership requests, to share content.

Exercise 5c
Share a map package

The following short exercise explains how to upload a map package to ArcGIS Online. You will publish the map you created in the previous exercise.

1. **Open the map you created in the previous exercise.**

As you recall, this map shows Chicago bike routes and bike rack locations.

2. **Sign in to your public ArcGIS Online account (File > Sign In).**

3. **From the File menu, click Share As > Map Package.**

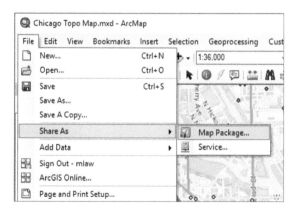

A **map package** contains the map document (.mxd) and the data referenced by the layers it contains in a portable file. Map packages can be saved locally or uploaded to ArcGIS Online and shared in a work group or organization.

4. Click the Analyze button.

If there are any issues that prevent map sharing, they will be reported at the bottom of the ArcMap window. (If necessary, move the Map Package window to view the Prepare panel.) Common issues are lack of a map summary or tags, both of which are required for publishing. You can edit the map summary, tags, and descriptions in the Item Description tab.

5. Click the Upload Package To My ArcGIS Online Account option. Replace the name with Chicago Bike Routes and Racks. Click the Share button. Click Yes when you are prompted to save the map document.

The packaging process begins. It may take several minutes.

6. When you get a message telling you the map has been successfully created online, click OK.

7. **In your web browser, go to www.arcgis.com and sign in. Go to your Content and click Chicago Bike Routes and Racks to see the map description.**

The map exists as a downloadable package that can be shared. If you belong to or own a group, you could share it with group members, providing an easy conduit for collaboration. You could also share it publicly, but we discourage doing so to prevent a great number of redundant maps residing in the public space.

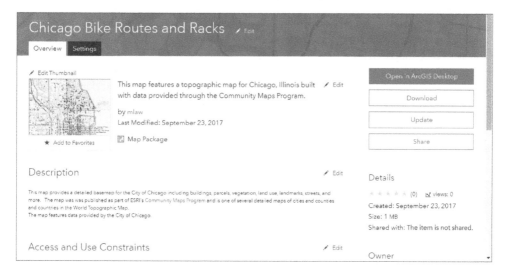

ABOUT SHARING CONTENT

You are encouraged to share maps and applications if they are useful to others and viewable in a web browser, mobile device, or ArcGIS application. However, you should not share any geographic information that is confidential, copyrighted, violates laws, or would cause a security risk if some-one gained access to it. When working with an organization, you may find it appropriate to share some content, but other content may be proprietary or business-sensitive data and should not be shared with people outside the organization. In addition, map creation and data collection are often works in progress and are not necessarily complete or run through final validation. However, once the content is finalized, feel free to share the data with the ArcGIS Online community and the public. For more information on best practices for sharing content, visit ArcGIS Online Help at https://doc.arcgis.com/en/arcgis-online/reference/best-practices-share.htm.

Summary

By now, you should have a good understanding of the fundamentals of using ArcGIS Online. You have learned how to register for ArcGIS Online, navigate the online mapping interface, and explore vast collections of maps and data available in the cloud. If you explored ArcGIS Online groups, you know how maps can be shared with users both publicly and privately.

The best way to get comfortable with ArcGIS Online is to use it. Use your free public account to search, view, and download data. Share your maps via ArcGIS Online. You may consider, now or in the future, getting a subscription account if you want to be able to use more sophisticated GIS tools with web maps. Remember that ArcGIS is a large, integrated system of Esri products—including ArcGIS Online—and that a great deal of information about ArcGIS is available online, including blogs, newsletters, and product demonstrations. If you are interested in learning more, go to the Esri website (esri.com) to learn about the complementary aspects of the ArcGIS system.

Terms

basemaps
groups
operational layers
mashup
layer package
map package

CHAPTER 6
Working with coordinate systems and projections

Examine coordinate systems
Project data
Define a map projection
Georeference a raster

Learning objectives

- Become familiar with geographic coordinate systems and projected coordinate systems
- Explore the details of coordinate systems
- Define a coordinate system
- Identify projection parameters
- Project data on the fly
- Align datasets with different coordinate systems
- Georeference a raster

Geographic information systems (GIS) depend on the proper use and definition of coordinate systems and projections. Using the correct coordinate system ensures geographic accuracy, whereas using an improper coordinate system will result in unusable or error-prone maps. Learning about coordinate systems and map projections can be challenging for beginners, but practice and experience will provide a solid foundation for practical use in ArcGIS applications.

A coordinate system is a reference framework that uses points, lines, and surfaces to determine positions in two- or three-dimensional space. In simpler terms, it helps you to find your way around a map. Defining a coordinate system is essential to every GIS mapping project, regardless of subject, projection, or scale. All coordinate systems are based on coordinate pairs that belong to a geographic grid, the location reference system for spatial features on the earth's surface. On a globe, locations on the earth's surface are defined in reference to lines of parallels and meridians, more commonly known as latitude and longitude, respectively. Latitude lines run parallel to the equator (at 0° latitude) and measure how far north or south you are from the

1
2
3
4
5
6
7
8
9
10
11
12
13
14
15
16
17
18
19
20

equator. Longitude lines run from pole to pole and measure how far east or west you are from the prime meridian (the meridian of 0° longitude that passes through Greenwich, England).

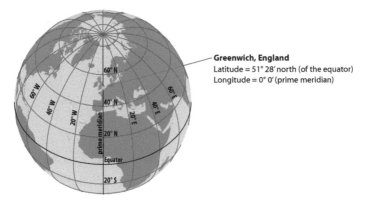

Greenwich, England
Latitude = 51° 28′ north (of the equator)
Longitude = 0° 0′ (prime meridian)

Lines of latitude and longitude. Data from Data and Maps for ArcGIS 2010, courtesy of ArcWorld supplement and Esri.

Latitude and longitude are measurements of angles, not of distance. Latitude is the angle between the equator and the parallel on which the point you are locating lies, as measured from the center of the earth. Longitude is the angle between the prime meridian and the meridian on which the point you are locating lies, as measured from the center of the earth.

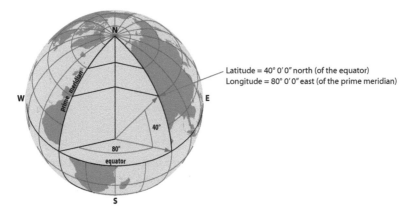

Latitude = 40° 0′ 0″ north (of the equator)
Longitude = 80° 0′ 0″ east (of the prime meridian)

Angles are used instead of distance, because as meridians converge toward the poles, the distance between them decreases. A degree of longitude is about 69 miles near the equator, about 49 miles at the 45th parallel, and zero at the North Pole. In other words, distance varies between meridians, and only the angle between them remains constant. Data from Data and Maps for ArcGIS 2010, courtesy of ArcWorld supplement and Esri.

The earth is divided into 360 degrees around the equator. Starting from the prime meridian, longitude values go to 180 degrees eastward and to −180 degrees westward. (The 180 degree and -180 degree meridians are the same.) Starting from the equator, latitude values go to 90 degrees at the North Pole and to -90 degrees at the South Pole.

Because latitude and longitude are angles, their values are commonly expressed in degrees-minutes-seconds (DMS) or sometimes as decimal degrees (DD). One degree equals 60 minutes, and 1 minute equals 60 seconds. For example, in DMS, Tokyo is located at 35° 41′ 6″ N (latitude) and 139° 45′ 5″ E (longitude). In DD, those values would convert to 35.685° and 139.7513889°, respectively.

Latitude and longitude are the basis of a geographic coordinate system, a system that defines locations on the curved surface of the earth. Because different estimates have been made of the earth's shape and size, there are a number of geographic coordinate systems in use. Although these systems are similar, the precise latitude-longitude coordinates assigned to locations vary from one system to the next.

Coordinate systems and projections are based on a defined spheroid and datum. A spheroid (also known as an ellipsoid) is a mathematical model that approximates the irregular shape of the earth using two axes. Because the earth bulges at the equator as a result of the earth's rotation, the model uses a major axis dissecting the equator and a minor axis dissecting the poles. The relationship between the shape of the earth and the shape of a spheroid is described using a datum, which essentially describes how a model fits into reality.

Left: Some geographic coordinate systems are based on the assumption that the earth is a sphere. *Right:* Most geographic coordinate systems are based on the assumption that the earth is a spheroid (exaggerated). The spheroid is a more accurate model because the earth bulges slightly at the equator and is flattened at the poles. Data from Data and Maps for ArcGIS 2010, courtesy of ArcWorld supplement and Esri.

Hundreds of datums are in use, and each one uses different methods to measure the earth. This poses a challenge to mapmakers, because each datum measures and accounts for different features, such as flatness ratio, circumference, or topographical differences. The goal of the datum is to provide the optimal fit between the geoid, a measure of the earth's radius, and the chosen spheroid used for a map.

Some commonly used datums in North America include NAD27 (North American Datum of 1927), WGS84 (World Geodetic System of 1984), and NAD83. NAD27 is based on the Clarke 1866 datum, a ground-measured spheroid. Technological advancements and satellites are now used to determine more precise spheroids and measurements. Many GIS data providers and web map producers are moving toward WGS84 as a common standard. However, a considerable amount of

existing data uses NAD27 and NAD83. So, it is important to be aware of the potential difference in feature positions between datums.

A map projection is used to represent the earth on a flat surface. In more technical terms, mathematical formulas are applied to a geographic coordinate system defined by a spheroid and datum to convert latitude (north-south orientation along the y-axis) and longitude (east-west orientation along the x-axis) values to a plane surface representing a geographic grid. The result is a two-dimensional (2D) surface that is composed of a grid of intersecting lines. One set of lines runs parallel to a horizontal x-axis, and the other set runs parallel to a vertical y-axis. The coordinates of any location or point are expressed as a distance value along the x-axis (longitude) and a distance value along the y-axis (latitude) from the intersection of the axes. Therefore, longitude and latitude values are expressed as x and y, respectively.

 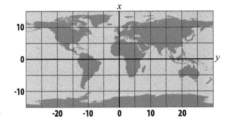

Left: A geographic grid with longitude values along the x-axis and latitude values along the y-axis. *Right:* On a flat map, coordinate values can represent any kind of linear unit. Data from Data and Maps for ArcGIS 2010, courtesy of ArcWorld supplement and Esri.

A projected coordinate system is a system that defines locations on a flat map based on x,y coordinates. The x,y coordinates of any given point depend on the map projection, on the units of measure (meters, feet, or whatever), and on which point the map is centered. If Cairo is made the center of the map, for instance, its x,y coordinates would be 0,0. Thus, the number of possible projected coordinate systems is infinite.

Keep in mind that not all data will be or must be projected, but it must have a geographic coordinate system. To understand how certain map projections work, you will also need to know how geometric objects are used to construct a flat map. Imagine a piece of paper being shaped to fit against a lighted globe in three configurations: a cylinder, a cone, and a flat plane (see the graphic). A cylindrical projection is possible when a cylinder is used, a conic projection when a cone is used, and an azimuthal projection when a flat plane is used.

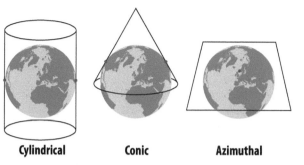

Cylindrical **Conic** **Azimuthal**

• Blue dot represents point of tangency

Common geometric shapes used to construct a map projection. Data from Data and Maps for ArcGIS 2010, courtesy of ArcWorld supplement and Esri.

The projections are possible when the geometric object makes contact with the globe at a line of tangency or at two lines of tangency, also called a secant. These lines of tangency are called standard parallels if they follow a parallel or standard meridians if they follow a meridian. Depending on the case of tangency and orientation of the geometric object, different projections can be produced, including secant conic, polar azimuthal, transverse cylindrical, and many more (see graphic). The distortion of each projection varies, but as a general rule, a projection is the least distorted along the tangent and grows more distorted leading away from it. Secant projections allow for greater accuracy in measurements, particularly on large-scale maps where both standard parallels run within the constraints of the map. Secant projections, however, are not as suitable for small-scale maps.

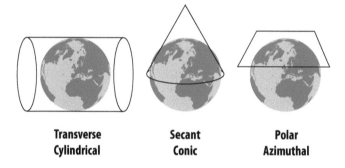

**Transverse Secant Polar
Cylindrical Conic Azimuthal**

Variations of common geometric shapes create different map projections. Data from Data and Maps for ArcGIS 2010, courtesy of ArcWorld supplement and Esri.

1
2
3
4
5
6
7
8
9
10
11
12
13
14
15
16
17
18
19
20

The transformation of the earth's surface to a flat surface is inherently not perfect and causes distortion to shape, area, distance, or direction, depending on the projection. To counter distortion, map projections can be grouped into four types of preserved properties:

- **Conformal projections**, which preserve shape
- **Equal-area projections**, which display an area at its proper relative size
- **Equidistant projections**, which preserve scale, and thus distance
- **Azimuthal**, or **true-direction**, projections, which maintain accurate directions

Choosing the correct map projection is an important consideration. It allows you to control the type of distortion in a map. In many cases, mapping at a larger scale (such as 1:24,000) requires a different map projection than mapping at a smaller scale (such as 1:1,000,000). For an area the size of a small country such as Switzerland, for example, the use of an appropriate projection will result in little to no distortion. For an area the size of a large continent, or even the world, however, the use of a compromise or specialized map projection can minimize distortion. In the end, distortion is inevitable, but using good judgment and an appropriate map projection will minimize distortion of your map.

The graphics shown here exhibit several examples of popular map projections.

The Mercator projection is a cylindrical projection that preserves shape but distorts area. Any straight line drawn with this projection represents an actual compass bearing. Data from Data and Maps for ArcGIS 2010, courtesy of ArcWorld supplement and Esri.

The World Winkel Tripel projection is used for world maps to minimize shape and area distortion. Data from Data and Maps for ArcGIS 2010, courtesy of ArcWorld supplement and Esri.

The Goode homolosine projection is a combination of the Mollweide and sinusoidal projections. There is no distortion along the central meridians of the interrupted lobes where areas are represented accurately. Data from Data and Maps for ArcGIS 2010, courtesy of ArcWorld supplement and Esri.

World Bonne is an equal-area projection that has true scale along the central meridian and all parallels. Shapes are distorted as you move away from these lines. Data from Data and Maps for ArcGIS 2010, courtesy of ArcWorld supplement and Esri.

For more information about coordinate systems and map projections, consult ArcGIS Desktop Help. On the Contents tab, browse to Map > Map projections > What are map projections?

Projecting data in ArcMap

ArcMap is able to handle an infinite number of map projections—both predefined and customized—to give you the most flexibility in working with your datasets. Your data will be either unprojected or projected. Unprojected data means that it is simply stored in a defined geographic coordinate system, such as WGS84, with latitude and longitude coordinate values shown in decimal degrees, or map units. You will find out how to look up the coordinate system of a dataset in the exercises in this chapter.

Projected data means that a map projection has been applied to your dataset. The definition of a projected coordinate system includes the projection it is based on, the measurement units, and other details. Each projected coordinate system is also associated with a particular geographic coordinate system. Typically, the map units are converted from latitude-longitude to x,y coordinates in meters or feet. Additional information that can be found with a defined projection includes the central meridian, false easting and false northing, standard parallel, and central parallel.

1
2
3
4
5
6
7
8
9
10
11
12
13
14
15
16
17
18
19
20

Common projections used in ArcMap are the state plane and Universal Transverse Mercator (UTM) systems. The State Plane Coordinate System (SPCS) divides each state into manageable regions for large-scale mapping and bases it on three projections: Lambert conformal conic, transverse Mercator, and oblique Mercator. The UTM system divides the world into 60 zones–each one is 6 degrees and designated by a number and a letter N (northern hemisphere) or a letter S (southern hemisphere). The UTM system works best for maps that include small areas that are entirely encompassed by a zone. It attempts to minimize distortion, is useful for mapping areas that extend in a north-south direction, and provides a standardized method for projecting data. Depending on the geographic location of your data, a suitable map projection is most likely available in ArcMap.

You can locate the coordinate system of a dataset in ArcCatalog, on the Description tab (to view its metadata). You can also find it in ArcCatalog on the feature Properties dialog box (right-click a feature class and click Properties), on the XY Coordinate System tab. In ArcMap, you can find it on the Layer Properties dialog box, on the Source tab. Alternatively, you can find coordinate system information for a data frame in ArcMap by clicking the Coordinate System tab on the Data Frame Properties dialog box.

This data frame has an Albers equal-area conic projected coordinate system. The projected coordinates were derived from its NAD83 geographic coordinate system.

The first layer added to a data frame in ArcMap sets a precedent for the coordinate system used in that data frame, meaning that additional layers will have to be projected to it. When the first layer is added, it is displayed according to the coordinate values of its features—geographic or projected. When a second layer is added, ArcMap reads its coordinate system and determines whether it matches the coordinate system of the data frame, based on the first layer. If it matches, there's no problem. If it doesn't, ArcMap attempts to automatically resolve the conflict with a map projection. Using a mathematical operation that recalculates geographic coordinates to projected coordinates, or vice versa, ArcMap determines which equation (forward or backward) is suitable for the projection. So if the second layer you add has a different coordinate system from the first, ArcMap recalculates the layer's coordinates to match the data frame. It is important to note that only the display position of the data is being changed, and not the data itself. This process is called on-the-fly projection.

On-the-fly projection doesn't permanently alter the feature coordinates of the dataset on disk or the dataset's coordinate system information. On-the-fly projection affects only a single data frame. It works best when layers are based on the same geographic coordinate system. If any layer added to the map has a different geographic coordinate system from the data frame, you will get a message such as the example in the graphic.

ArcGIS warning for conflicting geographic coordinate systems.

The message tells you that ArcMap can't convert one geographic coordinate system to another without additional parameters that you must set. It can display the layer, but the spatial alignment may be slightly off. This difference is usually too small to notice at the scale of a world map, or even a continental map. But if you are making a map that demands highly accurate feature positioning, you will need to perform a geographic transformation, or a datum transformation.

Projecting data on the fly is very useful and is an essential part of ArcMap. However, it may slow the redrawing of the map on screen because of the complex projection equations that are required. It is also not permanent, which may satisfy certain uses, but it can standardize and simplify workflows to have data in the appropriate projection from the start. ArcToolbox has all the tools required to project data from one coordinate system to another, and then save on disk.

Projections and Transformations tools in ArcToolbox.

Under the Data Management > Projections and Transformations toolset, the Project tool allows you to transform the map projection of a dataset from one coordinate system to another. It converts the x,y coordinates from one layer to another using transformation instructions defined in the software. This operation outputs a new layer, retaining the original dataset. You can use the Project tool in both ArcCatalog and ArcMap. You will learn more about how it works in the exercises in this chapter.

Lining Up Data in ArcGIS: A Guide to Map Projections, second edition, by Margaret M. Maher (Redlands, CA: Esri Press, 2013), is an excellent reference for any GIS user looking to learn techniques to solve data misalignment, dive deeper into data projections, and create custom projections to align data.

The Define Projection tool overwrites the coordinate system information (map projection and datum) stored within a dataset. It should be used only on datasets that have an unknown or incorrect coordinate system defined. Also, you must know the correct coordinate system of the dataset to successfully define a projection. This tool should not be confused with the Project tool.

Data

This chapter uses the following datasets:

- Canada_Mexico.shp–polygons of areas that make up Canada and Mexico *(Source: ArcWorld Supplement)*
- Cities.shp–points of cities across the United States *(Source: US Census)*
- ColoradoCounties.shp–polygons of areas that consist of counties in the state of Colorado *(Source: Esri, derived from Tele Atlas, US Census, Esri [Pop2010])*
- Lakes.shp–polygons of lakes in North America *(Source: ArcWorld)*
- Latlong.shp–lines of latitude and longitude *(Source: Esri)*
- USstates.shp–polygons of areas that consist of states in the United States *(Sources: ArcUSA, US Census, Esri [Pop2010])*

Exercise 6a
Examine coordinate systems

US Census Bureau statistics are updated every decade. For this exercise, create a map of the United States that shows the change in population per square mile between 2000 and 2010. First, use ArcCatalog to explore where the coordinate information is found and to gain some familiarity with the properties of map projections.

1. **Start ArcCatalog. In the Catalog tree, double-click Folder Connections and browse to the contents of your C:\EsriPress\GTKArcGIS\Chapter06\Data folder.**

2. **Click USstates.lyr, and click the Preview tab.**

If you don't see the file extension, on the menu bar, click Customize > ArcCatalog Options, and uncheck the Hide File Extensions check box.

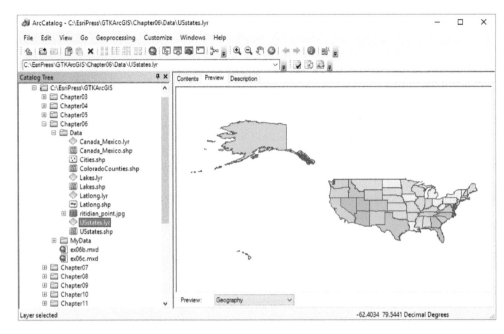

The US states layer is symbolized using a color ramp that shows the change in population (per square mile) in the last decade—light green for smaller increases and dark green for larger increases. You may notice that the appearance of the United States is distorted; its shape and area may look strange compared to other maps you may have seen in an atlas or in a published map. This is because the data is being displayed according to its geographic coordinate system and not a common US map projection such as Lambert conformal conic. Notice that the northern boundary of the United States, also known as the 49th parallel, is displayed as a straight line. Alaska has an east-west stretch that is not proportional, and Texas looks much smaller than its actual size.

3. Click the Preview arrow in the lower-left corner of the window, and click Table. Scroll to the right.

	POP2000	POP2010	POP00_SQMI	POP10_SQMI	SQMI	POP_CHANGE	
▶	1211537	1309580	110.8	119.8	10931	8.1	
	5894121	6756150	82.7	94.8	71297	14.6	
	902195	983932	6.1	6.7	147038	9.1	
	1274923	1338645	32.2	33.8	39555	5	
	642200	662194	9.1	9.4	70698	3.1	
	754844	827263	9.8	10.7	77115	9.6	
	493782	548154	5	5.6	97813	11	
	5363675	5741617	89.7	96	59822	7	
	1293953	1581697	15.5	18.9	83570	22.2	
	608827	626078	63.3	65.1	9614	2.8	
	4919479	5334772	56.7	61.4	86835	8.4	
	3421399	3865839	34.8	39.3	98381	13	
	1235786	1329915	132.2	142.2	9350	7.6	
	2926324	3057995	52	54.3	56272	4.5	
	6349097	6555736	622.1	642.3	10206	3.3	
	1711263	1822473	22.1	23.6	77352	6.5	

1 ▶ ▶I ▤ ▦ (of 51)

Preview: Table ▼

The table has 2000 and 2010 population figures for each state, including population per square mile (POP00_SQMI and POP10_SQMI) and the change in population per square mile (in percentages) represented by the POP_CHANGE attribute. Every state experienced population growth during the past decade. The layer is classified and symbolized based on the POP_CHANGE attribute.

*How was the POP_CHANGE attribute calculated? (**Hint:** It involves two attributes and a simple math formula.)*

Next, you will confirm that the US states shapefile has a geographic coordinate system.

4. **In the Catalog tree, right-click USstates.shp, and click Properties. On the Shapefile Properties dialog box, click the XY Coordinate System tab.**

The Shapefile Properties dialog box shows that the current coordinate system is called GCS_North_American_1983. This is a common geographic coordinate system for spatial data covering the United States and is used by many data providers. The datum it uses is called D_North_American_1983 (NAD 1983), and the prime meridian is defined at 0 degrees longitude. It is from this dialog box that the parameters of a coordinate system can be modified. However, you will not be modifying this particular layer.

What are the angular units of this coordinate system?

What is listed as the prime meridian?

There are four other shapefiles related to this exercise in the Data folder: Cities, Latlong, Canada_Mexico, and Lakes. The latter three shapefiles also have a corresponding layer file that contains a symbolization scheme and a path to the source data. You will view the properties of the first three layers. (If you want, you could explore the fourth, but only the first three are necessary to show the differences between them.)

5. **Click Cancel to close the USstates Shapefile Properties dialog box. In the Catalog tree, double-click Cities.shp to open its properties.**

The Shapefile Properties dialog box shows that this layer has a geographic coordinate system called GCS_WGS_1984 (WGS 1984). It is another popular geographic coordinate system that is in wide use. Since it is different from the coordinate system of the US states layer, it may not align exactly. ArcMap, however, is designed to make it work.

What spheroid does this layer use?

In exercise 6b, you will see how ArcMap reacts to these coordinate system inconsistencies, and how to solve the problem.

6. **Click Cancel to close the Cities Shapefile Properties dialog box. Open the properties for Latlong.shp.**

The Shapefile Properties dialog box shows that this layer is in the GCS_North_American_1983 coordinate system, like the US states layer. The data should align properly with the US states layer without any coordinate system transformation or projection on the fly.

7. Click Cancel to close the Latlong Shapefile Properties dialog box. In the Catalog tree, click Canada_Mexico.shp, and preview its geography. Then open its properties.

TIP

In ArcCatalog, the coordinate system information can be viewed in the data file properties but not in the related layer file (.lyr), if there is one.

Although a preview of Canada and Mexico can be seen, the Shapefile Properties dialog box shows that this layer is not in any geographic coordinate system and simply states that it is unknown. How is this possible? ArcCatalog still shows a preview of the actual data and reads the known feature coordinate values, even though a geographic coordinate system is not defined. If this file is added to ArcMap, it will be drawn, but it will not be projected because its geographic coordinate system is unknown. You will fix this situation in ArcCatalog.

In the Preview display, notice that the southern border of Canada has a long straight line similar to the northern border of the US states layer. Although it may not be the case in every situation, you have now been provided with a hint that the Canada and Mexico shapefile shares the same geographic coordinate system as the US states shapefile.

TIP

In many cases, trying to estimate the geographic coordinate system of a dataset that has none defined is based on trial and error, using best estimates. If the data is in the United States and shows a geographic extent in which the coordinates to the left of the decimal point are six, seven, or eight digits, the data is probably projected to the state plane or UTM coordinate system. You can view the feature extents in ArcCatalog (Shapefile Properties > Feature Extent tab) or in ArcMap (Layer Properties > Source tab). It is also possible to import the coordinate system of another feature layer that contains a defined state plane or UTM boundary and apply it to your shapefile or feature class with an unknown coordinate system to confirm whether it aligns properly.

8. On the Canada_Mexico Shapefile Properties dialog box, on the XY Coordinate System tab, click the Add Coordinate System button ⊕, and click Import. Browse to ..\Chapter06\Data, click USstates.shp, and click Add.

The Canada_Mexico shapefile now assumes the coordinate system information of the US states layer–GCS_North_American_1983. This means that when the shapefile is added to ArcMap, it will be recognized and will not have issues of alignment or need further geographic projection. If needed, view its coordinate system information again in its Shapefile Properties.

9. Click OK to accept the change.

10. To exit ArcCatalog, on the menu bar, click File > Exit.

It is common to find data in all sorts of coordinate systems and projections, or even data that is missing coordinate system information altogether. Learn more about how a shapefile or geodatabase stores coordinate information.

Now that you have a better understanding of the data and coordinate systems involved in this chapter, you are able to make proper decisions on how to handle them in exercise 6b.

TIP

You can also view feature class properties and coordinate system information from the Catalog window in ArcMap.

Exercise 6b
Project data

The map you create in this exercise consists of a shapefile of the United States displayed in three data frames, with a different projected coordinate system for each data frame (one for the mainland states, one for Alaska, and one for Hawaii). The data frames will also include other supporting layers: one that shows latitude and longitude lines and another that shows Canada and Mexico to complete the look of the map.

How is it possible to use one dataset in three data frames? Once data layers are added to ArcMap, they can be moved and copied between data frames. As you learned in chapter 3, multiple layers (even in different data frames) can be created from a single source dataset. This is more efficient than adding separate data layers with separate projected coordinate systems to the same map and is more practical than showing all 50 states in the same data frame. If that were the case, there would be many files to manage, and you would have to zoom to a map scale that would make it difficult to see the detail of all the states together.

In this exercise, you will use three versions of the Albers equal-area conic projection—one for each data frame. This is a popular projection of the United States because it maintains area and keeps distortion of the shapes of the states to a minimum.

1. **Start ArcMap and open the ex06b map document from your ..\GTKArcGIS\Chapter06 folder.**

 ArcMap opens a layout that has three empty data frames: Alaska, Mainland States, and Hawaii. Each data frame is identified by a label at its center. An active data frame is indicated in the table of contents by a data frame name in boldface and on the layout page by a dashed outline. A small legend that shows the percentage increase of population change is visible in the lower-right corner of the page.

 You will add the US states layer to each data frame and learn how map projections apply to each one. You will also learn how to add other supporting layers.

2. **Add USstates.lyr to the Mainland States data frame. (Use the Add Data button on the Standard toolbar or drag the layer file from the Catalog window.)**

1
2
3
4
5
6
7
8
9
10
11
12
13
14
15
16
17
18
19
20

TIP

When using the Catalog window, you may need to refresh the map display after dragging over data. To do so, click the Refresh button in the lower-left corner of the display area.

The US states layer is added to the Mainland States data frame. It is in the GCS_North_American_1983 geographic coordinate system that you previewed in ArcCatalog. By default, ArcMap includes the entire spatial extent of the layer. Next, you will add a few more layers before projecting the data and adjusting the map scale.

3. Add Canada_Mexico.lyr to the Mainland States data frame, and position it below the US states layer.

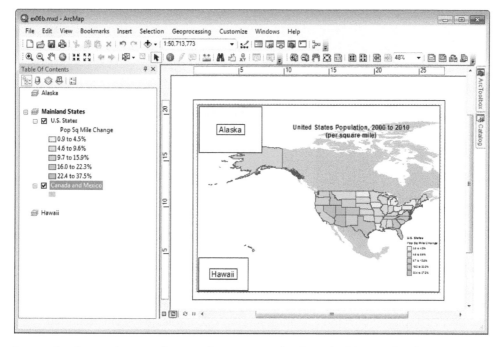

In exercise 6a, you imported a coordinate system for Canada_Mexico.shp. Because it is the same coordinate system as the US states layer, there are no alignment or accuracy issues.

TIP

If you didn't import a coordinate system for Canada_Mexico.shp, an Unknown Spatial Reference dialog box will appear because there is no coordinate system information. If this is the case, return to step 8 of exercise 6a.

The Canada and Mexico layer is styled in a light-gray fill color to act as a subtle background to the thematic US states layer. Next, you will add two more layers to the data frame.

4. Add Lakes.lyr to the Mainland States data frame. It should be placed above all the layers in the table of contents. The Lakes layer, symbolized using a white color, is used to add a visual aesthetic to the map.

5. Add Latlong.lyr, and drag it below all the other layers in the Mainland States data frame in the table of contents.

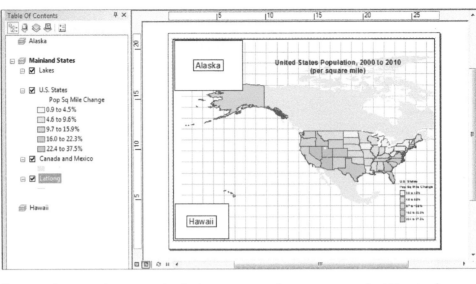

Because these two layers are also in the same coordinate system as the US states layer, everything lines up as it should. The Latlong layer provides reference for the geographic coordinates displayed on screen. It represents the x,y coordinate system from −180 degrees to 180 degrees longitude and −90 degrees to 90 degrees latitude, and the lines are perpendicular to each other at 90 degrees. However, viewing the world in this type of coordinate system shows significant distortion, especially in shape and area.

Use the Measure tool to measure the distance between two lines of latitude or two lines of longitude. What is the difference in degrees?

6. Add Cities.shp to the Mainland States data frame.

Geographic Coordinate Systems Warning ✕

The following data sources use a geographic coordinate system that is different from the one used by the data frame you are adding the data into:

Data Source	Geographic Coordinate System
Cities	GCS_WGS_1984

Alignment and accuracy problems may arise unless there is a correct transformation between geographic coordinate systems.

You can use this button to specify or modify the transformation(s) used by this data frame: [Transformations...]

The Transformations dialog can also be accessed from the Data Frame Properties dialog's Coordinate Systems tab after you have added the data.

☐ Don't warn me again in this session
☐ Don't warn me again ever

About the geographic coordinate systems warning [Close]

A warning appears when the layer is added. The message tells you that this layer has a different geographic coordinate system from the data frame. You previewed this layer in ArcCatalog and know that it is in the GCS_WGS_1984 geographic coordinate system. When you added the US states layer to the Mainland States data frame, the data frame was automatically set to GCS_North_American_1983. Even though the Cities layer has a different coordinate system, ArcMap can still display it, but the spatial alignment may be slightly off. At this map scale, it may not even be noticeable. However, it may cause further problems when it is projected.

7. **Read the messages on the warning dialog box, and click Close.**

TIP

If you don't see this warning, it's likely that you checked the Don't Warn Me Again Ever check box.

8. **In the table of contents, double-click Cities. On the Layer Properties dialog box, click the Source tab and look at the Data Source panel.**

This is a quick verification that the Cities layer is in the GCS_WGS_1984 geographic coordinate system. You will use an ArcToolbox tool to transform it from its current coordinate system to the same coordinate system as the US states layer, GCS_North_American_1983.

9. Click Cancel to close the Layer Properties dialog box.

10. On the Standard toolbar, click the ArcToolbox button.

11. In the ArcToolbox window, expand the Data Management Tools > Projections and Transformations toolset.

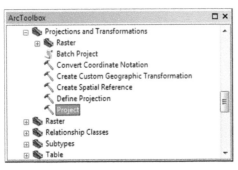

12. Double-click the Project tool to open its dialog box.

13. Click the Input Dataset or Feature Class arrow, and click Cities. Click the Browse button next to the Output Dataset or Feature Class box, and browse to your ..\GTKArcGIS\ Chapter06\MyData folder. In the Name box, type **Cities_GCS_NAD1983.shp**, and click Save.

TIP

The Input Dataset or Feature Class list contains layers that can be projected. Alternatively, the Browse button allows you to browse to the data source.

14. Click the Properties button next to the Output Coordinate System box.

On the Spatial Reference Properties dialog box, you can choose from coordinate systems defined by ArcGIS, import a coordinate system from another dataset, or create one. (The latter two options are found using the Add Coordinate System button.) You need to project the Cities layer to GCS_North_American_1983.

15. In the browse panel, expand the Geographic Coordinate Systems > North America > USA and territories folder, and scroll down and click NAD 1983. Click OK.

TIP

You can also choose the coordinate system from your Favorites list.

The Current coordinate system panel on the Spatial Reference Properties dialog box is updated with the new coordinate system details.

16. Click OK to close the Spatial Reference Properties dialog box.

ArcMap automatically populates the Project dialog box with an appropriate transformation using the coordinate system you selected. However, you might need to use your knowledge of the data (for example, are you focusing on the 48 contiguous states?) to choose an appropriate transformation code. The geographic transformation option, in this case, provides ArcMap with the correct transformation to convert WGS84 to NAD83.

The Geographic Transformation box is optional. The Project tool can project data from PCS to PCS (for example, state plane to UTM), transform GCS to GCS (NAD83 to NAD27), or PCS to GCS, and vice versa. In the event you are projecting data from PCS to PCS and they share the same datum, you are not transforming the data and therefore do not need a transformation code.

17. Click OK.

The Project tool runs, and when it is finished, a pop-up notification briefly appears in the notification area, at the far right of the taskbar, indicating the process was successful.

ArcMap adds the transformed Cities layer to the map. It may not look much different, but it is in a new coordinate system. It is important to remember that the new Cities layer now shares the same coordinate system as the other layers, and changing the coordinate system of the data frame will not yield the warning messages you saw before.

1
2
3
4
5
6
7
8
9
10
11
12
13
14
15
16
17
18
19
20

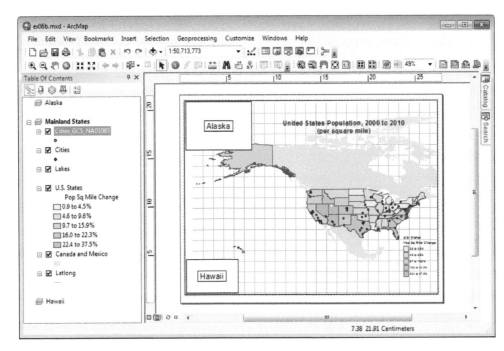

18. Close ArcToolbox or move it out of the way. In the table of contents, right-click Cities (not Cities_GCS_NAD1983), and click Remove.

This layer is not needed anymore because you now have a cities layer with the appropriate coordinate system to use. Next, you will change the coordinate system of the data frame to project layers on the fly.

19. In the table of contents, double-click the Mainland States data frame. On the Coordinate System tab, expand the Projected Coordinate Systems > Continental > North America folder.

A variety of North American map projections are listed.

20. Scroll down and click USA Contiguous Albers Equal Area Conic, and click OK.

The coordinate system of the data frame is now changed, and all the layers are projected on the fly to the USA Contiguous Albers Equal Area Conic projection. The lines of latitude and longitude exhibit a curvature similar to the shape of a cone. The overall shape of the US states layer has changed and may look more familiar to you.

*Which lines of latitude are defined as standard parallels for this projection? (**Hint:** View the data frame properties.)*

1
2
3
4
5
6
7
8
9
10
11
12
13
14
15
16
17
18
19
20

21. On the menu bar, click Bookmarks > Mainland States.

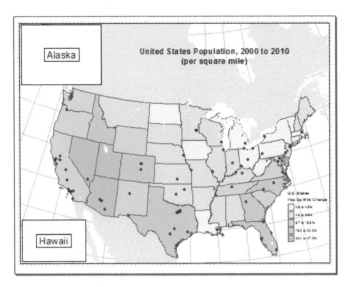

The map scale is changed to 1:20,000,000, and the layers are centered in the data frame. From here, you will begin the process of copying the data layers to the Hawaii and Alaska data frames.

22. In the table of contents, hold down Ctrl and click the Cities_GCS_NAD1983 layer, U.S. States layer, and the Canada_Mexico layer to select them. Right-click any selected layer, and click Copy. While still in the table of contents, right-click the Alaska data frame, and click Paste Layer(s).

23. In the Alaska data frame layer list, hold down Ctrl and click Canada_Mexico to remove it from the selected layers. Then, right-click either of the two selected layers, and click Copy. Finally, right-click the Hawaii data frame, and click Paste Layer(s) to move the Cities_GCS_NAD1983 and U.S. States layers to the Hawaii data frame.

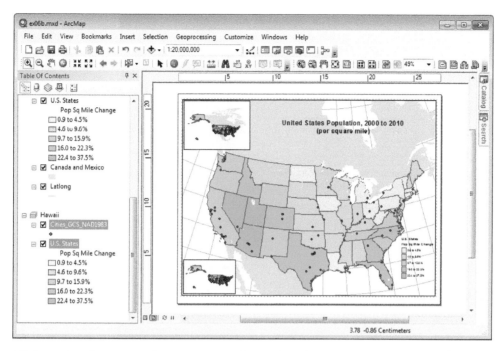

All three data frames now have content in them. The Alaska and Hawaii data frames have the data displayed in the GCS_North_American_1983 geographic coordinate system. Next, you will give these data frames map projections that are specifically designed for these areas of the United States.

24. In the table of contents, double-click the Hawaii data frame. On the Coordinate System tab, expand the Projected Coordinate Systems > Continental > North America folder, click Hawaii Albers Equal Area Conic, and click OK. A Geographic Coordinate system warning may appear here—dismiss it and continue the exercise.

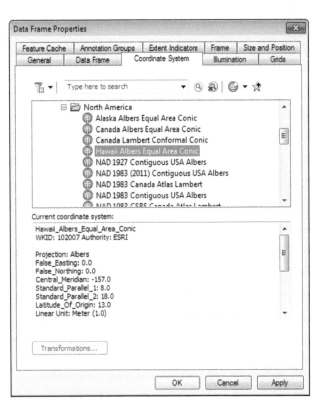

The layers are projected on the fly. The mainland states look as if they are rotated. This projection minimizes distortion, specifically for Hawaii. The Canada_Mexico layer was not included in this data frame because you wouldn't be able to see it once the map scale and position of the data frame are adjusted.

25. Activate the Hawaii data frame (right-click it in the table of contents, and click Activate). On the menu bar, click Bookmarks > Hawaii.

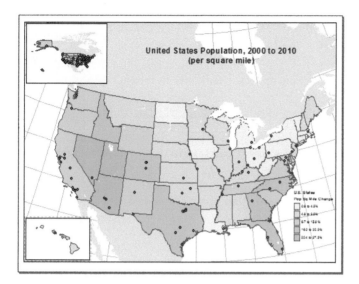

26. Repeat the same steps to project the Alaska data frame to Alaska Albers Equal Area Conic. Activate the Alaska data frame, and zoom to Alaska using the bookmark.

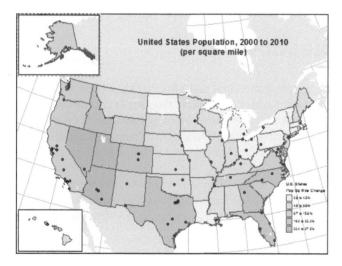

Each of the three data frames has a variation of the Albers Equal Area projected coordinate system. Data frame properties can be adjusted at any time. It is also possible to work on individual data frames in Data View mode. To change the map display, activate a data frame.

1
2
3
4
5
6
7
8
9
10
11
12
13
14
15
16
17
18
19
20

ON YOUR OWN

Examine other projected coordinate systems that could be applied to the Mainland States data frame. Change the map projection, and see how it affects the four map properties: shape, area, distance, and direction.

27. To save your work, rename your map document and save to your MyData folder for chapter 6.

28. If you are continuing to exercise 6c, leave ArcMap open. Otherwise, exit the application and click No if prompted to save your changes.

Exercise 6c
Define a map projection

Gathering data from various sources can turn into a large task very quickly. It is even more challenging when you realize some of the data does not include the proper coordinate system information. For shapefiles, this information is stored in a separate projection file with the extension.prj. For geodatabases, coordinate system information is stored as a spatial reference, which is set as a property of a stand-alone feature class or of a feature dataset.

ArcMap reads feature coordinate values when a layer is added. When the coordinate system information is missing, it attempts to display the layer based on these x,y coordinate values. If the x-values have a range of −180 to 180 and the y-values have a range of −90 to 90, it can be assumed that the layer is in a geographic coordinate system.

Unfortunately, ArcMap won't know which one, so you will have to assign one. If you cannot determine the correct coordinate system by researching the metadata or questioning the data source, you will have to make an educated guess, and then compare the results with another dataset to see whether it aligns properly. Alternatively, if the x,y coordinates are six-, seven-, or eight-digit numbers, the layer could be in a projected coordinate system, but again, the software can't determine which one. This leads ArcMap to display a warning message saying that the data sources you added are missing spatial reference information.

ArcMap can display the layer, but it cannot project its coordinate system. This means the layer will be displayed incorrectly, and its units will be unknown. As is usually the case when you use trial and error, the results are unpredictable. If a layer is added to a data frame set to a projected coordinate system, the layer may be displayed, but it may be seriously misaligned with the data that is in the correct location on the map. Other times, misalignment may be only minimal, a result that might be acceptable depending on the accuracy required.

In ArcToolbox, the tool used to assign coordinate system information to a dataset is called Define Projection. This tool overwrites any existing coordinate system information (map projection and datum) that is stored and should be used only if the coordinate system information is unknown or incorrect.

The most important point is, you must know the correct coordinate system of the dataset before using this tool. To find the missing coordinate system information, you often have to contact the data provider and look through the metadata and supporting documentation.

1. **Start ArcMap. Open the ex06c map document from your ..\GTKArcGIS\Chapter06 folder.**

 The map is open in layout view. There are two data frames, each in a projected coordinate system: state plane and Universal Transverse Mercator (UTM). The State Plane data frame is active. Both frames include a U.S. states layer, zoomed to the state of Colorado. Notice the slight difference in curvature of the state boundaries. The left data frame is projected to the Colorado North State Plane Coordinate System, and the right data frame is projected to the UTM Zone 13 Coordinate System.

1
2
3
4
5
6
7
8
9
10
11
12
13
14
15
16
17
18
19
20

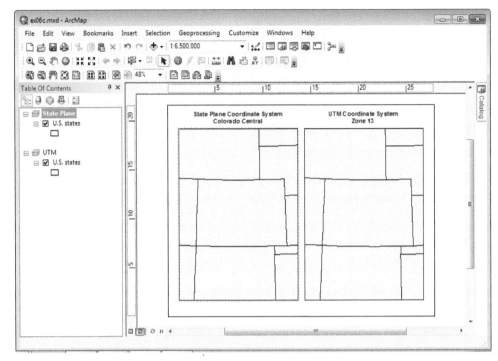

The State Plane Coordinate System (SPCS) was developed in the 1930s by the US Coast and Geodetic Survey of the National Oceanic and Atmospheric Administration to provide a conformal mapping system for the country. It is commonly used for county and statewide surveying and data applications. It is not recommended, however, for use across state lines. To maintain accuracy and minimize distortion, larger states have multiple zones.

The UTM divides the earth into 60 zones, each 6 degrees of longitude wide. There is almost no distortion at the center of each zone and gradually increasing distortion outward from the center. UTM grid coordinates are expressed as a distance in meters east of the central meridian, known as "easting," with an assigned value of 500,000 meters, and a distance in meters north of the equator, known as "northing," with an assigned northing value of 0 meters.

ON YOUR OWN

Explore more about the differences and uses of the state plane and UTM coordinate systems.

2. Add ColoradoCounties.shp from your ..\GTKArcGIS\Chapter06\Data folder to the State Plane data frame.

ArcMap displays a warning that the layer is missing spatial reference information.

TIP

If you don't see the warning, you may have dismissed these types of warnings in a prior ArcMap session. To reset the default behavior of coordinate system warnings, close ArcMap and browse to C:\Program Files\ArcGIS\Desktop10.8\Utilities. Run the AdvancedArcMapSettings.exe file. On the ArcMap Advanced Settings Utility dialog box, click the TOC/Data tab and make sure the Skip Datum Check check box is not checked.

3. Click OK to close the warning message.

You may notice quickly that something is wrong: ArcMap added the layer (you can see it in the table of contents), but it is not visible in the State Plane data frame. Because the Colorado counties layer did not have defined coordinate system information, ArcMap does not know where to properly place it.

4. Right-click the Colorado counties layer, and click Zoom To Layer. On the Tools toolbar, click the Fixed Zoom Out button several times until you can see surrounding states.

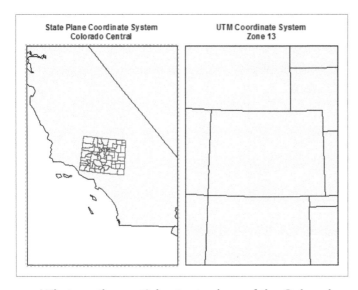

What are the spatial extent values of the Colorado counties layer?
*(**Hint:** View the Source tab on the layer properties dialog box.)*

Since ArcMap had no information to use, it drew the Colorado counties layer using a projection with latitude and longitude coordinates as if they were x,y coordinates in the state plane coordinate system. As a result, the layer is being placed in California, which is obviously incorrect. In a situation such as this, it is up to you to find the appropriate coordinate system information. For this exercise, you are given the information that the coordinate system of Colorado counties is Albers equal-area conic. Next, you will define the correct coordinate system information for the Colorado counties layer using ArcToolbox.

5. On the Standard toolbar, click the ArcToolbox button.

6. In the ArcToolbox window, expand the Data Management Tools > Projections and Transformations toolset. Double-click the Define Projection tool.

7. Click the Input Dataset or Feature Class arrow, and click Colorado counties.

> ## TIP
>
> The Define Projection tool differs from the (Feature) Project tool used in exercise 6b. The Project tool projects spatial data from one coordinate system to another. The Define Projection tool overwrites the coordinate system information (map projection and datum) stored with a dataset and should be used only for datasets that have an unknown or incorrectly defined coordinate system. If this tool is used on data that already has a defined coordinate system, you will be warned that running the tool will change the coordinate system.

The coordinate system is listed as unknown. So next, you will browse for the coordinate system information to apply to the layer.

8. Click the Properties button next to the Coordinate System box.

9. On the Spatial Reference Properties dialog box, expand the Projected Coordinate Systems > Continental > North America folder. Scroll to and click USA Contiguous Albers Equal Area Conic.

The Current coordinate system panel is updated. Notice that the geographic coordinate system used is GCS_North_American_1983. (You may have to scroll down to see it.) This is the same system as in the State Plane data frame.

10. Click OK.

This coordinate system information will be saved to a projection file (.prj) associated with the Colorado counties shapefile.

11. Click OK to run the Define Projection tool.

While the Define Projection tool is running, ArcMap locks the layer so it cannot be accidentally modified or moved. A notification indicates the process was successful. ArcMap projects the Colorado counties layer to the Colorado North State Plane Coordinate System projection. The State Plane data frame is updated, and the Colorado counties layer disappears from the map display.

TIP

If you still see ColoradoCounties being drawn in California, click the small Refresh button at the bottom of the display area.

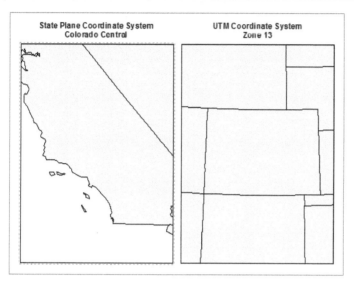

12. On the menu bar, click Bookmarks > Colorado State.

The counties are now aligned with the U.S. states layer.

13. In the table of contents, drag the Colorado counties layer to the UTM data frame, above the US states layer.

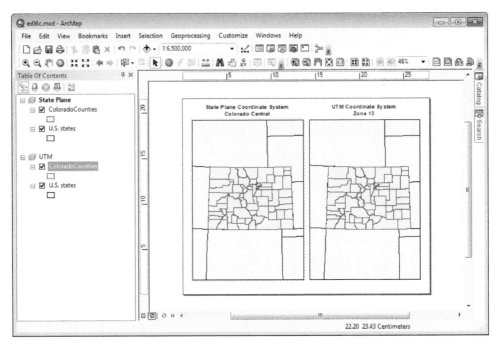

The Colorado counties layer is copied to the UTM data frame and is projected to UTM Zone 13. This is possible because the projected UTM coordinate system is also based on the GCS_North_American_1983 geographic coordinate system.

How does mapping at different map scales using SPCS and UTM coordinate systems affect the overall distortion and accuracy of your map?

14. To save your work, give it a new name and save to your MyData folder for chapter 6.

15. If you are continuing to exercise 6d, leave ArcMap open. Otherwise, exit the application and click No if prompted to save your changes.

Exercise 6d
Georeference a raster

Georeferencing is a GIS task frequently performed on scanned paper maps (raster maps) to align them to a known coordinate system. After a map is georeferenced, it can be used for further data creation (digitizing features) or simply used as another GIS raster layer (commonly as a base layer).

Before using ArcMap georeferencing tools, you must first check the map for available metadata including information about the coordinate system or projection and lines of longitude and latitude. Second, you must identify several usable ground control points on the coordinate grid that can be used to calculate x,y coordinates based on the map projection. The information you find will help provide every pixel in the scanned image a geographical dimension. For example, ArcMap can assign a pixel that has an x,y coordinate of 543, 237, the real-world coordinates of 12° 27′ E and 41° 54′ N.

1. **Open a new blank map in ArcMap. Open the Catalog window and browse to the GTKArcGIS\Chapter06\Data folder. Right-click ritidian_point.jpg, and click Item Description.**

 The Item Description dialog box appears.

2. **Click the Preview tab. Use the navigation tools to explore the map. (ArcMap may prompt you to create pyramids. Create the pyramids and continue.)**

TIP

Using pyramids can speed up the display of raster data by retrieving only the data at a specified resolution required for the display.

An image of a scanned topographic map of Ritidian Point, Guam, is shown on the Preview tab. As it sits, the raster does not have any geospatial information. If it was imported into ArcMap, it would simply have a Cartesian plane coordinate system applied based on the pixel dimension of the image. To use such an image in a GIS, you must first georeference it.

You will want to assign the Ritidian Point image the correct projection system that it represents, and thus change the coordinate system to represent this projection. This involves (1) specifying the image's coordinate system in Catalog, (2) specifying the coordinate system for the image by digitizing specific control points, and (3) rectifying or "stretching" the scanned image to fit the new projection.

3. Zoom in to view the text in the lower-left corner of the map.

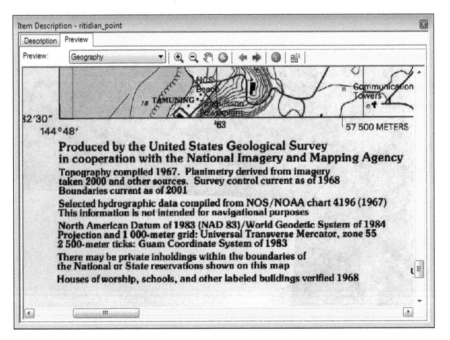

In the fourth paragraph, information is given about the map's coordinate system and projection. It is important to know this information before georeferencing the image. Also, scroll to the right to see the map's scale and other information.

What coordinate system and projection is the map in?

4. Close the Item Description window. In the Catalog window, right-click ritidian_point. jpg, and click Properties.

5. On the Raster Dataset Properties dialog box, locate the Spatial Reference property and click the corresponding Edit button.

6. On the Spatial Reference Properties dialog box, expand Projected Coordinate Systems > UTM > WGS 1984 > Northern Hemisphere, and click WGS 1984 UTM Zone 55N.

7. Click OK to close the Spatial Reference Properties dialog box.

The Spatial Reference property is updated with the chosen projected coordinate system information. Now that ArcGIS knows what projection the image is in, the next step is to georeference the image.

8. Click OK to close the Raster Dataset Properties dialog box.

9. From the Catalog window, drag ritidian_point.jpg to the data view. (Click Yes if prompted to create pyramids.)

10. On the main menu bar, click Customize > Toolbars > Georeferencing.

The Georeferencing toolbar has a collection of drop-down items and interactive tools to allow you to georeference raster datasets. To georeference your raster, you will identify a series of ground control points–known x,y coordinates–that link locations on the raster with locations in the spatially referenced data. Ground control points are locations that can be accurately identified on the raster dataset and in real-world coordinates. Once several control points have been identified, ArcMap will calculate–using a polynomial transformation–and shift the raster dataset from its existing location to the spatially correct location.

11. On the Georeferencing toolbar, click the Viewer button 🖾, and click the View Link Table button 🖽. Resize both windows so that you can still see the main map.

12. Using the graphic as a reference, use the Image Viewer and zoom in to the data frame, and find the intersection of the grid lines indicated by label 1. (The labels are just for reference and don't appear on the actual raster.) On the Georeferencing toolbar, click the Add Control Points button ⚹.

13. Using either the Image Viewer or the Data View of the main ArcMap window, click the intersection of the grid lines once to add the ground control point, and immediately click again in the same position. (If you moved the pointer to another position, right-click and click Cancel Point, and try again.)

The second click links the ground control point to the real-world coordinate. Notice in the Link window that the first link has been created and that the X and Y Source values are the same (or very close to the same) as X and Y Map values. This happens because ArcMap does not yet know the real-world coordinates indicated on the map. You will change the X and Y Map values using information from the raster.

14. On the Image Viewer, follow the grid line and pan to the bottom boundary of the map. Note the grid line value listed. Pan back to the control point, and pan to the left boundary of the map. Note the grid line value listed.

The bottom boundary grid line value is listed as 265–in UTM units, this means it is 265,000 meters. The left boundary grid line value is listed as 1500–or 1,500,000 meters.

15. **In the Link window, select the first link, click X Map, and replace the existing value with 265000. Click Y Map and enter 1500000.**

Don't worry if your X Source and Y Source values don't match the graphics exactly– slight discrepancies are the result of human input and are expected.

The first link has been completed. The main map will disappear from the data view–this is normal. At any time, click Full Extent on the Tools toolbar to see the raster. While adding more control points, the map will look stretched as it transforms to the new coordinate information.

TIP

When you use the Pan or Zoom tools in the Image Viewer, you will have to again click the Add Control Points tool from the Georeferencing toolbar before creating another control point.

16. **Create three more links using the graphic in step 11 as a reference. The X and Y Map values you will enter are as follows:**

 - **Control point 2 (265000, 1511000)**
 - **Control point 3 (271000, 1511000)**
 - **Control point 4 (271000, 1500000)**

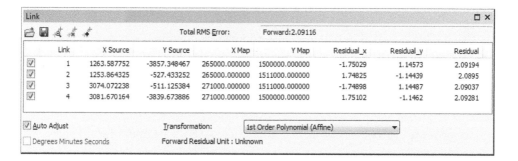

	Link	X Source	Y Source	X Map	Y Map	Residual_x	Residual_y	Residual
✓	1	1263.587752	-3857.348467	265000.000000	1500000.000000	-1.75029	1.14573	2.09194
✓	2	1253.864325	-527.433252	265000.000000	1511000.000000	1.74825	-1.14439	2.0895
✓	3	3074.072238	-511.125384	271000.000000	1511000.000000	-1.74898	1.14487	2.09037
✓	4	3081.670164	-3839.673886	271000.000000	1500000.000000	1.75102	-1.1462	2.09281

Total RMS Error: Forward: 2.09116

☑ Auto Adjust Transformation: 1st Order Polynomial (Affine)

☐ Degrees Minutes Seconds Forward Residual Unit : Unknown

Once the links are created, a measure of the error—the residual error—will be calculated and displayed. The error is the difference between where the source point ended up as opposed to the actual location that was specified. All residual errors closer to zero are considered more accurate. Although an RMS (root mean square) of nearly zero is achievable, this does not mean that the raster will be perfectly georeferenced. If you notice significant error in a link, remove it and place the control point again. This can be tedious but can lead to a more accurately georeferenced image. In a practical application, it is best to choose more control points that are evenly distributed around the raster.

17. **On the Georeferencing toolbar, click the drop-down menu and click Rectify. On the Save As dialog box, choose the Output Location as ..\GTKArcGIS\Chapter06\MyData. Then change the Name to georef_ritidianpt.**

1
2
3
4
5
6
7
8
9
10
11
12
13
14
15
16
17
18
19
20

18. **Click Save. Close the georeferencing windows. Notice the new raster in your MyData folder.**

Rectifying will create a raster dataset that is georeferenced using the map coordinates and the spatial reference. You are not required to permanently transform your raster dataset to display it with other spatial data; however, you should do so if you plan to use it in other projects, perform analysis, or share with others who use different software packages.

19. **To save your work, give it a new name and save to your MyData folder for chapter 6.**

20. **If you are continuing to the next chapter, keep ArcMap open. Otherwise, exit the application and click No if prompted to save your changes.**

Summary

In this chapter, you learned some of the fundamentals of projections and coordinate systems. You should have a good understanding of how latitude and longitude form the basis of a geographic coordinate system and how this affects locations on the curved surface of the earth. Similarly, you should have learned that map projections represent the earth on a 2D, flat surface. Every map projection introduces some level of distortion to shape, area, direction, or distance.

It is important to know that a spheroid is a mathematical model that approximates the irregular shape of the earth. A datum is used to provide an optimal fit between the geoid and the chosen spheroid used for a map. Projected coordinate systems such as state plane and UTM are widely used in the United States. Each system provides projections that are used for local and state-level mapping.

You have learned how to use ArcMap and ArcCatalog to explore and manage the coordinate system information about data. ArcMap can hold multiple data frames, and data can be moved between them. The data is projected on the fly: the dataset's coordinates are recalculated, and the display position of the data is changed to match the data frame's coordinate system.

Geoprocessing tools can be used to project data permanently as well as to define a projection for GIS data that may have information for an unknown coordinate system.

You also learned how to georeference a raster using information available on the raster to link ground control points to real-world coordinates. There are many practical uses for georeferencing. Sometimes aerial photography and satellite imagery are missing spatial information; thus, georeferencing the raster to a coordinate system will allow you to align other layers to it.

Review the learning objectives of this chapter and see if you are comfortable working with projections and coordinate systems in ArcGIS. If not, review some sections of this chapter as a reminder. In the beginning, working with projections that are unfamiliar to you may be a matter of trial and error. That's OK—it's often the best way to learn. Just be sure to make a copy of any original data as you experiment and learn.

Terms

coordinate systems
latitude and longitude
degrees-minutes-seconds
decimal degrees
geographic coordinate system
spheroid
datum
geoid
map projection
projected coordinate system
standard parallels
standard meridians
map units
false easting
false northing
on-the-fly projection
datum transformation
georeferencing

CHAPTER 7
Symbolizing features

Create custom symbology
Symbolize features by categorical attributes
Use styles and create layer files
Symbolize rasters

Learning objectives

- Change the default ArcMap symbology for points, lines, and polygons
- Reorder layers to influence effects of drawing order
- Symbolize features using the same symbol for each feature
- Change the full-extent specifications of a map
- Symbolize features based on descriptive attributes
- Create a definition query to limit the features being displayed
- Select symbols from a predefined style
- Create layer files
- Choose a new color scheme for a raster layer
- Use layer transparency effectively

Symbolizing features means assigning colors, markers, sizes, widths, angles, patterns, transparency, and other properties that will make them recognizable on a map. Symbols often look like the objects they represent—for example, a lake polygon that is blue and fire hydrants that are marked by icons that look like fire hydrants. Sometimes the relationship is less straightforward. On a street map, varying line thicknesses may be used to show whether a street is a local road, an arterial, or a highway without implying that the widths on the map are proportional to the widths of the actual streets. Symbology may also be purely conventional. Cities, for example, are commonly symbolized as circles, even though city boundaries are seldom round.

By varying symbol properties, you convey information about features.

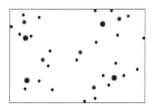

Left: Fill patterns and colors differentiate perennial lakes, intermittent lakes, and salt pans. *Center*: Line thickness, color, and solidity show a hiking trail and main, secondary, and utility roads. *Right*: Unique icons mark major and minor cities.

Symbology is also influenced by scale. A city may be a point symbol on one map and a polygon on another.

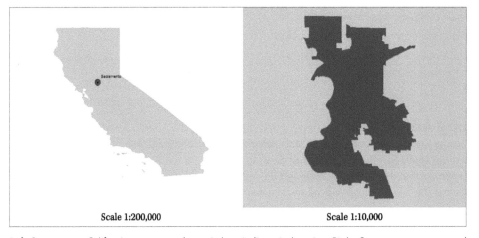

Left: Sacramento, California, represented as a circle to indicate its location. *Right*: Sacramento represented as a polygon to indicate its boundaries. Data from Data and Maps for ArcGIS 2010, courtesy of ArcWorld, ArcUSA, US Census, Esri (Pop2010 field), and the City of Sacramento, CA.

ArcGIS has thousands of symbols for common map features. Symbols are organized in the Symbol Selector by style (Environmental, Transportation, Weather, and so on) to make it easy to find the ones you need. You can also create your own symbols.

When you add spatial data to ArcMap as a layer, all features in the layer have the same symbology. Points are displayed with small circles, for instance, and polygons are blocks of color in the shape of the objects they represent. Line features are displayed using thin, simple lines. The colors for points, lines, and polygons are randomly chosen.

To change the default symbology, you use information from a chosen field in the layer attribute table.

1
2
3
4
5
6
7
8
9
10
11
12
13
14
15
16
17
18
19
20

Left: Terrestrial ecoregions in Madagascar. By default, all features (polygons, in this case) have the same symbology. Right: Each feature has a different fill color indicating its ECO_NAME attribute value. Courtesy of World Wildlife Fund (WWF). Olson, D. M. E. et al. 2001. "Terrestrial Ecoregions of the World: A New Map of Life on Earth." BioScience 51: 933-938.

You can save a layer's symbology by making a layer file (.lyr). Then, when you add that layer file to a map, the features are already symbolized the way you want them.

Symbolizing raster layers is similar to symbolizing feature layers but with fewer options. The only property that can be controlled is the color of the cells that make up the raster.

In this chapter, you will create the symbols for a Florida tourism map, and in chapter 9, you will create the labels. The completed map will look like the map shown next.

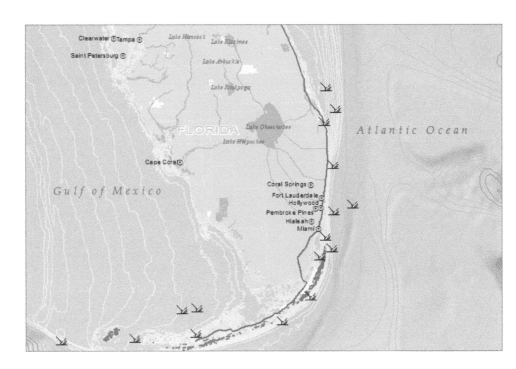

The Florida tourism map that you'll complete, including symbols, labels, and styles. Data courtesy of the Florida Geographic Data Library, the Office of Coast Survey's Automated Wreck and Obstruction Information System, FWC-FWRI, NOAA, US Census Bureau, USGS, Florida Department of Transportation, Florida Department of Environmental Protection, and Natural Earth.

Data

The data used in the following exercises includes ten feature classes (stored in the Florida geodatabase) and three raster datasets, as follows:

- Bathymetry—line features that represent changes in ocean depth *(Sources: Florida Fish and Wildlife Conservation Commission's Fish and Wildlife Research Institute and National Oceanic and Atmospheric Administration)*
- Cities—major cities of Florida *(Source: National Atlas of the United States, US Geological Survey)*
- Coral—polygons representing coral populations off the coast of Florida *(Source: Florida Fish and Wildlife Conservation Commission's Fish and Wildlife Research Institute)*
- Counties—polygons representing Florida county boundaries *(Source: US Census Bureau)*
- Hwy1—a single line feature depicting a major state highway that traverses the Atlantic coastline and Florida Keys *(Source: Florida Department of Transportation)*
- Lakes—polygons representing Florida's major lakes *(Source: Florida Department of Environmental Protection)*

- Rivers—line features representing Florida's major rivers *(Source: Florida Department of Environmental Protection)*
- Shipwrecks—a selection of historical shipwrecks along the southern Atlantic coast of Florida and the Florida Keys *(Source: National Oceanic and Atmospheric Administration Office of Coast Survey's Automated Wreck and Obstruction Information System)*
- Southeast—a polygon layer that includes demographics *(Source: ArcUSA, US Census, Esri)*
- StParks—polygons representing Florida state parks *(Source: Florida Department of Environmental Protection)*
- florida_coast—a raster image of seafloor topography *(Source: Natural Earth)*
- shdrl—a hillshade raster showing continuous elevation values *(Source: University of Florida GeoPlan Center)*
- usgsdem—a digital elevation model raster for the southern portion of Florida *(Source: US Geological Survey)*

Exercise 7a
Create custom symbology

For this exercise, you are a graphic designer creating a marketing product for the Florida Commission on Tourism. Your map, highlighting the southern Atlantic coast of Florida as well as the Florida Keys, will be printed in a brochure for prospective coastal tourists and scuba divers.

To symbolize, it is best to have your ArcMap window maximized if possible.

1. **Start ArcMap and open the ex07a map document in your C:\EsriPress\GTKArcGIS\ Chapter07 folder.**

1
2
3
4
5
6
7
8
9
10
11
12
13
14
15
16
17
18
19
20

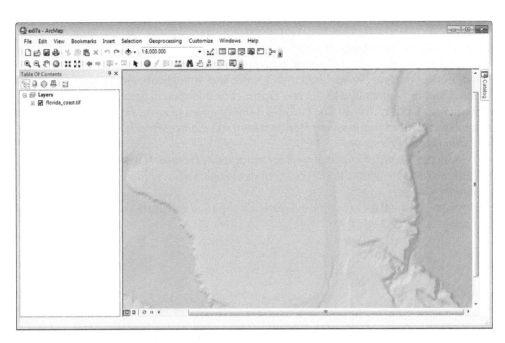

At this point, the map includes only a raster dataset representing the ocean floor terrain along the Florida coast.

2. Click the Add Data button , and add all of the datasets stored in Florida.gdb to the ..\GTKArcGIS\Chapter07\Data folder. (Press Shift to select all feature classes.)

Your map will look a little different from the graphic, because ArcMap randomly assigns symbol colors when data is added to a map.

You've got some work to do to make your map look presentable.

3. In the ArcMap table of contents, click anywhere in the white area to clear the selected layers.

Next, you will change some layer names for clarity.

4. Recall what you learned in earlier chapters, and rename the following layers:

- Shipwrecks to Historical Shipwrecks
- Hwy 1 to U.S. Route 1
- Rivers to Major Rivers
- Lakes to Major Lakes
- StParks to State Parks
- florida_coast.tif to Ocean Background

Remember, you are not renaming the source data files but only the map layer names as they exist in the current map document.

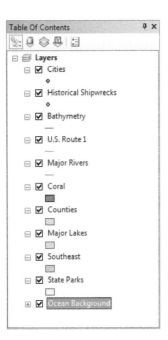

5. Uncheck the Cities and Historical Shipwrecks check boxes to turn the layers off. You will come back to these layers later.

6. Click the line symbol under Bathymetry to open the Symbol Selector.

Bathymetry lines indicate changes in ocean depth, such as the elevation lines on a topographical map.

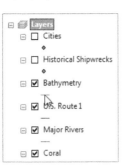

7. In the Symbol Selector, click the Color arrow. In the color palette, click Sugilite Sky, and click OK.

TIP

Pausing your mouse pointer over the color chips will reveal the color names.

The new color is fine, but you may notice that the bathymetry lines somewhat obscure the coast boundaries. To remedy this, you will change the default drawing order.

8. In the table of contents, drag Bathymetry below State Parks.

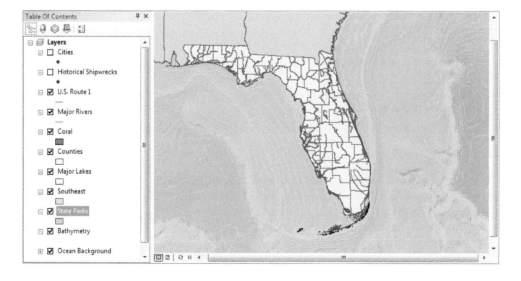

TIP

You can reorder layers only when the table of contents is in the default List By Drawing Order view.

Although lines are typically placed on top of polygons, why does it make sense to put bathymetry lines below polygons?

Here is a shortcut for performing simple color changes.

9. Right-click the symbol for the Major Rivers layer to open the color palette. In the palette, click Oxide Blue.

You want the same color for lakes, but because lakes are polygon features, there are two color options—one for the outline and one for the fill. For more complex symbology such as this, you can't use the shortcut method you just used for Major Rivers.

10. Click the symbol below Major Lakes to open the Symbol Selector. For Fill Color, click the color chip, and click Oxide Blue. For Outline Color, click No Color. Click OK to apply the color scheme.

You can't see lakes because the lake features are obscured by another polygon layer. As you work on your map, a more logical layering order becomes obvious. You want lakes to be the same color as rivers, and you also want them close together in the table of contents.

11. Drag **Major Lakes** just below **Major Rivers**. Then drag **State Parks** above **Counties.**

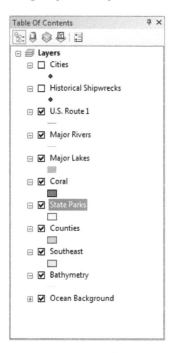

Next, you will modify the width of a line layer.

1
2
3
4
5
6
7
8
9
10
11
12
13
14
15
16
17
18
19
20

12. Open the Symbol Selector for U.S. Route 1. For Color, choose Mars Red. For Width, highlight 1.00 and replace it with **1.5**. Click OK to apply the changes.

ON YOUR OWN

Symbolize the remaining layers as follows:
- Coral:
 - Fill Color: Medium Coral
 - Outline Width: 1.5
 - Outline Color: Medium Coral
- State Parks:
 - Fill Color: Macaw Green
 - No outline
- Counties:
 - No fill
 - Outline Width: 0.40 (default)
 - Outline Color: Gray 10%
- Southeast:
 - Fill Color: Light Apple
 - Outline Width: 0.40 (default)
 - Outline Color: Gray 20%

13. Turn on Historical Shipwrecks.

The wrecks are located on the southern Atlantic coast and throughout the Florida Keys. This will be the focus of your map.

14. In the table of contents, right-click Historical Shipwrecks, and click Zoom To Layer.

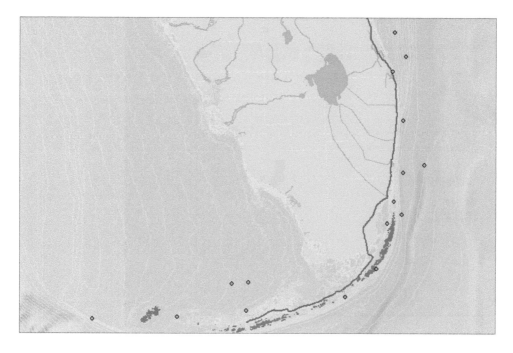

You will save this extent as the full extent of your map. That means whenever you click the Full Extent button, you will be brought back to this view.

ON YOUR OWN

Fine-tune the display: Change the scale to exactly 1:3,000,000, and use the Pan tool to drag the landmass to the center of the display. Be sure to maintain the visibility of the coral reefs in the Keys.

15. Right-click Layers (the data frame name), and click Properties. Click the Data Frame tab. In the Extent Used By Full Extent Command frame, click Other.

16. Click Specify Extent. On the Full Extent dialog box, click the Current Visible Extent option, and click OK. Click OK again to complete the task.

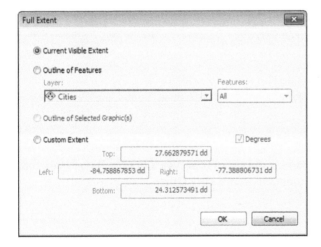

TIP

An alternative for opening the data frame properties is to go to the menu bar and click View > Data Frame Properties. Or double-click the data frame name.

17. Finally, change the data frame name from Layers to **South Florida Recreation.**

Notice that the data frame title is in boldface in the table of contents to show that it is active.

18. To save your work, rename your map document and save to your MyData folder for chapter 7.

19. If you are continuing to the next exercise, keep ArcMap open. Otherwise, exit the application, and click No if prompted to save your changes.

Exercise 7b
Symbolize features by categorical attributes

In exercise 7a, all rivers had the same marker symbol, and all state parks were drawn in the same color. You can make the symbology of a feature class more informative by assigning a different symbol to each unique value in an attribute field or to ranges of values. For example, you might symbolize a polygon layer of countries according to their names. Each feature would have a unique color, because every country has a different name. You might also symbolize countries by their political systems. Features with the same system would have the same color: republics might be blue, constitutional monarchies purple, and communist states red.

Attributes that are names or descriptions are called categorical attributes (or qualitative or descriptive attributes). They are usually text but may be numbers if the numbers are codes standing for descriptions or if they are ranges of numbers that compose a category, such as population ranges (rather than a specific population count). Attributes that are measurements or counts of features are called quantitative attributes. Using the previous example, a country's population is a count, and its area in square kilometers is a measurement.

In this exercise, you will symbolize features by categorical attributes. You will symbolize by quantitative attributes in chapter 8.

In some cases, you will want to limit the number of features displayed on the map without editing the source data. To do this, you would create a definition query, which you will also do in this exercise.

1. In ArcMap, open the ex07b map document from your Chapter07 folder.

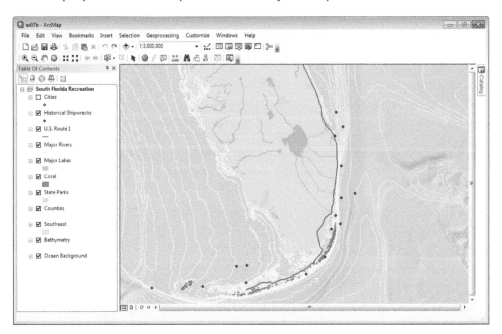

This map resembles the one you created in exercise 7a.

In this exercise, you want to make the symbology for State Parks more helpful. It would be nice if map readers could quickly determine whether a fee is required to visit the park.

2. In the table of contents, right-click State Parks, and click Open Attribute Table. Scroll to the right, and examine the Public_Acc (public access) field.

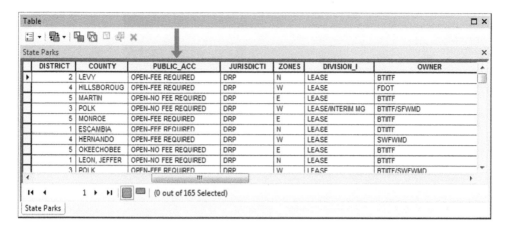

There are two public access categories: Open-Fee Required and Open-No Fee Required. You will choose a different symbol for each one so the categories can be easily discerned on the map.

3. Close the table, and open the layer properties of State Parks. Click the Symbology tab.

4. Under Show, click Categories. Unique values is selected by default.

5. In the Value Field list, click PUBLIC_ACC.

6. Click Add All Values. Uncheck the <all other values> check box.

ADDING VALUES

You may want to focus on certain features in a layer and deemphasize or ignore others. The Add Values button (next to Add All Values) allows you to select specific attribute values and assign symbology to features that have these values. Other features are assigned a single symbol–the one designated for <all other values>. If the <all other values> check box is not checked, these features are not displayed.

7. Double-click the symbol to the left of OPEN-FEE REQUIRED. In the Symbol Selector, you have default Esri symbol styles. Click Sun in the first row of color chips. Set the outline color to No Color, and click OK to close the Symbol Selector.

8. **Double-click the symbol next to OPEN-NO FEE REQUIRED. Click the Green color chip from the Esri style choices, set the outline color to No Color, and click OK to close the Symbol Selector. Click OK to apply the new color scheme.**

9. **In the table of contents, click PUBLIC_ACC twice slowly to make it editable. Rename the label Public Access.**

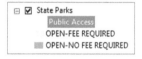

Now you know how to give features different symbols based on categorical attribute values. Next, you will limit the number of symbols used in the Cities layer.

ON YOUR OWN

Explore the other polygon symbols in the default Esri symbol styles: for example, give Major Lakes the symbol for Swamp or Glacier. Use either the Symbology tab on the Layer Properties dialog box or just click the symbol in the table of contents to open the Symbol Selector. When you are done experimenting, change the symbol back to Oxide Blue fill, no outline.

10. **Turn on Cities.**

The city symbols clutter the map. Because this map is a marketing tool for tourism, you want to display only the cities that are large enough to accommodate visitors and recreation enthusiasts. You will create a definition query to limit the display of cities to only those with a population of 100,000 or greater. This affects only the features shown on this map; it will not permanently delete features from the dataset. With a definition query, you can limit the display based on any attribute, whether categorical (such as population range) or quantitative (such as population).

11. **Open the properties of the Cities layer, and click the Definition Query tab.**

Layer Properties
General

Definition Query:

[Query Builder...]

OK Cancel Apply

12. **Click Query Builder.**

13. On the upper portion of the Query Builder dialog box, double-click POP_RANGE to add it to the query expression box.

14. Click the equals sign (=), click Get Unique Values, and double-click '100,000 - 249,999'. Then click the Or button.

15. Again double-click POP_RANGE, click the equals sign, and double-click '250,000 - 499,999'. Click the Or button. Then double-click POP_RANGE, click the equals sign, and double-click '500,000 - 999,999'. (There are no cities over 1 million.) Your query should match the graphic.

1
2
3
4
5
6
7
8
9
10
11
12
13
14
15
16
17
18
19
20

16. Click OK and then OK again to close the Query Builder and Layer Properties dialog boxes.

TIP

If any part of the expression is missing, you will see an error message. If this happens, go back to the Definition Query tab to ensure the expression is correct.

The number of cities has been dramatically reduced.

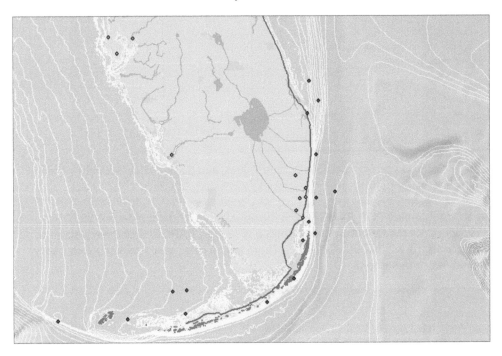

TIP

Alternatively, you could have created this simpler query: POP_2000 ≥ 100,000. You were given a more complex approach to show how to base your display on categories, as opposed to quantitative attributes.

17. Change the symbol for Cities. Use the Esri style Circle 3, and change the size to **10**. Click OK to close the Symbol Selector.

18. In the table of contents, right-click Cities and click Label Features.

This is a fast way to label features, but you need to make sure the labeling field is set to the correct attribute on the Layer Properties dialog box, on the Labels tab. ArcMap will choose a default field, but it may not always be the one you want. In this case, the default choice was correct. In chapter 9, you will modify label properties to maximize readability.

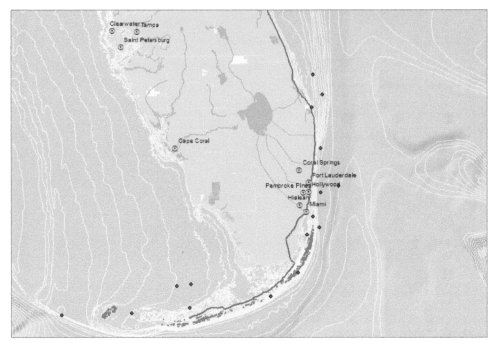

19. Finally, rename the Cities layer **Cities Pop 100,000+.**

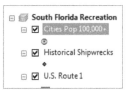

20. To save your work, rename your map document, and save to your MyData folder for chapter 7.

21. If you are continuing to the next exercise, leave ArcMap open. Otherwise, exit the application, and click No if asked to save your changes.

In exercise 7c, you will learn how to use other symbology styles and save your symbology to layer files.

Exercise 7c
Use styles and create layer files

In exercises 7a and 7b, you modified colors, sizes, and outlines of point, line, and polygon symbols. You also picked predefined symbols and colors from the default options provided in the Symbol Selector–these symbols and colors belong to the Esri style library. A **style** is a collection of predefined symbols, colors, and other map elements such as labels, north arrows, scale bars, and borders. ArcGIS has more than 20 style libraries, and you can make new ones that contain often used symbols from other styles or custom symbols you create.

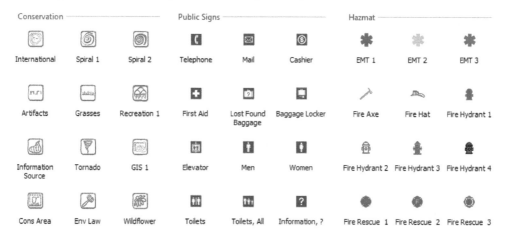

Left to right: Some symbols used in the Esri Conservation, Public Signs, and Hazmat style libraries.

Having invested time in symbolizing a layer, you may want to save it as a **layer file** (a file with a.lyr extension). Layer files store symbology information for a specific spatial dataset–they do not contain the actual data.

In this exercise, you will use a shipwreck symbol from the Esri Civic style library to represent historical shipwrecks on the Atlantic coast of Florida and the Florida Keys. Then you will save the symbology to a layer file.

The ships on this map all went down from 1942 to 1944, during World War II. They represent only a sample of the numerous wrecks in these waters. Mapping the location of wrecks is important to navigation, historical studies, and marine recreation. Shipwreck diving is part of Florida's tourism economy.

1
2
3
4
5
6
7
8
9
10
11
12
13
14
15
16
17
18
19
20

1. In ArcMap, open the ex07c map document from your Chapter07 folder.

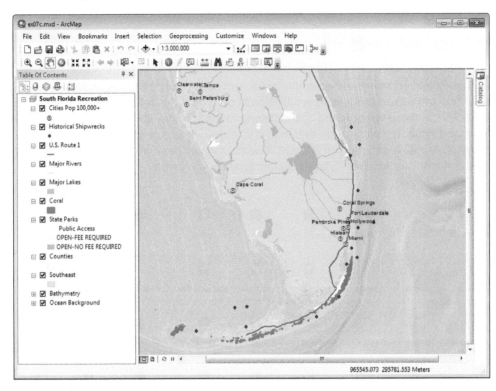

The South Florida map has the same symbology updates you made in exercises 7a and 7b.

2. In the table of contents, click the point symbol below Historical Shipwrecks to open the Symbol Selector. If you scroll through the list of symbols, you will find there is nothing that looks like a shipwreck.

3. On the lower right of the dialog box, click Style References to display the list of styles.

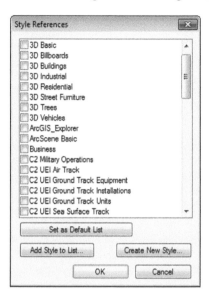

When a style is loaded, it has a check mark next to it. By default, only your personal style library and the Esri style library are loaded. (Your personal style library is empty unless you have added symbols to it.)

There are many styles collected in style libraries that are intended for specific types of maps. If you know you are looking for a specific symbol but aren't sure where to find it, you can search for it.

4. Click Cancel to close the Style References dialog box.

5. In the Symbol Selector, ensure that All Styles is selected, and in the search box, type **wreck** and press Enter.

ON YOUR OWN

To see all the symbols in a style library, select the style on the Style References dialog box to load the complete set to the Symbol Selector.

6. Scroll down to the Civic style library, and click the second Wreck symbol. Click OK to implement the change.

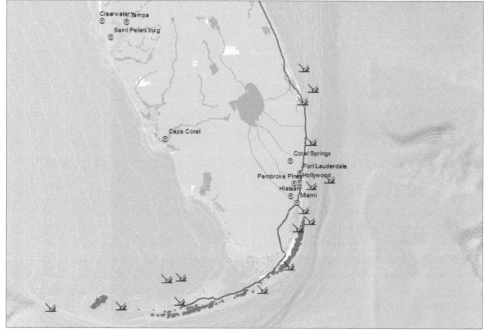

ON YOUR OWN

Use the Identify tool and click some of the shipwrecks to see their attributes. Keep in mind that when using graphic symbols, the centerpoint is not always easy to find. With this symbol, you must click right at the bottom of the V formed by the mast and the ship body, as shown in the graphic.

If you wanted to reuse the symbolized layer in other map documents—or send it to a colleague—you would create a layer file. Layer files allow you to store symbology information—the colors, shapes, and sizes you choose for features—so you never have to re-create it. Just as a person can wear many different outfits, a spatial dataset can have many different layer files.

7. In the table of contents, right-click Historical Shipwrecks, and click Save As Layer File.

8. On the Save Layer dialog box, browse to your ..\GTKArcGIS\Chapter07\MyData folder. Accept the default file name of Historical Shipwrecks.lyr, and click Save.

Layer files can be given unique names—they don't have to share the same name as their source data.

9. Repeat step 7 to save the State Parks symbology to a layer file. Use the default name and save to your MyData folder for chapter 7.

The Historical Shipwrecks and State Parks layer files can now be added to any map. Because they reference their source feature classes (Shipwrecks and StParks), the feature classes must also be accessible on disk or across a network. If you send a layer file to someone else, you must also send the data source that it references.

ON YOUR OWN

Can you figure out how to symbolize the bathymetry layer using graduated colors based on depth measurements in meters? Shallower waters should be light blue, and deeper waters, dark blue. Your results will resemble the graphic. If you succeed, you are ahead of the game—symbolizing features based on quantitative attributes is the focus of chapter 8. (**Hint:** Use the Layer Properties dialog box, not the Symbol Selector. You will have to change the sample size to 25,000 [Classify > Sampling].)

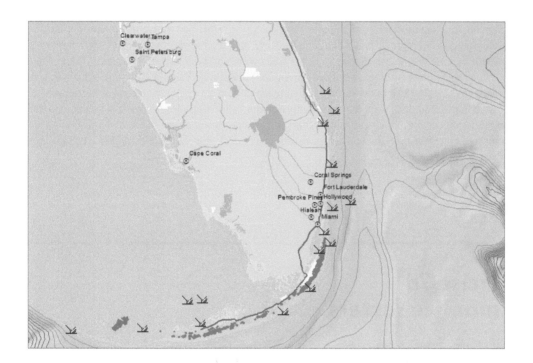

ON YOUR OWN

Optionally, create layer files for the other datasets in your map. Save the layer files to your MyData folder for chapter 7.

10. To save your work, rename your map document and save to your MyData folder for chapter 7.

11. If you are continuing to the next exercise, leave ArcMap open. Otherwise, exit the application, and click No if prompted to save your changes.

1
2
3
4
5
6
7
8
9
10
11
12
13
14
15
16
17
18
19
20

CARTOGRAPHIC REPRESENTATIONS

In ArcGIS Desktop, you can use the power of the geodatabase to store cartographic representations, often referred to simply as representations.

You can convert any geodatabase feature class to a representation class, which stores symbology rules for features. The Representation toolbar provides special tools that allow mapmakers to improve the appearance of features. You can also override the appearance or shape of features stored as representations, or remove them from the map display, without changing the actual source data.

For example, imagine that you are responsible for creating trail maps for all the state parks in Florida. The maps should share the same symbology scheme. You have a trails feature class that you symbolize with gray dashed lines and then convert to a representation class. Each park map uses a definition query to display only the trails within its boundaries. A few state parks have unique needs, however. Some have equestrian trails that should be differentiated on the map with a unique color. Sometimes parks have to delete a trail from a map or change its course because of maintenance. You don't want to permanently edit the trails database, so instead you select the trail symbol that requires modification and change its properties or shape or remove it from the map display. Later, if you need to, you can go back to the map's original appearance because the source data stays the same.

Exercise 7d
Symbolize rasters

In chapter 1, you learned that a raster is a matrix of identically sized square cells. Each cell in a raster stores a value, usually a quantity of something such as elevation, rainfall, or temperature that has been measured at the location represented by the cell.

When you symbolize a raster, you assign colors to cell values or ranges of cell values. Raster values usually lie on a continuous scale, such as the scale of integers or real numbers, and are symbolized by color ramps. Rasters can have a single band or multiple bands of data. Most satellite imagery has multiple bands, whereas raster surface layers typically have one band of data.

The shades of red in this color ramp become darker as the values increase from 0 to 100.

Rasters can represent many kinds of data and be symbolized using different color ramps.

Left to right: Rasters of elevation, hillshade, and slope for the same area. Raster images based on the National Elevation Dataset courtesy of the US Geological Survey.

A hillshade raster shows an elevation surface in relief. It can be mathematically derived from an elevation raster by setting a certain angle and altitude for a light source.

Satellite images and aerial photographs are rasters in which the cell values are measurements of reflected light. Scanned maps are rasters in which the cell values are measurements made by the scanning device.

An air photo (*left*) and a scanned map (*right*) of the area around San Diego's Mission Bay. If you zoom in close enough on any raster, its cell structure will be revealed, as shown by the inset in the orange box on the air photo. Satellite image and map courtesy of the US Geological Survey.

You can display raster data in ArcGIS Desktop, but to create it, you need one of the ArcGIS extensions for working with raster data—ArcGIS Spatial Analyst™, ArcGIS 3D Analyst™, or ArcGIS Geostatistical Analyst™. Raster data is available in many formats, on the internet and from commercial vendors.

Multiple rasters can be organized in a raster catalog. If you wanted to view multiple adjacent raster images at a time, you would store them in a mosaic dataset. Both raster catalogs and mosaic datasets can be created in a geodatabase, if desired.

This chapter provides a first look at some types of raster data. In chapter 20, you will create raster surfaces using ArcGIS Spatial Analyst.

In this exercise, you will symbolize an elevation raster and a hillshade raster to show the topography of southern Florida.

1. **In ArcMap, open the ex07d map document from your Chapter07 folder.**

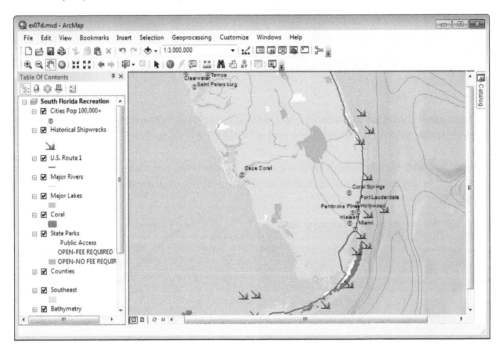

This map contains the familiar Florida layers with which you have been working.

2. **In the table of contents, expand the Ocean Background legend.**

You can see that the Ocean Background image is a three-band raster, symbolized by a red-green-blue (RGB) value for each band.

Next, you will add elevation and hillshade raster layers to the map.

3. **Click the Catalog tab to expand the Catalog window. (If the tab is not visible, click the Catalog button on the Standard toolbar.)**

4. Under the Home location, expand the Data folder, and click the shdrl raster.

5. Drag shdrl from the Catalog window to the ArcMap display area. If prompted to build pyramids, click Yes.

6. Now drag usgsdem from the Catalog window to the ArcMap display area. If prompted to build pyramids, click Yes.

A pyramid is a reduced-resolution raster layer that copies the original data in decreasing levels of resolution to enhance performance. The coarsest level of resolution is used to draw the entire dataset quickly. As the display zooms in, layers with finer resolution are drawn. Drawing speed is maintained because fewer pixels are needed to represent the successively smaller areas. You need to build pyramids only once for a raster dataset. They are stored with the data as a file with the extension.rrd.

The raster layers are added near the bottom of the table of contents, above the Ocean Background layer. Next, you will rename the raster layers.

7. In the table of contents, rename the usgsdem layer **Elevation**.

1

2

3

4

5

6

7

8

9

10

11

12

13

14

15

16

17

18

19

20

8. Rename shdrl **Hillshade.**

To see the rasters, you will need to change their position in the table of contents.

9. In the table of contents, click the Elevation layer. Hold down the Ctrl key and click Hillshade. Both rasters should be highlighted. Drag the layers below the Counties layer but above the Southeast layer.

The raster layers are repositioned in the table of contents and displayed on the map.

By default, rasters are drawn in shades of gray. Next, you will choose a different color ramp for the Elevation layer.

10. **In the table of contents, double-click the Elevation layer. On the Layer Properties dialog box, click the Symbology tab.**

There are four raster display methods in the Show panel. In the Stretched method (the default), ArcMap makes subtle transitions along the selected color ramp but doesn't give you precise information about which data values are associated with which shades of color. In the Classified method, color transitions are less subtle, but you can see exactly which value ranges correspond to which shades. In the Unique Values method, a different color is assigned to each data value in the raster. In the Discrete Color method, each unique value is symbolized randomly from a set of colors. This method is useful when many unique values are present.

1
2
3
4
5
6
7
8
9
10
11
12
13
14
15
16
17
18
19
20

11. Maintain the Stretched method. In the Color Ramp list, right-click the colors (not the drop-down arrow), and click Graphic View to turn it off.

12. Click the Color Ramp arrow, and click the Yellow-Green Light to Dark color ramp. Check the Invert check box. Click OK to close the Layer Properties dialog box.

The color ramp is applied to the layer. The colors range from dark olive green at lower elevations through lighter shades of olive green at higher elevations.

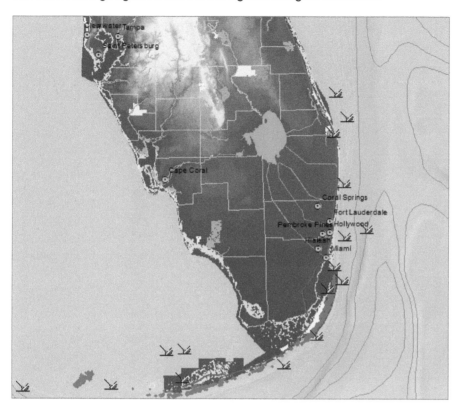

How would you describe the topography of southern Florida?

13. Click the Identify tool, and click any land area of the map. In the Identify window, in the upper portion, change the Identify from value from <Top-most layer> to Elevation. Then click the map again, in several locations, to see the elevation value.

14. Close the Identify window. Zoom in closely (approximately 1:60,000) on an area with varied elevation. Notice that the raster is made up of an array of square cells, or pixels.

To see surface relief on your map, you will next make the Elevation layer partially transparent so that the Hillshade layer shows through.

1
2
3
4
5
6
7
8
9
10
11
12
13
14
15
16
17
18
19
20

15. Zoom to the full extent of the map . Open the Layer Properties dialog box for the Elevation layer, and click the Display tab. In the Transparency text box, highlight the default value of 0 and type **50**. The Elevation layer will be 50 percent transparent. Click OK.

TIP

Transparency can also be used effectively on vector layers. Transparency for all features in vector layers is set on the Display tab. Alternatively, click the Advanced button on the Symbology tab to set varying transparency according to field values.

1
2
3
4
5
6
7
8
9
10
11
12
13
14
15
16
17
18
19
20

The Hillshade layer is now visible below the Elevation layer, providing a realistic impression of the landscape.

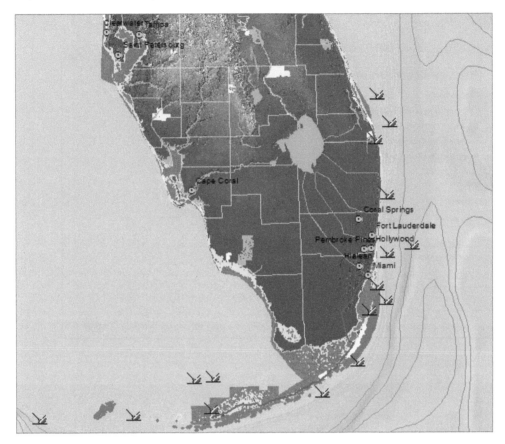

16. Again, zoom in to better see the hillshading effect.

17. To save your work, rename your map document, and save it to your MyData folder for chapter 7.

18. If you are continuing to the next chapter, leave ArcMap open. Otherwise, exit the application, and click No if prompted to save your changes.

In chapter 8, you will symbolize layers by quantitative attributes rather than categorical ones. You will also learn how to classify data—that is, adjust the value ranges for applying symbols. And you will work with symbology methods specifically for numeric data.

In chapter 9, you will revisit the Florida data and add labels and annotation to complete your map. In chapter 10, you will learn how to create a nice-looking map layout so you can print your map and present it to others.

1
2
3
4
5
6
7
8
9
10
11
12
13
14
15
16
17
18
19
20

Summary

In this chapter, you gained experience with the first step in mapmaking: symbolizing vector and raster layers. Realize that map symbology can get sophisticated—consider the work of professional cartographers. Once you have mastered the basics, you can explore the Symbol Selector editor and advanced cartography tools to create more detailed symbology. Remember, to reuse or share symbology, create a layer file. The layer file stores the symbology and also points to the source data file, so when sharing, be sure to provide the source data as well as the layer file.

Terms

style
layer file
hillshade
pyramid

CHAPTER 8
Classifying features

Classify features by standard methods
Map density
Use graduated and chart symbols

Learning objectives

- Know what to consider when classifying quantitative attribute data
- Become familiar with classification methods
- Choose symbology based on classification type
- Create dot density maps
- Create graduated-color maps

An important aspect of using ArcMap is to decide what data to map and how to map it. You just worked with qualitative data in chapter 7. In this chapter, you will work with quantitative data. Common examples of quantitative attributes are the population of a country, the area of a city block, and the length of hydroelectric transmission lines.

Quantitative maps show numeric values in relation to one another on a continuous scale. Although you can map individual values, it is often more useful to group or classify these values. This will help your readers to separate the data visually, so it can be understood more easily. Deciding how many classes are appropriate for the data and the method of classification used is called classifying data. Depending on the message you are trying to convey, ArcMap can create several types of quantitative maps: graduated color, graduated symbol, proportional symbol, dot density, and charts.

A graduated-color map uses a change in color to represent data classifications. A ramp with a single color (called monochromatic) is suited to data that moves from small to large values. Sequential schemes such as population increase can be shown with light colors for smaller amounts of change and dark colors for larger amounts of change. A ramp with two colors can be used to represent a divergent scheme, such as population increase *and* decrease. The extreme values are shown in dark colors, and the middle values are shown in light colors.

A graduated-symbol map uses a symbol that varies in size to represent numeric attribute values. Points that increase in size can represent a city's growing population. Different line widths can be used to represent the increasing flow of a pipe or a river. Generally, three to seven graduated colors are used. If more than seven colors are used, the difference between colors may not be distinguishable by the human eye.

A proportional-symbol map is similar to a graduated-symbol map, except that the symbols are drawn proportionally in size. A country with a population of 50 million people would have a symbol five times larger than a country with a population of 10 million people. These symbols are most appropriate for data that has a relatively narrow range of values.

A dot density map represents quantities using randomly placed dots to form a density pattern inside a polygon. Each dot is equal to a predetermined attribute value—the more dots, the greater the value of the polygon. However, the density pattern may be different from the actual distribution of values, and a common technique is to use a masking layer to exclude certain areas, such as a neighborhood or a zip code boundary, to create a more realistic depiction of where dots should be placed.

A chart map can be used to represent multiple attribute values for a single feature. ArcMap can create pie, bar, and stacked charts that can be similar in size or vary based on proportionality.

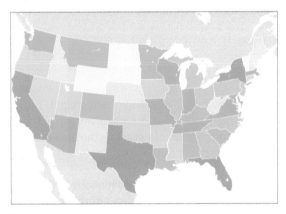

Graduated-color map. Data from Data and Maps for ArcGIS 2010, courtesy of ArcUSA, US Census, Esri (Pop2010 field).

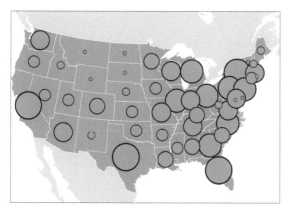

Graduated-symbol map. Data from Data and Maps for ArcGIS 2010, courtesy of ArcUSA, US Census, Esri (Pop2010 field).

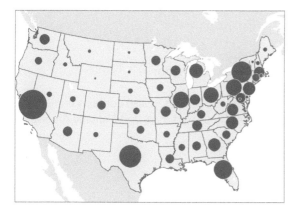

Proportional-symbol map. Data from Data and Maps for ArcGIS 2010, courtesy of ArcUSA, US Census, Esri (Pop2010 field).

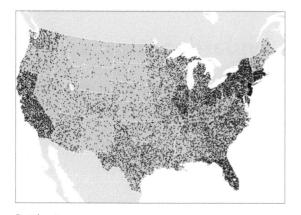

Dot density map. Data from Data and Maps for ArcGIS 2010, courtesy of ArcUSA, US Census, Esri (Pop2010 field).

1
2
3
4
5
6
7
8
9
10
11
12
13
14
15
16
17
18
19
20

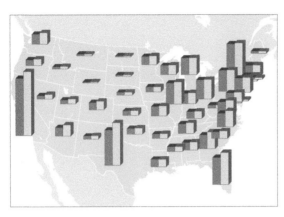

Map using charts. Data from Data and Maps for ArcGIS 2010, courtesy of ArcUSA, US Census, Esri (Pop2010 field).

Data

This chapter uses the following dataset, stored in TexasCounty.gdb:

- Counties–polygons of the counties within the state of Texas. It also contains crime data for the years 1994 and 2002. *(Sources: ArcUSA, US Census, Esri, and National Archive of Criminal Justice Data)*

Exercise 8a
Classify features by standard methods

When mapping quantitative data, you want to classify the data appropriately so the information is not skewed or misleading. ArcGIS has seven classification methods: manual classification, equal interval, defined interval, quantile, natural breaks, geometrical interval, and standard deviation.

Manual classification

With manual classification, you can set class breaks manually by typing them or by adjusting the histogram break bars in ArcMap.

Equal interval

In equal interval classification, the entire range is equally divided by the number of classes you choose. If the range of values is 1 to 1,000 and there are five classes, this method will create classes of 1–200, 201–400, 401–600, 601–800, and 801–1,000. This method is useful for highlighting changes in the extremes and is most appropriate for familiar data ranges, such as percentages or temperatures. With this method, it is also possible to have classes that contain no values.

Defined interval

A defined interval is similar to an equal interval. However, you get to define the interval size to determine how many classes there are. If the students in your class have a range of test scores from 50 percent to 90 percent, for example, and the interval size is 10, this method will create four classes, 50–60, 61–70, 71–80, and 81–90.

Quantile

With quantiles, all classes have the same number of features. This method is appropriate when data is linearly distributed. It can create a balanced-looking map because no classes have too many, too few, or no values; however, the resultant map can be misleading. The key is to increase the number of classes to prevent data from being placed in a different class. When an appropriate number of classes are used, the data can be visually represented more accurately.

Natural breaks (Jenks)

Natural-breaks classes are based on natural groupings inherent in the data, and boundaries are set where there are relatively large gaps between values. Developed by Professor George Jenks of the University of Kansas, this method is useful for classifying unevenly distributed data such as population.

Geometrical interval

The geometrical-interval classification scheme creates class breaks based on class intervals that have a geometrical series, such as a logarithmic distribution. This method was specifically designed to accommodate continuous data and minimize variance within classes to ensure that each class range has approximately the same number of values and that the change between intervals is fairly consistent.

1
2
3
4
5
6
7
8
9
10
11
12
13
14
15
16
17
18
19
20

Standard deviation

Data that you know is normally distributed (in a bell-shaped curve) can benefit from standard-deviation classification. This method creates classes according to a specified number of standard deviations from the mean value. We recommend a divergent color scheme for this type of classification.

In this exercise, you will create thematic maps for a crime-mapping project by applying different classification methods and different numbers of classes.

1. **Start ArcMap. Open the ex08a map document from your C:\EsriPress\GTKArcGIS\ Chapter08 folder.**

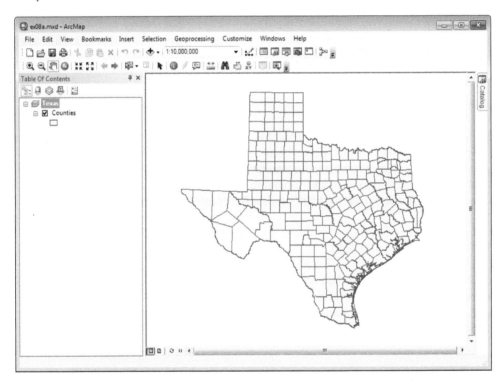

The map is open in data view. It contains a single data frame called Texas that contains the Counties layer. You will explore the layer's attributes to gain an understanding of what you will be mapping.

2. **In the table of contents, right-click the Counties layer, and click Open Attribute Table. Scroll to the right to see more of the table column headings.**

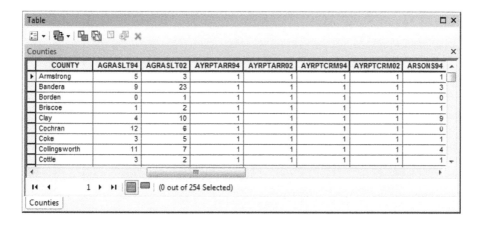

The table contains many attributes such as county name, population, and crime data from 1994 and 2002. You will be using a few of these attributes for the mapping tasks in this chapter.

3. Close the attribute table. In the table of contents, double-click the Counties layer name to open the Layer Properties dialog box. Click the Symbology tab.

4. Under Show, click Quantities.

The Graduated colors option of the Quantities method is highlighted, and the dialog box is changed. The Value list shows the available attributes—that is, numeric fields only.

5. In the Fields frame, click the Value arrow, and click BURG02.

The BURG02 attribute represents burglaries for each county in 2002. By default, the values in the BURG02 field are grouped into five classes using the natural-breaks (Jenks) classification method.

6. If your color ramp is different from the one in the previous graphic, click the Color Ramp arrow, and click the Blue Light to Dark ramp. (To find it by name, right-click the color ramp color bar, and click Graphic View to turn off that view.) Click OK to close the Layer Properties dialog box.

The graduated-color symbology is applied to the map, and the table of contents is updated to show the values represented by each color. The light colors represent fewer burglaries, and the dark colors represent more burglaries. This may be useful to crime analysts who want to quickly see where the most burglaries are occurring. It doesn't account, however, for how many people live in a county or how many square miles are in a county. You will explore these situations in exercise 8b. For now, you will see how other classification methods affect the map.

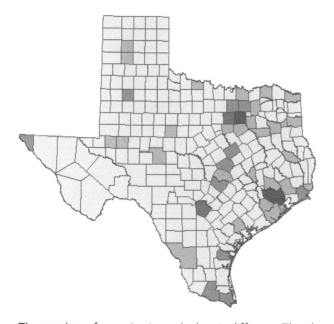

The number of counties in each class is different. The class with the highest values has just two members: Dallas and Harris Counties. The next highest class also has two members: Tarrant and Bexar Counties. The remaining counties belong to the three lowest classes.

ON YOUR OWN
Use the Identify tool to look at the attribute information of various counties.

7. In the table of contents, open the properties of the Counties layer, and click the Symbology tab if it is not already active.

8. In the Classification frame, click the Classes arrow, and click 3. The number of classes and their value ranges are adjusted. Click Apply, and move the Layer Properties dialog box away from the map display.

1
2
3
4
5
6
7
8
9
10
11
12
13
14
15
16
17
18
19
20

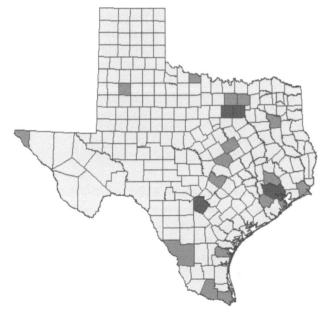

The map reflects the new classification. The highest burglaries class now contains four counties. Many counties that previously belonged to separate classes are now grouped together.

Why might these counties have a higher incidence of burglary than surrounding counties?

9. On the Symbology tab, in the Classification frame, click Classify.

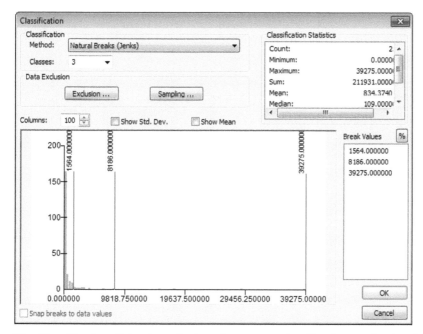

The Classification dialog box displays the current classification method and the number of classes. Under Classification Statistics, you can see the count, minimum, maximum, sum, mean, median, and standard deviation of the BURG02 field. (You may need to scroll to the right to see the values.) In the Break Values box, you can see the upper threshold value that defines each class. The percentage (%) button displays decimals as percentages.

The large window contains a histogram, or frequency distribution chart. The x-axis (horizontal) shows the range of values in the BURG02 field. The y-axis (vertical) is a count of features. Vertical blue lines are class breaks (also shown in the Break Values box). The gray columns represent the number of features that fall within a specific span of the total range of values. By default, 100 columns are drawn in the histogram, so each column represents 1 percent of the total range. In this case, values range from 0 to 39,275, so the total range of the histogram is 39,275. (This wouldn't be the case if the number of burglaries began with a number other than zero. Here, 39,275 divided by 100 is rounded to 393, so each column shows the number of features that fall within each span of 393 values.)

10. In the Break Values box, click the first break, at 1564.000000.

1
2
3
4
5
6
7
8
9
10
11
12
13
14
15
16
17
18
19
20

1

2

3

4

5

6

7

8

9

10

11

12

13

14

15

16

-17

18

19

20

Many of the gray columns are grouped within the first class. This means there is a high count of counties with fewer than 1,564 burglaries—exactly 228 counties. (The count is shown below the histogram window.) Next, you will take a closer look at the columns in the histogram.

How many elements (or features) are in the other classes?

11. Right-click the histogram near the origin of 0,0 (where the x-axis and y-axis intersect), and click Zoom In. Repeat the zoom in. If you zoomed in on another spot in the histogram by mistake, right-click and click Reset Zoom, and try again.

(Your zoom results may be a little different from the graphic.)

By zooming in on the histogram, you can get a better sense of how the values are distributed. Many of the county burglaries actually fall within the bottom 0.5 percent of the value range (below an x-axis value of 196.5). The data is skewed toward the low end of the histogram, meaning that there are relatively few burglaries in these counties. (The mean number of burglaries is 138.) A gray column will not be drawn if no features fall within the value range it would represent. The number of columns can be set to anything from 10 to 100. With 10 columns, for instance, each would represent 10 percent of the value range.

12. In the Columns box, change the default value of 100 to **25**.

Now each gray column represents one twenty-fifth (or 4 percent) of the value range.

13. In the Classification frame, click the Method arrow, and click **Equal Interval**.

The class breaks are now equally divided (still in three classes). This method places breaks so that each class has the same range of values. Almost all the values fall into the first class. Unlike the natural-breaks method, with equal-interval classification, it is possible to have classes that have no values.

ON YOUR OWN

Try increasing the number of classes to see how the breaks are placed. When you are finished, return the number of classes to 3.

14. In the Classification frame, click the Method arrow, and click Quantile.

The quantile method creates classes that have an equal number of features.

ON YOUR OWN

Zoom in on the histogram twice to get a better picture of three classes with the same number of features. How many counties fall within each class?

15. Click OK to close the Classification dialog box. The current classification settings are updated to reflect the quantile method. Click Apply, and move the Layer Properties dialog box away from the map display.

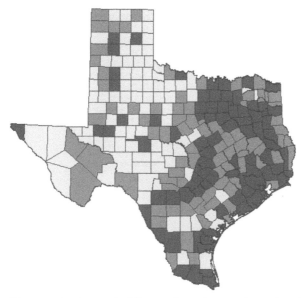

The map looks much different from the previous classification methods. Many more counties now fall within the highest burglaries class. As this example shows, the way you choose to depict data can provide different impressions, some of which can be misleading. You will have to adjust the classification method again to use natural breaks.

16. On the Symbology tab, click Classify. On the Classification dialog box, for Method, click Natural Breaks (Jenks) and change the number of classes to 7. Click OK.

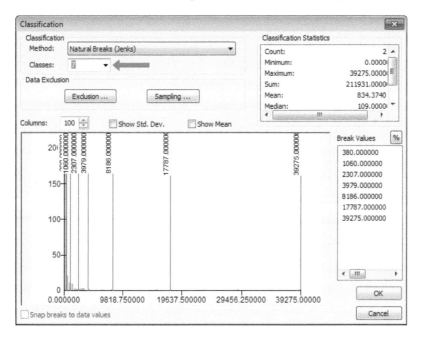

Before finalizing the map, you will make some small changes to the class labels to make them look better.

17. **In the lower section on the Symbology tab, click the Label column heading, and click Format Labels.**

The Number Format dialog box has options to edit the look of the numeric labels. This does not affect the classification values but only the way they appear. You can control the number of decimal places and significant digits, alignment, and whether to include comma separators. You can also choose special formats listed in the Category box.

18. **On the Number Format dialog box, change the Rounding number of decimal places to 0. Also, check the Show Thousands Separators check box.**

1
2
3
4
5
6
7
8
9
10
11
12
13
14
15
16
17
18
19
20

19. Click OK to close the dialog box. See how your label changes are reflected on the Symbology tab, and click OK.

20. On the menu bar, click View > Layout View.

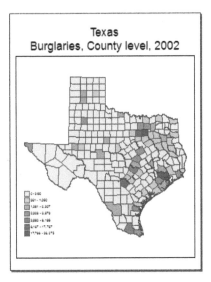

21. On the Tools toolbar, click the Full Extent button.

The layout includes a legend that shows the seven classes created using natural breaks.

ON YOUR OWN

Try classifying some of the other attributes using different methods to see how the data is distributed. Use the Identify tool to look at the COUNTY field in relation to the BURG02 field.

22. To save your work, rename your map document and save to your MyData folder for chapter 8.

23. If you are continuing to the next exercise, leave ArcMap open. Otherwise, exit the application, and click No if prompted to save your changes.

Exercise 8b
Map density

In this exercise, you will continue to create thematic maps for a crime-mapping project. In exercise 8a, you saw how crime can be mapped to show areas of high and low burglary incidence by applying different classification methods. However, you know that not all counties have the same population—for example, Dallas County has over 2.2 million people, and El Paso County has just over 650,000. But which of the two is more densely populated? To answer that question, you would need to know the number of people per square unit of area. As long as you have both a population attribute and an area attribute, you can do this. Simply divide the population by area. How about calculating crime density per capita? Similarly, you divide a crime attribute by population. Dividing one attribute by another to find the ratio is called normalization.

1. Start ArcMap. Open the ex08b map document from your Chapter08 folder.

1 2 3 4 5 6 7 8 9 10 11 12 13 14 15 16 17 18 19 20

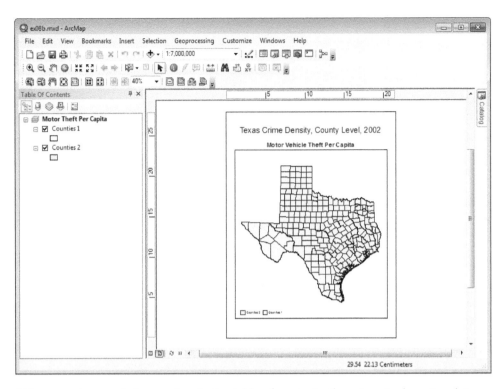

This opens the map in layout view. In the table of contents, there is a single active data frame called Motor Theft Per Capita. The Motor Theft Per Capita data frame has two identical Counties layers (coming from the same data source). You will make a population density map that is overlaid by a dot density map of motor vehicle thefts.

2. On the menu bar, click **View > Data View.**

3. Open the Counties 2 layer attribute table.

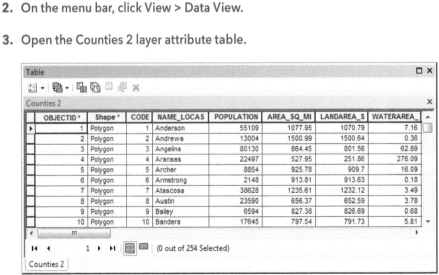

The POPULATION attribute contains the population count of each county. The AREA_SQ_MI attribute contains the size of each county in square miles. Next, you will divide the POPULATION attribute by the AREA_SQ_MI attribute to create a ratio of population density for each county.

4. **Close the attribute table. Double-click the Counties 2 layer, and click the Symbology tab. Under Show, click Quantities. (Graduated colors is automatically selected.) In the Fields frame, click the Value arrow, and click POPULATION. Click the Normalization arrow, and click AREA_SQ_MI. Set the color ramp to Red Light to Dark.**

The natural-breaks (Jenks) classification method is chosen, and five classes are defined. In the Label column, the values are rounded, but the decimal places are not consistent.

5. **Click the Label column heading, and click Format Labels. On the Number Format dialog box, select the Number of decimal places option and change the number to 2. Click OK to close the dialog box.**

6. Click OK to close the Layer Properties dialog box. Select the check box next to Counties 1 to hide the layer.

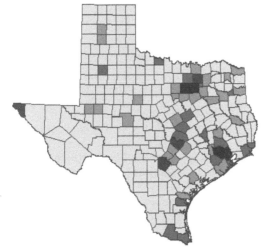

The class with the greatest population density has three counties: Tarrant, Harris, and Dallas. (You can verify this using the Identify tool.) The next highest class contains four other counties. The remaining counties of Texas have low population density. Graduated colors are one way to represent density. Another way is to use a dot density map.

7. Select the check box next to Counties 1 to make the layer visible again.

8. Open the Counties 1 properties. On the Symbology tab, click Quantities > Dot density. In the Field Selection list, scroll down, and click MVTHEFT02. Click the right-facing arrow (>) button to add it to the list on the right.

TIP

If you added the wrong attribute or added too many attributes, click the left-facing arrow (<) button to move an attribute back to the list on the left.

The box on the right in the Field Selection frame indicates that the incidence of motor vehicle theft in 2002 will be represented by the dot symbol. Next, you will change the color of the dot so that it has higher contrast and can stand out from the population density graduated-color map.

9. **Double-click the symbol next to MVTHEFT02 to open the Symbol Selector. Click the Color chip and click Black, and click OK.**

10. **In the Background frame, click the Line button. In the Symbol Selector, click the Color chip,** and click No Color. Then click OK.

This will prevent the county lines from appearing.

11. **In the Densities Calculated frame, click the horizontal slider and move it left or right. The preview gives you an idea of what the density will look like. In the Dot Value box, change the value to 100, and click OK.**

A dot value of 100 means that each dot represents 100 motor vehicle thefts.

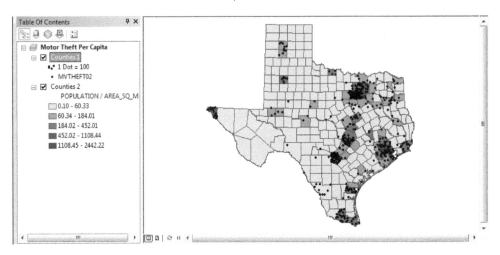

In ArcMap, you can also build query expressions to exclude dot density points or use other layers to act as a mask to exclude dots or place them in the specified area only. In the next few steps, you will use a masking layer to hide dot density points.

12. From your chapter 8 Data folder, add the **HarrisAdjacentCounties** feature class from the **TexasCounty** geodatabase.

13. Zoom in on the newly added layer. It is located in the southeastern area of the state. (Your feature class may be colored differently from the one in the graphic.)

14. Open the Counties 1 properties. On the Symbology tab, click the Properties button. On the Dot Density Symbol Properties dialog box, check the Use Masking check box, select HarrisAdjacentCounties in the Control Layer list, and click the Place Dots Only In These Areas option. Then click OK.

The options on this dialog box allow you to use a mask to either hide or show dot density points. When using a mask to exclude, no points will be visible in the defined area. When using a mask to include, points only in the defined area will be visible.

15. Click OK again to close the Layer Properties dialog box. In the table of contents, turn off the HarrisAdjacentCounties layer.

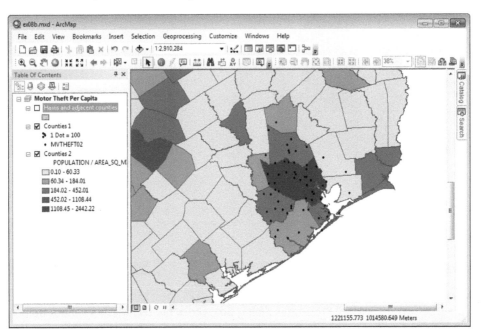

When creating dot density maps, consider the area units (such as census tracts or block groups), dot size, dot value, and dot placement. The distribution of data can create variations in the pattern of a dot density map. The random distribution of dots can often be misleading. For example, the dots in a county are randomly distributed. However, in real life, the majority of the population is centered in urban areas. No exact method exists for choosing the most appropriate dot value and size, so you will need to experiment until you produce a good result. Generally, you should choose a medium-size dot. Small dots produce a widely dispersed dot pattern and may convey disproportionate map accuracy. Large dots produce overly dense dot maps and hide any noticeable patterns.

ON YOUR OWN

Try changing the dot value and the dot size to see how the density pattern changes. To turn off the masking layer, return to the Dot Density Symbol Properties dialog box, and uncheck the Use Masking check box.

16. To save your work, rename your map document, and save it to your chapter 8 MyData folder.

17. If you are continuing to the next exercise, leave ArcMap open. Otherwise, exit the application, and click No if prompted to save your changes.

Exercise 8c
Use graduated and chart symbols

Graduated-symbol maps use proportionately sized symbols such as circles or squares to represent different ranges of values. This type of map allows you to visualize classifications by means of differently sized symbols, making an effective visual tool.

Charts are also an effective visual communication tool. A pie chart can show multiple sets of quantitative data at the same time: its overall size can be proportionate to an attribute value, and its divisions (or slices) can show the composition of several attribute values. For example, you might use it to show how much of a country's population is divided into different age groups.

Bar charts are useful in side-by-side comparisons, especially in cases in which attribute values do not make up a whole. For example, you might use bar charts to compare census tract household income in 2000 versus 2010. Another variation of the bar chart is a stacked bar chart representing the sum of attributes. For example, if you wanted to display the marital status in a county, you could symbolize the values for three attributes—single, married, and divorced—using a stacked bar chart.

 In this exercise, you will create three more maps for this crime-mapping project. The first will use graduated symbols to map aggravated assault. The second will use bar charts to compare two years of larceny (theft). And the third will use pie charts to show the makeup of all the crime attributes included in the dataset.

1. In ArcMap, open the ex08c map document from your Chapter08 folder.

1
2
3
4
5
6
7
8
9
10
11
12
13
14
15
16
17
18
19
20

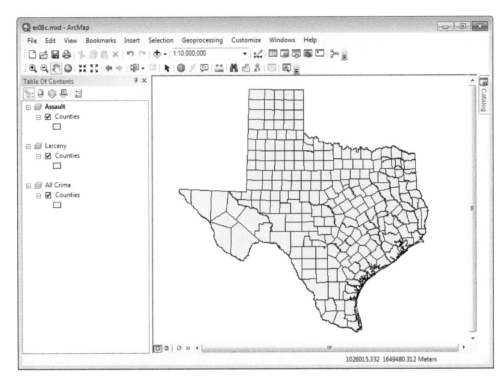

The map is open in data view. In the table of contents, there are three data frames: Assault, Larceny, and All Crime. Each data frame has a Counties layer (all from the same data source). The Assault data frame is active.

2. In the table of contents, open the attribute table of the Counties layer in the Assault data frame. Scroll to the right to the AGRASLT02 column heading. Right-click the AGRASLT02 column heading, and click Sort Descending.

TIP

You may need to widen the column to see the entire field name.

The AGRASLT02 attribute represents the number of aggravated assaults for each of the 254 counties in 2002. You will use a natural-breaks classification to symbolize the data with five classes.

3. Close the table. In the table of contents, double-click the same Counties layer to open its properties. On the Symbology tab, click Quantities > Graduated symbols. In the Fields frame, click the Value arrow, and click AGRASLT02.

In the Symbol column, five differently sized symbols are shown representing five ranges. You will change the symbol template to something more appropriate.

4. Click the Template button to open the Symbol Selector.

5. In the Symbol Selector, click the Circle 1 symbol. Click the Color chip, and click Poinsettia Red. Click OK to close the dialog box.

Next, you will change the background color.

6. Click the Background color chip to open the Symbol Selector.

7. In the Symbol Selector, scroll down the style library and click the Med Yellow symbol. Then click OK.

Next, you will adjust the symbol size to make the difference between the smallest and largest symbols more apparent.

8. Change the Symbol Size from 4 and 18 to **3** and **20**, respectively.

9. In the Label column, click 0-198 and type **Less than 199**. For the remaining labels, change the hyphen to the word **to**. For example, change 199-604 to **199 to 604**. Then click OK.

The map now shows graduated symbols for each county. The table of contents shows the graduated-symbol settings you made. Similar to the density maps you created in exercise 8b, a greater number of incidents are located in counties with a higher population.

You will continue by creating bar charts to compare larceny between 1994 and 2002.

10. In the table of contents, right-click the Larceny data frame, and click **Activate**. Open the Counties attribute table. Scroll to the right to the **LARCY94** and **LARCY02** column headings.

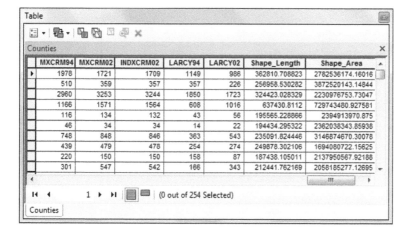

MXCRM94	MXCRM02	INDXCRM02	LARCY94	LARCY02	Shape_Length	Shape_Area
1978	1721	1709	1149	986	362810.708823	2782536174.16016
510	359	357	357	226	256958.530282	3872520143.14844
2960	3253	3244	1850	1723	324423.028329	2230976753.73047
1166	1571	1564	608	1016	637430.8112	729743480.927581
116	134	132	43	56	195565.228866	2394913970.875
46	34	34	14	22	194434.295322	2362038343.85938
748	848	846	363	543	235091.824446	3146874670.30078
439	479	478	254	274	249878.302106	1694080722.15625
220	150	150	158	87	187438.105011	2137950567.92188
301	547	542	166	343	212441.762169	2058185277.12695

(0 out of 254 Selected)

Counties

These two layer attributes represent the number of larcenies for each of the 254 counties in 1994 and 2002. Next, you will look at the attribute statistics to get a better sense of how the data is distributed.

11. Right-click the LARCY02 column heading, and click Statistics.

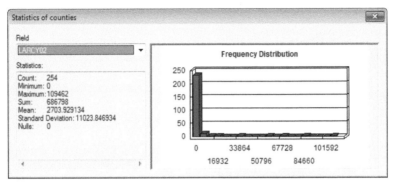

Many of the counties have low counts of larceny (the tallest blue bar in the Frequency Distribution chart). As you saw earlier, a majority of the incidents occur in highly populated counties. Instead of creating bar charts for each county, you will exclude counties that have a low population.

12. Close both the Statistics of counties window and the attribute table.

13. In the table of contents, open the properties for the Counties layer (for the Larceny data frame). On the Symbology tab, under Show, click Charts > Bar/Column. Scroll down the Field Selection list, click LARCY94, and click the right-facing arrow (>) button to add it to the list on the right. Add LARCY02 as well.

Your bar chart colors are randomly selected and may be different from the ones in the graphic.

14. Click the Exclusion button to open the Data Exclusion Properties dialog box. Click the Query tab, and in the Exclude clause list, double-click "POPULATION". Click the less than (<) button, press the spacebar, and type **500000** in the expression box.

TIP

Do not use comma (,) separators—for instance, to separate digits—in query expressions. Commas will invalidate the query. The Verify button can confirm whether a query is valid.

You just constructed a query that *excludes* counties with a population less than 500,000.

TIP

Querying data is a fundamental technique when working with GIS. You will learn more on how to query data in part 5, "Getting information about features."

15. On the Data Exclusion Properties dialog box, click the Legend tab. Check the Show Symbol For Excluded Data check box. Click the Symbol color chip to open the Symbol Selector.

16. In the Symbol Selector, click the Fill Color chip, click Gray 20%, and click OK.

17. In the Label box, type **Population less than 500,000** and click OK.

The "excluded" symbol and label will be added to the table of contents after you confirm your changes on the Layer Properties dialog box in step 19.

But first, you will make a slight adjustment to the size of the bars in the chart.

18. Click the Properties button to open the Chart Symbol Editor dialog box. In the Bars frame, adjust the width to a value of **10**, and click OK.

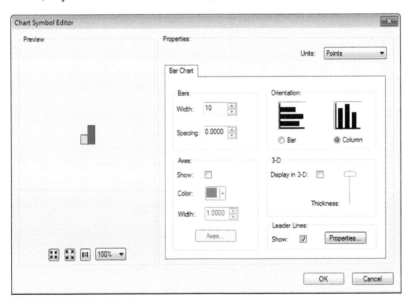

19. Click OK to close the Layer Properties dialog box.

The bar charts are drawn on the map for the counties that were not excluded (those with a population greater than 500,000). These counties include Bexar, Travis, El Paso, Tarrant, Dallas, Harris, and Hidalgo.

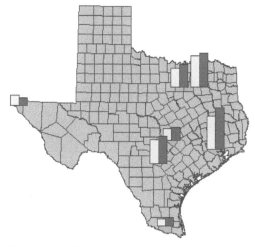

Next, you will create the final map to show all the crime attributes included in the 2002 dataset.

20. Activate the All Crime data frame. Open the properties of the data frame's Counties layer. On the Symbology tab, under Show, click Charts. The Pie option is highlighted. In the Field Selection list, click AGRASLT02, and click the right-facing arrow (>) button to add it to the list on the right. Continue to add the following attributes: ARSONS02, RAPE02, MVTHEFT02, MURD02, ROBB02, BURG02, and LARCY02.

1
2
3
4
5
6
7
8
9
10
11
12
13
14
15
16
17
18
19
20

21. Right-click the Color Scheme ramp, and click **Graphic View** to turn off that view. Click the Color Scheme arrow, and click **Enamel**.

A symbol color is randomly assigned to each attribute using the selected color scheme. Similar to the bar charts, you will exclude some values and change the size of the pie charts.

22. Click the Exclusion button to open the Data Exclusion Properties dialog box. Click the Query tab, and in the Exclude clause list, double-click "POPULATION". Click the less than (<) button, press the spacebar, and type **500000** in the expression box.

ON YOUR OWN
Change the color scheme of the excluded data to your liking.

23. Click the Legend tab. Check the Show Symbol For Excluded Data check box. Click the Symbol color chip to open the Symbol Selector. Click the Fill Color chip, click **Gray 20%**, and click OK.

24. In the Label box, type **Population less than 500,000**, and click OK.

25. Next, click the Size button. On the Pie Chart Size dialog box, click the Vary Size Using The Sum Of The Field Values option. Adjust the symbol size to **20**, select the Appearance Compensation (Flannery) option, and click OK.

The size of each pie chart will be based on several factors: the numeric size (in points), the variation due to the sum of the attribute values, and the appearance compensation option. When compensation is activated, ArcMap scales up proportional circles (in this case, pie charts) so that the larger the circle, the greater the amount of scaling. The compensation of proportional circles was the major research focus of James Flannery, a professor of geography specializing in cartography, in the 1950s.

1
2
3
4
5
6
7
8
9
10
11
12
13
14
15
16
17
18
19
20

26. Click OK to accept your settings. A different color scheme from the one in the graphic may be applied to your map.

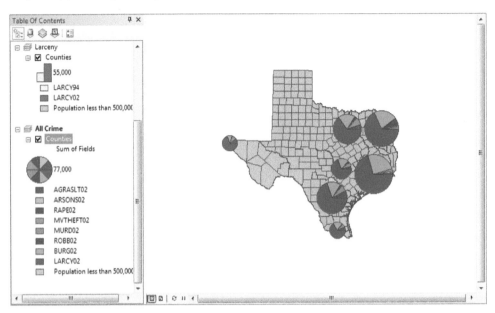

27. On the menu bar, click View > Layout View.

Your map scales may not be exact. On the page layout, click a data frame to activate it, click the Bookmarks menu, and use the bookmark provided for each data frame.

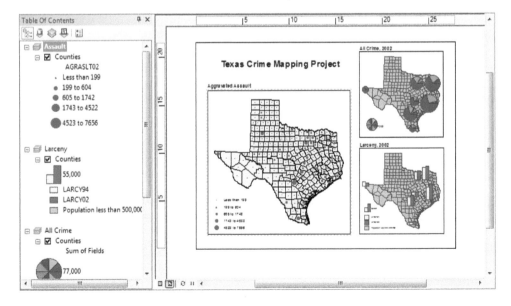

ON YOUR OWN
Create layer files for several of your data layers.

28. If you like, rename your map document, and save it to your MyData folder for chapter 8.

29. If you are continuing to the next chapter, leave ArcMap open. Otherwise, exit the application, and click No if prompted to save your changes.

Summary

In this chapter, you learned what to consider when classifying quantitative attributes. This was useful in determining which ArcMap classification method to use and when to view a data histogram to see how data is distributed. You also learned how to create a visual summary with bar and pie charts. These are some of the most useful tools for displaying information. By now, you should be comfortable manipulating data to achieve different views using graduated colors, graduated symbols, proportional symbols, and dot density effects.

Terms

graduated-color map
graduated-symbol map
proportional-symbol map
dot density map
manual classification
equal interval
defined interval
quantile
natural breaks
geometrical interval
standard deviation
histogram
normalization

1
2
3
4
5
6
7
8
9
10
11
12
13
14
15
16
17
18
19
20

CHAPTER 9
Labeling features

Use dynamic labels
Set rules for label placement
Create graphic labels
Convert dynamic labels to annotation

Learning objectives

- Create dynamic labels
- Modify the default appearance of dynamic labels
- Use predefined styles for labels
- Fine-tune the placement properties of dynamic labels
- Prioritize placement of dynamic labels
- Set a scale dependency for label display
- Create graphic labels using the text tools on the Draw toolbar
- Convert labels to annotation and modify placement

Broadly speaking, a label is any text that names or describes a feature on a map, including proper names (Haiti, Buckingham Palace, Interstate 10), generic names (Lake, Tundra, Hospital), descriptions (Residential, Commercial, Hazardous), or numbers (7,000 placed on a mountain to indicate its elevation). In ArcMap, labels specifically represent values in a layer attribute table. You can add other bits of text to a map using the Draw toolbar, but, strictly speaking, these are not labels and do not behave the same way. Labels, for example, stay close to the features they are associated with as you zoom in or out. Text is not similarly sensitive. Its position may change markedly in relation to the features if you zoom or pan the map.

Default labels (*top*) are often sufficient, but you can modify the default labels (*bottom*) to improve the map's appearance by changing the font, color, size, character width, and spacing; adding special effects such as halos; or specifying placement settings, such as curving river labels along the lines of features. You can also choose from many predefined label styles offered in ArcMap. Data from Data and Maps for ArcGIS 2010, courtesy of ArcWorld and ArcWorld Supplement.

As seen in chapters 3 and 7, you can label all the features in a layer with a single click. This process, called dynamic labeling, is fast, but it has its limitations. For one thing, ArcMap chooses the label positions. (You can set guidelines, but the actual placement is out of your control.) If the scale of your map changes, or if different layers are added or removed, ArcMap may change the position of labels. Still, dynamic labels may be useful for interactive internet map publishing.

An alternative to dynamic labels is annotation. Annotation is static text or graphics that you can individually select, position, and modify. You can enter annotation manually or convert existing labels to annotation. Annotation features are stored as part of a map; in an annotation group the same way as dynamic labels; or in a geodatabase, as an annotation feature class. If you are creating a printed map and want significant control over the appearance and placement of labels, annotation is your best option.

TIP

In the following exercises, your map display will be redrawn many times as ArcMap adds and moves labels. If your map fails to redraw as it should, click the Refresh button 🔄 in the lower-left corner of the map display.

Data

In this chapter, you will use most of the Florida data you used in chapter 7. The feature classes, stored in the Florida geodatabase, are as follows:

- Bathymetry–line features that represent changes in ocean depth *(Sources: Florida Fish and Wildlife Conservation Commission's Fish and Wildlife Research Institute and National Oceanic Atmospheric Administration)*
- Cities–major cities of Florida *(Source: National Atlas of the United States)*
- Coral–polygons representing coral populations off the coast of Florida *(Source: Florida Fish and Wildlife Conservation Commission's Fish and Wildlife Research Institute)*
- Counties–polygons representing Florida county boundaries *(Source: US Census Bureau)*
- Hwy1–a single line feature depicting a major state highway that traverses the Atlantic coastline and Florida Keys *(Source: Florida Department of Transportation)*
- Lakes–polygons representing Florida's major lakes *(Source: Florida Department of Environmental Protection)*
- Rivers–line features representing Florida's major rivers *(Source: Florida Department of Environmental Protection)*
- Shipwrecks–a selection of historical shipwrecks along the southern Atlantic coast of Florida and the Florida Keys *(Source: National Oceanic and Atmospheric Administration Office of Coast Survey's Automated Wreck and Construction Information System)*
- Southeast–a polygon layer that includes demographics *(Sources: ArcUSA, US Census, Esri)*
- StParks–polygons representing Florida state parks *(Source: Florida Department of Environmental Protection)*
- florida_coast–an image of seafloor topography *(Source: Natural Earth)*

Exercise 9a
Use dynamic labels

Dynamic labels are easy to work with because they behave as a group. When you turn them on or off, change their symbol, or change the attribute values they express, it affects all the labels in the layer. The drawback to dynamic labeling is that ArcMap doesn't always put the labels where you want them. For this reason, labels are often converted to annotation in the late stages of map production.

The position ArcMap chooses for dynamic labels depends on many factors, including the size of your application window. Your results may look slightly different from the graphics in the exercises for this chapter. For labeling, it is best to maximize your ArcMap window if possible.

In the exercises in this chapter, you continue to fine-tune a Florida tourism map that will be printed in brochures.

1. **In ArcMap, open the ex09a map document from your C:\EsriPress\GTKArcGIS\ Chapter09 folder.**

 The cities already have default labels, but you need additional labels to make the map more helpful.

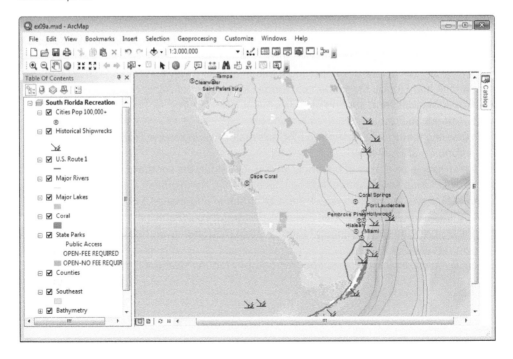

First, you will label the state of Florida and its water features.

2. **In the table of contents, right-click the Southeast layer, and click Label Features.**

Each feature in the layer is labeled, but because this map shows only one state, it has only one label, for Florida, in the center of the map display. ArcMap tries to put the labels in the center of each polygon. Next, you will change the font and color of the label.

3. **Open the properties for the Southeast layer, and click the Labels tab.**

In the Text String frame, the Label Field list shows which field in the attribute table is being used for labels (STATE_NAME). Below this, you have the current label symbology. The labels for this layer are drawn in 8-point black Arial.

4. **Click the color chip. On the color palette, click Gray 20%.**

5. In the Text Symbol frame, in the size list, replace 8 with **14**.

6. Click the Symbol button. In the Symbol Selector, click the Edit Symbol button.

7. On the Editor dialog box, click the Formatted Text tab. Next to Character Width, highlight 100.00 and enter **130**. Under Text Case, click All Caps.

8. Click the Mask tab. Under Style, click Halo. Change the size to **1.**

9. Click OK to close the Editor dialog box. Click OK to close the Symbol Selector. Click OK to close the Layer Properties dialog box.

You have successfully created a label for the state of Florida that is well suited for its purpose: the halo ensures the text isn't lost on the map, but the light-gray color makes it visually less prominent, so it doesn't compete with other features or labels.

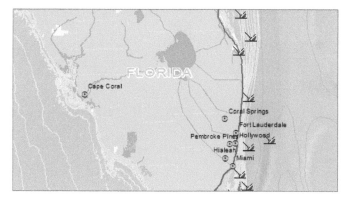

You will use a predefined style to label lakes.

10. Open the properties of the Major Lakes layer, and click the Labels tab. Check the Label Features In This Layer check box. Click the Symbol button to open the Symbol Selector.

11. In the Symbol Selector, scroll through the Esri style library, and click the label style River. Then click OK. (No unique style for lake labels exists, so you will use the River style instead.)

12. Click OK to apply the changes, and OK to close the Layer Properties dialog box.

The lakes on the map are now labeled.

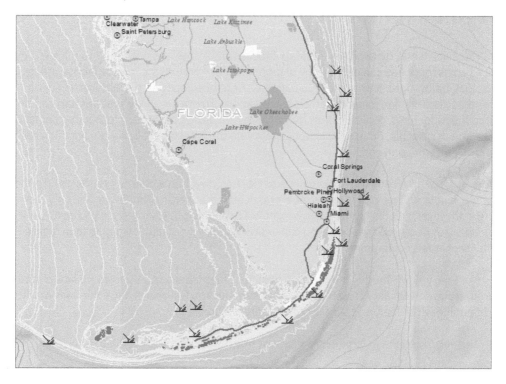

The Florida label moved position once the lakes were labeled. Dynamic labeling means labels will change position depending on the placement of other features and other labels and the size of the map scale.

13. To save your work, rename your map document and save to your MyData folder for chapter 9.

14. If you are continuing to the next exercise, leave ArcMap open. Otherwise, exit the application, and click No if prompted to save your changes.

ArcMap chooses label positions to minimize conflicts with other text and features on the map, but you can control the positioning somewhat, as you will see in exercise 9b. If you want absolute control of the label positions, you must either create graphic labels, which you will do in exercise 9c, or convert dynamic labels to annotation, which you will do in exercise 9d.

Exercise 9b
Set rules for label placement

When you place labels dynamically, ArcMap chooses label positions according to rules you can modify. Labels for points can occupy any of eight positions around the feature (above, below, above right, below right, and so on). You can rank these positions and prohibit certain ones as desired. Labels for line features can be placed above, below, or on the features and can be made to follow features' curves. Labels for polygons are always placed as close to the center of the polygon as possible, but you can choose to rotate the label vertically or remove the label if it doesn't fit inside the polygon.

In addition, label and feature weights can be set for each layer on a map to prioritize and handle labeling conflicts. If there isn't room to label two nearby features in different layers, only the feature in the layer with the higher label weight will be labeled. Additionally, if you don't want certain features obscured by labels, you can assign them higher feature weights to minimize the possibility of labels overlapping them.

In this exercise, you will modify the properties of the Cities labels, and then create labels for the Major Rivers layer with distinct rules for label placement. You will also change the layer priorities that determine which feature is labeled when there is a conflict.

1. In ArcMap, open the ex09b map document from your Chapter09 folder.

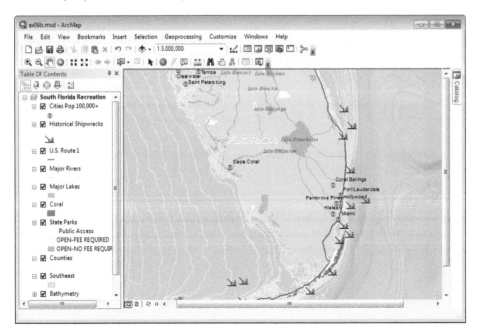

The map looks as it did at the end of exercise 9a. If the cities labels were restricted to the left of the points, there would be less clutter in the southern Atlantic coast area.

2. **Open the properties of Cities Pop 100,000+, and click the Labels tab.**

3. **Under Other Options, click Placement Properties. On the Placement Properties dialog box, make sure the Placement tab is active.**

When labels are offset horizontally around points (the default), each possible position can be ranked. The graphic in the Point Settings frame shows the current setting. The feature is represented by the white square in the middle, and the numbers in the surrounding squares rank each position–1 is the most desirable and 3 the least. The number 0 means a label cannot occupy that position. In the current setting, ArcMap places labels above and to the right of features as long as they do not overlap other labels in any layer. All positions are allowed, although positions below and to the left of the feature are discouraged with this setting.

In this exercise, you want to center the labels to the left of the features.

4. On the Placement Properties dialog box, click Change Location to open the Initial point placement dialog box.

5. On the Initial point placement dialog box, scroll down and click the setting for Prefer Center Left, all allowed. Click OK.

6. Click OK to close **the Placement Properties dialog box, and click OK again to close the La**yer Properties dialog box.

This is better. Most of the city names should be located at the left side of the corresponding point. However, at this scale, many of the labels overlap the point symbols, and it is not clear which point is Hollywood and which is Pembroke Pines. Depending on the size of the ArcMap window, some city labels may still be in conflict with shipwreck features. You will make these final adjustments later, when you convert the labels to annotation.

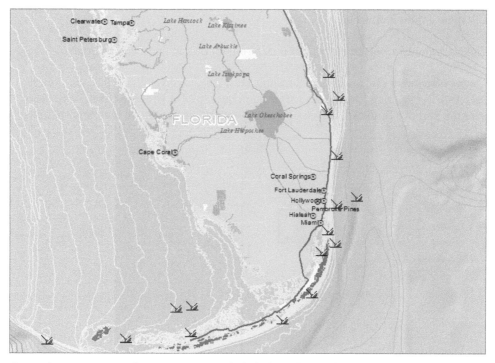

7. Open the properties of Major Rivers, and click the Labels tab. Turn on the labels for the layer.

8. Choose the Rivers symbol style, and click Apply.

9. Click Placement Properties.

 The Placement Properties dialog box is different for points, lines, and polygons.

10. On the Placement tab, under Line Settings, click the Curved option if it is not already selected. For Position, choose Above (and uncheck the On The Line check box). Make sure the Remove Duplicate Labels option is selected, and maintain the other defaults.

When might you choose to place a label directly on a line feature?

11. Click OK to close the Placement Properties dialog box, and click OK again to close the Layer Properties.

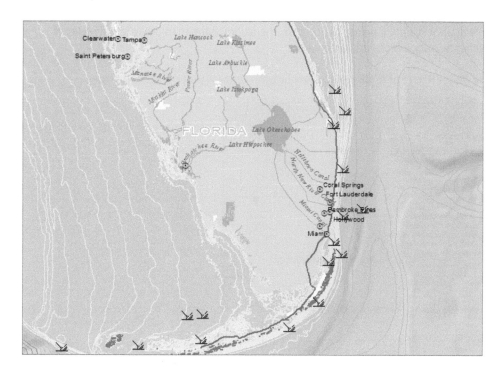

ArcMap tries to label as many features as possible without overlapping labels. You want to label most of the rivers and canals, but doing so makes the map look cluttered, and your city labels look wrong again. You will downgrade the priority level of rivers to help clean up the map.

ALLOWING LABELS TO OVERLAP

By default, ArcMap does not put labels on top of each other. This means that, depending on the scale and the part of the map you are looking at, some features in a layer may not be labeled. To be sure every feature in a layer is labeled, you can tell ArcMap to place overlapping labels. You can also control label-on-label and label-on-feature overlaps for individual layers.

12. On the Customize menu, click Toolbars > Labeling to turn on the Labeling toolbar.

The Labeling toolbar, with several buttons for managing labels, appears in the ArcMap application window.

13. On the Labeling toolbar, click the Label Priority Ranking button.

On the Label Priority Ranking dialog box, the layers of the map are listed in order of labeling priority, which you can change.

14. Click Major Rivers to highlight it, and click the Bottom (not the Down) arrow.

15. Move the Label Priority Ranking dialog box out of the way, and click Apply.

Notice how the map changes. There are fewer labels for rivers and canals, and the map is easier to read.

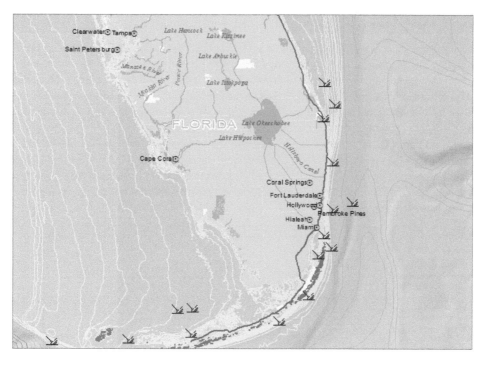

16. Click OK to close the Label Priority Ranking dialog box.

ON YOUR OWN

Read the ToolTips for the other buttons on the Labeling toolbar.

1
2
3
4
5
6
7
8
9
10
11
12
13
14
15
16
17
18
19
20

LABEL MANAGER AND SETTING LABEL CLASSES

You can use the Label Manager dialog box to set label properties and scale dependencies for all layers using one window rather than opening each layer's properties separately. You can also use the Label Manager dialog box to create different classes of labels based on symbology categories or other differentiating attributes. For instance, in the Florida map, you could choose one type of label for the "No fee" state parks and a different type of label (or no label) for the "Fee required" parks. Or if you have a roads feature class whose attribute table contains a field for road type, you could use a different label for each class–for example, symbol markers for highways, black text for main roads, and gray text for local roads. Creating label classes also allows you to have more than one label per feature–for example, if you wanted counties to be labeled with both their name and population, you could use two different labels. Once you create label classes, you can set the label properties for each class separately.

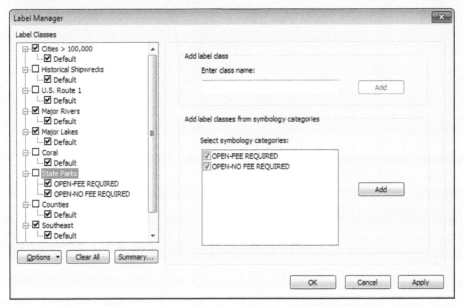

MAPLEX

For advanced label placement options, you can turn on the Maplex Label Engine. Maplex allows you to modify the density of labels and allows more control over label positioning for different data types. For more information on Maplex, consult ArcGIS Desktop Help.

Another way to reduce label clutter is to implement scale-dependent labels (as done in chapter 3).

17. Open the properties of Major Rivers. On the Labels tab, under Other Options, click Scale Range.

18. On the Scale Range dialog box, click the Don't Show Labels When Zoomed option, and from the Out beyond menu, choose 1:1,000,000. Click OK to close the Scale Range dialog box. Click OK again to close the Layer Properties dialog box.

The Major Rivers labels will appear only at scales larger than 1:1,000,000. If the tourism brochure uses a large-scale local map in addition to the main map, rivers and canals will be labeled.

19. On the menu bar, click Bookmarks > Canals to see the scale-dependent labels.

1
2
3
4
5
6
7
8
9
10
11
12
13
14
15
16
17
18
19
20

20. Zoom to the full extent of the map, and close the Labeling toolbar.

21. To save your work, rename your map document, and save to your MyData folder for chapter 9.

22. If you are continuing to the next exercise, leave ArcMap open. Otherwise, exit the application, and click No if prompted to save your changes.

Exercise 9c
Create graphic labels

If you want to add stand-alone text to your map or if you just want to label a few features interactively, you can use the tools on the Draw toolbar. Most of the tools on the Draw toolbar are for creating stand-alone text, but the Label tool allows you to choose a label that is connected to a feature's attributes. The labels or text boxes created from the Draw toolbar act like graphics–they can be selected and moved wherever you like. In this exercise, you will add text to mark the Gulf of Mexico and the Atlantic Ocean, and then create a graphic label for the US Route 1 feature.

This technique is meant for cases in which you have only a few features to label or if you want to label base layers that don't have attributes.

1. In ArcMap, open the ex09c map document from your Chapter09 folder.

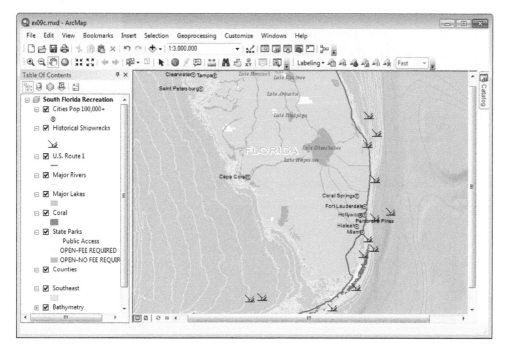

As before, some of your dynamic labels may be in slightly different locations from in the graphic.

Map readers would be better oriented if you labeled the Gulf of Mexico and the Atlantic Ocean. However, you don't have an ocean layer with attributes to which you can attach labels. (The ocean background is a raster layer, which doesn't have text attributes.) Because there are no distinct Gulf of Mexico or Atlantic Ocean features, there is nothing to label. Instead, you will add a graphic label using the Text tool. Graphic labels are selectable and stand alone on the map—they are not connected to any feature.

2. If it is not already available, display the Draw toolbar. (On the Customize menu, click Toolbars > Draw.)

3. On the Draw toolbar, click the Text tool ⓐ. Move the mouse pointer over the map. The pointer changes to crosshairs with an A (for text). Click in the Gulf of Mexico to add a text box.

4. In the text box, type **Gulf of Mexico**, and press Enter.

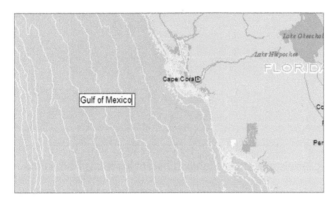

The text is drawn in 10-point black Arial, as you can see on the Draw toolbar. The bright blue outline indicates the text is selected. Next, you will increase the font size.

5. Right-click the Gulf of Mexico label. On the shortcut menu, click Properties. On the Properties dialog box, on the Text tab, click Change Symbol.

This opens the Symbol Selector. Next, you will select a style and modify it according to your preferences.

6. In the Symbol Selector, scroll down until you get to the predefined label style for Ocean, and click it to select the color and spacing determined by that style. Next, click the font arrow, and click Book Antiqua (if it is installed on your computer) or another font of your preference.

7. Click OK to close the Symbol Selector. Click OK again to apply the new properties to the text.

 Now that it is larger, the text may have to be moved.

8. Move the mouse pointer over the text. The pointer becomes a four-headed arrow. Use the pointer to drag the text over the gulf, and release the mouse button. Click in an empty blue area to clear the text selection.

1
2
3
4
5
6
7
8
9
10
11
12
13
14
15
16
17
18
19
20

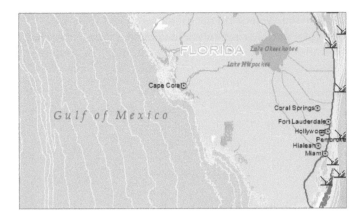

ON YOUR OWN

Repeat steps 3 through 8, but this time label the Atlantic Ocean

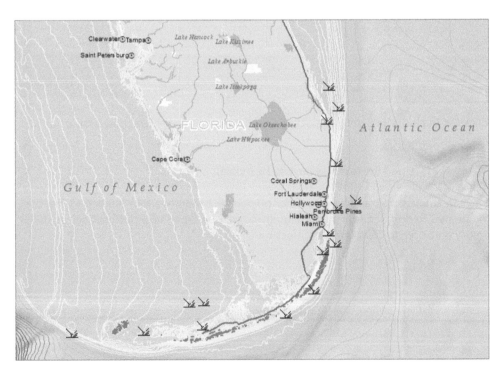

Finally, you will label US Route 1 using the Label tool.

9. On the Draw toolbar, click the down arrow next to the Text tool to see the other text tools available. Click the Label tool.

TIP

When you use any text tool, it will become the default tool shown on the Draw toolbar until you choose a different tool.

10. On the Label Tool Options dialog box, click the Place Label At Position Clicked and the Choose A Style options. In the style library, click the first symbol, for U.S. Route.

11. Move your mouse pointer over the U.S. Route 1 line feature (the red line running down the Atlantic coast), just east of Lake Okeechobee. When you get to the MapTip US1, click to place the label.

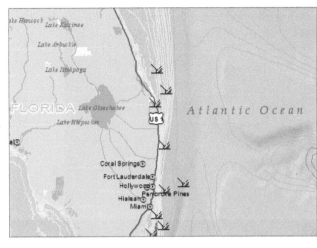

12. With the label selected—it should be outlined in bright blue—right-click it, and click Properties. Change the symbol size to 6 pt black Arial in the Symbol Selector.

13. Close any open dialog boxes, click the Select Elements tool (on either the Tools toolbar or the Draw toolbar) ▶, and click anywhere on the map to clear the highway symbol selection.

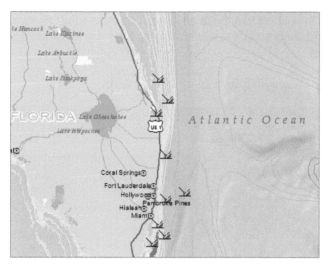

14. To save your work, rename your map document, and save to your MyData folder for chapter 9.

15. If you are continuing to the next exercise, leave ArcMap open. Otherwise, exit the application, and click No if prompted to save your changes.

Exercise 9d
Convert dynamic labels to annotation

Although you have considerable control over dynamically placed labels, you can't specify exact positions or symbolize them individually. To do this, you must convert the labels to annotation, which means, essentially, that you tell ArcMap to turn over full responsibility for label placement and appearance to you. Annotation can be selected, moved, and modified piece by piece, like the text boxes you created in exercise 9c. The difference is, you can quickly create annotation for all features in a layer rather than one at a time. Typically, you convert labels to annotation when your map is nearly finished and you want to make final adjustments to the labels.

Annotation can be saved within your map document in an annotation group, the same way as dynamic labels, or to an annotation feature class within a geodatabase. It then becomes much like any other spatial dataset—it can be copied, edited, and added to any map document.

For more information, consult ArcGIS Desktop Help.

In this exercise, you will convert the cities labels to annotation so you can make adjustments to the label positions. Once converted to annotation, the labels will keep the same size and location relative to surrounding features.

1. **In ArcMap, open the ex09d map document from your Chapter09 folder.**

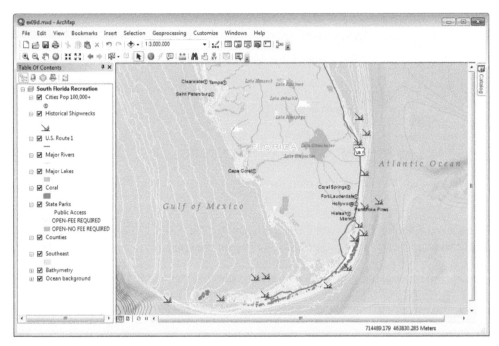

2. **In the table of contents, right-click the Cities Pop 100,000+ layer and click Convert Labels to Annotation. This opens the Convert Labels to Annotation dialog box.**

The default option for storing annotation is to store it in a geodatabase. However, since you want to modify only a handful of labels, you will store the annotation as an annotation group in the map document.

MAP ANNOTATION OR GEODATABASE ANNOTATION?

If you have a large number of labels, it is best to use geodatabase annotation, because it is faster and can handle a greater number of annotation features more efficiently. Creating annotation often takes a lot of time, so saving to a geodatabase can be desirable, allowing the work to be reused and shared. Geodatabase annotation can be either stand alone or feature linked. Stand-alone annotation is used for text that is not associated with a map feature. In feature-linked annotation, the labels are associated with another feature class, so if a feature is moved or deleted, the annotation is moved or deleted as well. You will learn more about geodatabases in chapter 11.

3. **Click the option to store annotation in the map. Click Convert.**

Convert Labels to Annotation

Store Annotation Reference Scale

○ In a database ● In the map 1:3,000,000

Create Annotation For

● All features ○ Features in current extent ○ Selected features

Feature Layer	Annotation Group
Cities Pop 100,000+	CitiesPop 100000+ Anno

Destination: Annotation group stored in the map document

☑ Convert unplaced labels to unplaced annotation Convert Cancel

The labels are converted to annotation. In the process, their positions may have changed slightly.

Next, you will move the labels for Hollywood and Pembroke Pines so it is easy to see which feature goes with which label. You will also move the labels for the other cities just a bit, so the text doesn't touch the corresponding point symbols.

4. **Zoom to the Canals bookmark.**

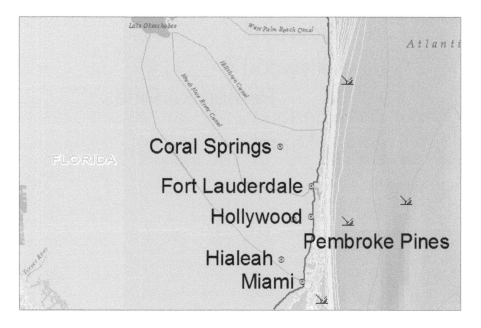

Notice how large the annotation labels are on the map display compared with the dynamic labels for lakes and rivers. Unlike dynamic labels, annotation labels are static, meaning they act like a feature–just as a lake grows in appearance as you zoom in on a map, so does the annotation.

TIP

If you want your dynamic labels to act like static annotation when zooming in, go to the full extent of the map and set a reference scale for the data frame. Right-click the data frame in the table of contents, and click Reference Scale > Set Reference Scale.

ON YOUR OWN

Use the Identify tool to determine which point represents Hollywood.

5. On the Tools toolbar, click the Select Elements tool. Click the Hollywood label to select it.

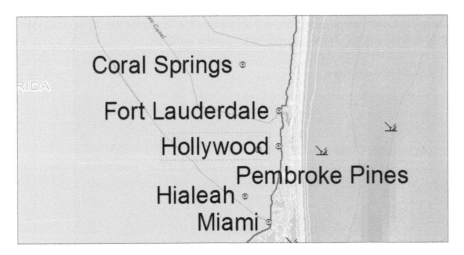

6. Drag the Hollywood label slightly north.

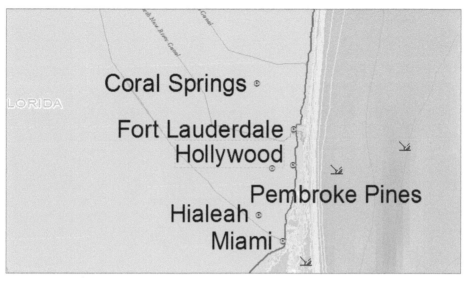

7. Click the Pembroke Pines label, and drag it to the southwest of the point feature, as shown in the graphic. Click somewhere away from the label to clear the selection. Click Refresh.

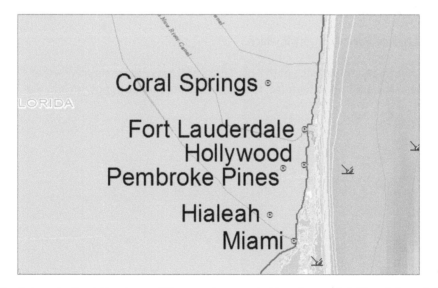

8. Return to the full extent of the map to see whether the readability of these labels has improved.

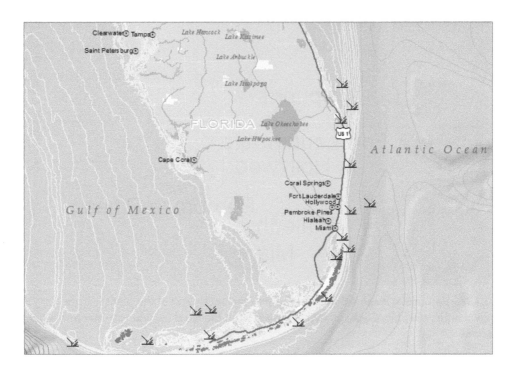

Much better—it is clear now which city is which.

9. **Zooming in as necessary, select each of the city labels and move them slightly farther away from their associated point features, to ensure that the text does not run into the point symbols.**

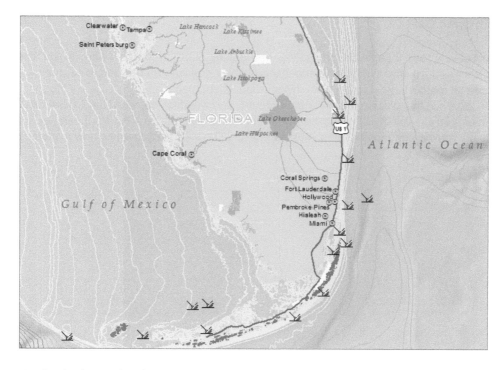

You finished your Florida tourism map—good job!

10. To save your work, rename your map document, and save to your MyData folder for chapter 9.

ON YOUR OWN

As you learned how to do in exercise 5c, share the map package to your ArcGIS Online account.

11. If you are continuing to the next chapter, leave ArcMap open. Otherwise, exit the application, and click No if prompted to save your changes.

Summary

You now have a good foundation for creating nice-looking maps: in chapters 7 and 8, you learned how to create informative symbology; and in this chapter, you learned how to create labels using several methods. Remember, with dynamic labels, ArcMap places the labels (although you can influence that placement by setting priorities and weights), but with stand-alone text or annotation, you can select and move individual labels and give them a unique appearance. In chapter 10, you will continue building your mapmaking knowledge by learning how to create a layout.

Terms

label
annotation

CHAPTER 10
Making maps for presentation

Create a layout
Add titles and text
Add standard map elements
Add final touches and set print options

Learning objectives

- Become familiar with page layout tools
- Create effective map layouts
- Add text to a layout
- Create map elements such as a north arrow, scale bar, and legend
- Become familiar with print settings and options

A map is not complete without several critical features and a well-composed page layout. Any map that is destined to be printed or used online (or on-screen) should contain a certain collection of map elements laid out and organized on the page. These elements include data frames (containing map layers), a legend, a scale bar, a north arrow, a map title, descriptive text, and additional graphics. Cartographic design should be informed by both aesthetics and the ability to communicate. Cartographic elements need to accurately convey information about the map without diverting attention from the map.

The measure of a good layout is how well it conveys information to its readers. A good layout should not only provide the facts, but also persuade opinion. When creating a layout, you should ask yourself some key questions:

- What is the purpose and intended use of the map?
- Who is the target audience?
- What physical (or digital) size and page orientation will it have?
- Will the map stand on its own, or will it be part of a map series with maps that share a template?

- How will the map elements be organized on the page?
- Will the layout contain more than the common map elements, such as graphs or reports?
- What is the intended map scale, and how large will the data frames in the layout be?

A layout created in ArcMap that shows the main map, inset maps, scale bars, and a legend. 2012-2013 Lawrence Transit System Map courtesy of City of Lawrence, KS. Data courtesy of City of Lawrence and University of Kansas.

You may also want to think about how the layout of data frames and map elements affects the balance and composition of the layout. A well-planned page layout with properly sized data frames and a clear visual hierarchy of surrounding elements will allow readers to explore the map on a logical path. Readers will generally focus on the map itself; however, haphazardly placed text, elements, or graphical objects may draw their attention from the intended focal point. Finding the right balance between elements that should be in the foreground and those that are best suited to the background is one of the keys to good mapmaking.

In this chapter, you will create the layout shown here of Queensland, Australia, bioregions and bio-subregions, which are researched areas of diverse biology and geographic landscape.

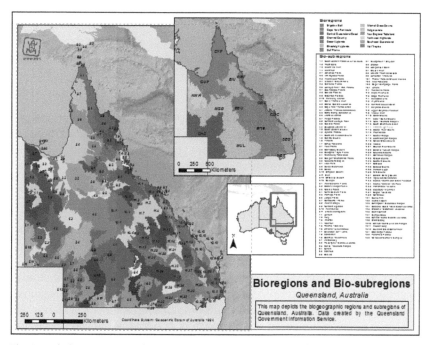

The Australia bioregions map that you'll complete including layout, labels, legend, inset maps, and surrounding map elements. Data from Data and Maps for ArcGIS 2010, courtesy of ArcWorld Supplement, State of Queensland (Department of Science, Information Technology, Innovation and the Arts) 2010, 3.0 Australia (CC BY 3.0).

Data

The exercises in this chapter use the following datasets:

- australia.shp–polygons of the provinces of Australia *(Source: Data from Data and Maps for ArcGIS 2010, courtesy of ArcWorld Supplement)*
- bioregions.shp–polygons of bioregions in the province of Queensland, Australia *(Source: Courtesy of State of Queensland [Department of Science, Information Technology, Innovation and the Arts] 2010, 3.0 Australia [CC BY 3.0])*
- biosubregions.shp–polygons of bio-subregions in the province of Queensland, Australia *(Source: Courtesy of State of Queensland [Department of Science, Information Technology, Innovation and the Arts] 2010, 3.0 Australia [CC BY 3.0])*
- country_background.shp–polygons of surrounding countries, used as a background layer *(Source: ArcWorld Supplement)*
- ocean_background.tif–raster of the ocean, used as a background layer *(Source: Natural Earth)*

Exercise 10a
Create a layout

The amount of space on which elements can be arranged may be anything from a letter-size sheet of paper to a wall poster. The orientation may be vertical (portrait) or horizontal (landscape). Choosing the dimensions and orientation of the map before you start makes the layout process easier. It will influence your decision on factors such as data frame size, map scale, text size, and other map elements.

ArcMap has rulers, guides, and a grid to help you arrange map elements on a page. You can also align, nudge (move incrementally), distribute (space evenly), order, rotate, and resize selected elements to place them where you want them.

For this exercise, you are a GIS student researching the biodiversity of Queensland, Australia, and have just finished mapping its 13 biogeographic regions and 133 biogeographic subregions (also known as bioregions and bio-subregions, respectively). Bioregions represent broad landscape patterns that are the result of the interplay among such factors as geology, climate, and biota. Bio-subregions are based on finer differences in geology, vegetation, and other biophysical attributes and are the basis for determining the major regional ecosystems. You will create a page layout as part of a presentation used to provide useful information to other researchers, students, and people interested in studying biodiversity.

1. **Start ArcMap, and open the ex10a map document from your C:\EsriPress\GTKArcGIS\ Chapter10 folder.**

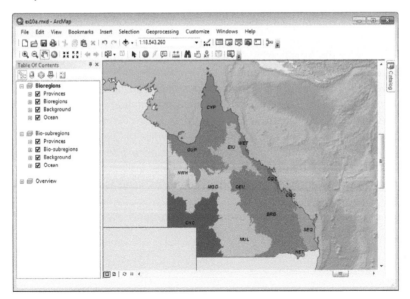

The map document contains three data frames: Bioregions, Bio-subregions, and Overview. Data frames allow multiple maps to be shown on the page layout. The Bioregions data frame is currently active. Bioregions are classified and labeled based on their abbreviated names (for instance, Brigalow Belt is labeled BRB) and are symbolized using a color ramp with various shades of orange.

2. **In the table of contents, expand the Bioregions layer to see the bioregion names and corresponding feature colors. When you are finished examining the feature symbols, collapse the layer.**

 What type of classification is used to classify these bioregions: categorical or quantitative?

 All three data frames will be part of your page layout. You will focus the attention on the bio-subregions map, and position the bioregions map next to it. For reference, you will include a map of Australia with an extent indicator showing where the state of Queensland is located. In ArcMap, an extent indicator is simply a rectangle that is placed in a data frame to reference the specific geographic location of another data frame.

3. **On the menu bar, click View > Layout View.**

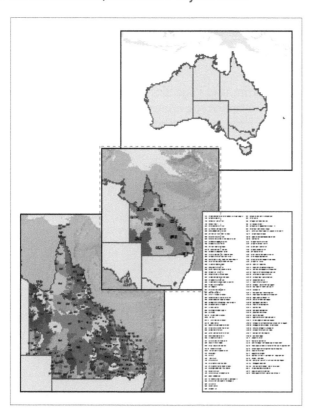

In layout view, you see how the printed map will look. The data frames are stacked on the virtual page according to their order in the table of contents (although they can be reordered any way you want). Also included is an index of bio-subregions that consists of two text boxes and a rectangle. It will be easier to see what you are doing if you make the ArcMap application window as large as possible.

4. Maximize the ArcMap window. (Click the Maximize button on the right end of the title bar.) We recommend that you maintain a large ArcMap window size for all the exercises in this chapter.

Notice that in layout view, the Layout toolbar automatically appears (if it wasn't already turned on) and is active. It may be floating or already docked to your ArcMap window. If it is docked, you will not see the title of the toolbar. To see it, you will have to undock the toolbar.

5. On the Layout toolbar, click the Zoom Whole Page button.

The virtual page is enlarged within the layout window. At the moment, the page orientation is portrait. Next, you will adjust the page settings.

6. On the menu bar, click File > Page and Print Setup.

7. On the Page and Print Setup dialog box, in the Map Page Size frame (not to be confused with the Paper frame, in which the Landscape option will not change the orientation of the layout), set the orientation to Landscape, as shown in the graphic. Uncheck the Use Printer Paper Settings check box if it is checked. Click OK.

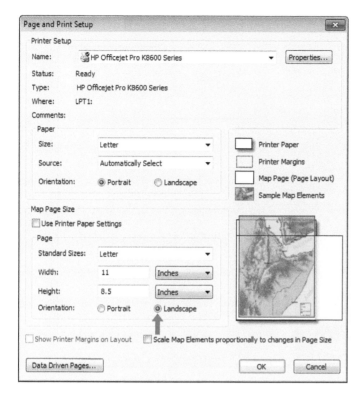

The orientation of the layout is changed. You also need to change the orientation of the printer paper, but you will do this later in the chapter. Don't worry about the position of the overview map—you will fix that next.

To help you place data frames exactly, you will turn on page guides. Page guides are cyan-colored lines originating from arrows on the vertical and horizontal layout rulers. These guides help you align and position data frames and map elements but do not appear on the printed map. In this exercise, some guides have already been set for you.

8. **On the menu bar, click View > Guides.**

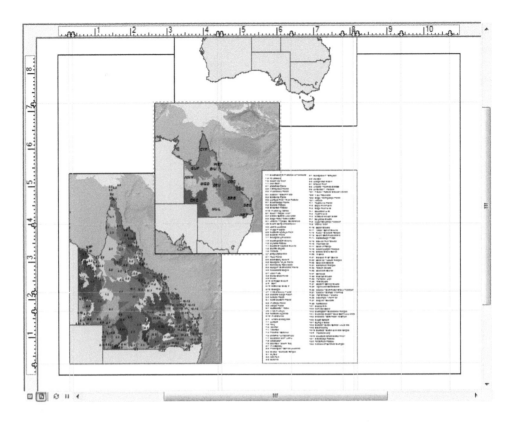

The guides are displayed on the layout.

9. Place the mouse pointer over the vertical ruler on the left side of the display. Click to add a guide at 8 inches and another one at 8.2 inches, as shown in the graphic.

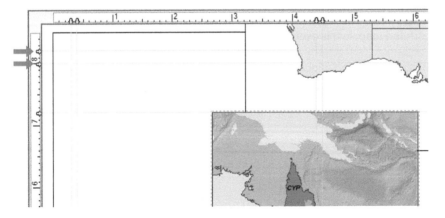

TIP

Don't worry if you didn't click exactly the right spot on the ruler. To move the guide, drag it to the correct position. The measurement is indicated in the upper-left corner of the display area.

The guides will be more useful if map elements snap (automatically align) to them.

10. **Right-click an empty part of the page. On the shortcut menu, click Guides > Snap to Guides.**

Next, you will resize and arrange the data frames on the page, beginning with Bio-subregions (the largest data frame).

11. **In the display area, click the Bio-subregions data frame to select it.**

Once selected, it is outlined with a cyan dashed line and marked with cyan selection handles.

12. **Move the mouse pointer over the data frame. When the pointer becomes a four-headed arrow, use the pointer to drag the frame so its upper-left corner snaps to the guides at 8 inches on the vertical (left) ruler and 0.3 inch on the horizontal (top) ruler.**

Next, you will make the Bio-subregion's data frame bigger by dragging a selection handle. When you resize a data frame, ArcMap automatically adjusts the map extent and the display scale.

TIP

To make sure the map extent or the display scale stays the same, you can set a fixed extent or scale on the Data Frame tab on the Data Frame Properties dialog box.

13. Place the mouse pointer over the selection handle in the lower-right corner of the data frame. When the pointer becomes a double-headed arrow, use the pointer to drag the corner of the data frame until it snaps to the guides at 0.3 inch (vertical ruler) and 6.4 inches (horizontal ruler).

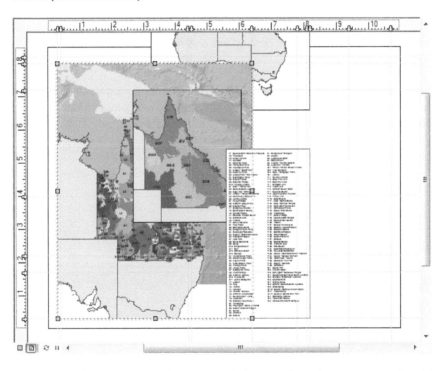

14. On the Standard toolbar, in the Map Scale box, replace the existing scale with 1:13,000,000, and press Enter.

The data frame is updated with an appropriate map scale to show the entire Bio-subregions layer.

Next, you will move the Bioregions data frame.

15. On the virtual page, click the Bioregions data frame to select it. Drag the frame so its upper-left corner snaps to the guides at 8.2 inches (vertical ruler) and 4.4 inches (horizontal ruler). On the Standard toolbar in the Map Scale box, replace the existing scale with **1:25,000,000**, and press Enter.

Next, you will reposition the Bio-subregions index.

16. In the display area, click the index to select it. Drag the index so its lower-right corner snaps to the guides at 2.2 inches (vertical ruler) and 10.7 inches (horizontal ruler).

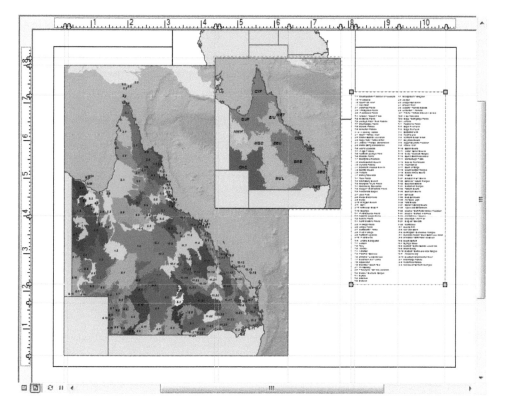

Next, you will reposition, rescale, and change the layer order of the Overview data frame.

17. In the display area, click the Overview data frame to select it, and right-click it. On the shortcut menu, click Order > Bring to Front.

18. Open the properties of the Overview data frame. On the Size and Position tab, in the Size frame, enter **2** in the Width box and **1.8** in the Height box. Then click OK.

General	Data Frame	Coordinate System	Illumination	Grids
Feature Cache	Annotation Groups	Extent Indicators	Frame	Size and Position

Position
X: 3.2342 in
Y: 6.5829 in

☐ As Offset Distance

Anchor Point:

Size
Width: 2 in
Height: 1.8 in

☐ As Percentage

☐ Preserve Aspect Ratio

Element Name

Overview

Don't worry about the appearance or map scale of the overview map. You will fix this in a moment.

19. Drag the Overview data frame so its lower-right corner snaps to the guides at 2.2 inches (vertical ruler) and 7.8 inches (horizontal ruler).

20. On the Standard toolbar, change the scale to **1:120,000,000**, and press Enter.

The data frames now have their final sizes and positions. Next, you will finish the Overview data frame by creating an extent indicator.

21. Open the properties of the Overview data frame. On the Extent Indicators tab, click Bio-subregions in the Other data frames list, and click the right-facing arrow (>) button to move the item to the Show Extent Indicator For These Data Frames list. Click OK.

An extent indicator framed in red is drawn to show the spatial extent of the Bio-subregions data frame inside the Overview data frame.

TIP

> If your maps look slightly different, use the Pan tool on the Tools toolbar to fix the position of the maps.

22. To save your work, rename your map document, and save to your MyData folder for chapter 10.

23. If you are continuing to the next exercise, leave ArcMap open. Otherwise, exit the application, and click No if prompted to save your changes.

In exercise 10b, you will add a title and additional text to the map.

USING MAP TEMPLATES

One of the easiest ways to keep your maps consistent is to use a map template. Any map document (.mxd) can be made into a map template. Templates are layouts that contain layers, data frames, page layouts, and other common map elements. Using a template to create a map series can save time because you don't have to re-create each element of the map. Like other parts of ArcMap, templates can be shared to increase productivity within an organization.

Use a map template

When you open ArcMap or create a document, the Getting Started dialog box appears and provides options to help you choose a recent map or a map template. When you create a map document from a template, an untitled map appears based on the selected template. Provided templates are listed under the Templates node under New Maps. Types of templates range from blank, standard page-size templates to traditional layouts that include a basemap.

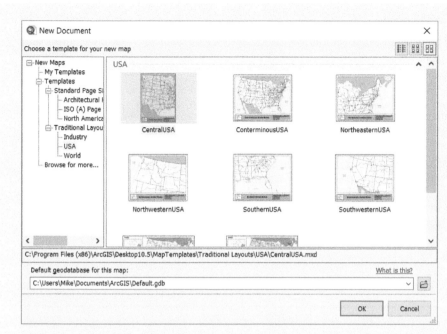

While working in a map document, you can change the current template to another one. On the Layout toolbar, click the Change Layout button, and on the Select Template dialog box, choose a new template.

1

2

3

4

5

6

7

8

9

10

11

12

13

14

15

16

17

18

19

20

Create a map template

To create a template that only you can see, save or copy a map document to a folder in your user profile (C:\Users\[user]\AppData\Roaming\Esri\Desktop10.x\ArcMap\Templates). You can create folders to organize templates within this user profile folder. Templates stored in this location are listed under My Templates in New Maps on the Getting Started dialog box.

Alternatively, to create a template that all users of your system can use, you will have to save or copy a map document to the default location in the ArcGIS installation directory (C:\Program Files (x86)\ArcGIS\Desktop10.x\MapTemplates). The default location can be changed in the System Paths tab, using ArcMapAdvancedSettings.exe located in the Utilities folder of the install directory. Similarly, categories (folders) can be created to help you organize your templates.

Categories will appear on the Getting Started, New Map, and Change Layout dialog boxes only if a map document is present in the folder under the Templates or My Templates headings.

Exercise 10b
Add titles and text

All maps have titles, and many have subtitles as well. A good title helps the reader understand what to look for on the map. Additional text might be a short description to further explain what the map is about.

Now that you have laid out the three data frames and index for the Queensland, Australia, bioregions and bio-subregions map in exercise 10a, you will add a title, subtitle, and description.

1. In ArcMap, open the ex10b map document from your Chapter10 folder. Maximize the ArcMap window, and click the Zoom Whole Page button on the Layout toolbar.

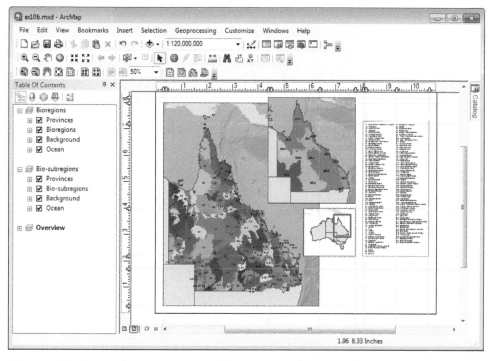

The map is in layout view. It looks as it did at the end of exercise 10a.

Next, you will insert the title and position it in the space in the lower-right corner of the page.

2. On the menu bar, click Insert > Title. On the Insert Title dialog box, in the text box, type **Bioregions and Bio-subregions**, and click OK.

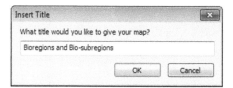

TIP

If you make a mistake and need to change the text, double-click the title to open its text properties. You may notice that the Properties text box contains the following text: <dyn type="document" property="title"/>. This code dynamically calls the title entered, which is then populated on the Map Document Properties dialog box, in the Title text box (on the menu bar, go to File > Map Document Properties).

The title is centered near the top of the page. However, it is too small. You can change its font, size, style, or color on the Draw toolbar.

3. **Make sure the title is selected, indicated by a cyan box surrounding it. Drag the title to the empty white area in the lower-right corner of the layout.**

4. **To turn on the Draw toolbar, on the Customize menu, click Toolbars > Draw. The toolbar may be floating or docked. On the Draw toolbar, for font size, highlight the current value and replace it with 20, and press Enter. Click the Bold button.**

The title font is changed to 20-point Bold Arial.

5. **Snap the upper-right corner of the title to the guides at 1.7 inches (vertical ruler) and 10.7 inches (horizontal ruler).**

Next, you will add a subtitle, and position it below the title.

6. On the menu bar, click Insert > Title. A duplicate of the title is created at the top of the layout. Drag the new title just below the existing title in the lower-right corner of the map. Double-click the text box. On the Properties dialog box, replace the existing text with **Queensland, Australia.**

Next, you will change the font and style of the subtitle.

7. On the Properties dialog box, click Change Symbol. In the Symbol Selector, change the font size to **14,** click the Italic button, and click OK. Click OK again to close the Properties dialog box.

Next, you will add descriptive text.

8. On the Draw toolbar, click the Text button arrow, and click Rectangle Text.

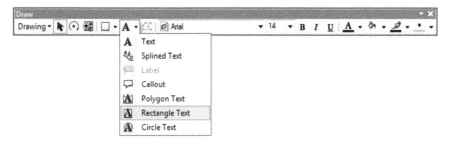

9. Press and drag the tool to draw a rectangle below the subtitle. Double-click the new text box. Delete the selected text and type **This map depicts the biogeographic regions and subregions of Queensland, Australia. Data created by the Queensland Government Information Service.** Next, click the Left Justify button.

10. On the Properties dialog box, click the Size and Position tab. In the Size frame, change the width to **4 in** and the height to **0.65 in**, and click OK.

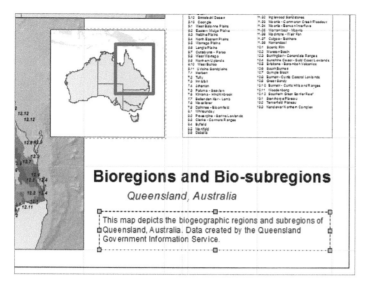

Next, you will fine-tune the positions of the three elements and group them together.

The subtitle needs to be very close to the title to fit in the space prepared for it.

11. Select the subtitle, and right-click it. On the shortcut menu, click Nudge > Nudge Up. (You can also use the Up Arrow key to nudge it.) Keep nudging until the subtitle is as close to the title as possible without touching it.

If you go too far, you can use Nudge Down (or press the Down Arrow key) to nudge the subtitle down. At any time, use the Layout Zoom In tool to zoom in if you have difficulty seeing these elements. You may also drag them into position using the Select tool.

12. Similarly, nudge the description rectangle up.

13. With the description rectangle selected, hold down Ctrl and click the title and the subtitle. All three elements are selected. Right-click any element, and on the shortcut menu, click Align > Align Center.

14. With the elements still selected, right-click any element and, on the shortcut menu, click Group.

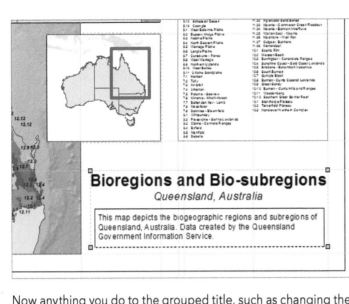

Now anything you do to the grouped title, such as changing the font size or the color, will affect all the elements that are grouped together. If the right side of your grouped objects moved away from the vertical guide, simply nudge it or snap it back to the guide.

TIP

To ungroup an element, right-click the element, and on the shortcut menu, click Ungroup.

Next, add a background rectangle to frame all the text elements.

15. On the Draw toolbar, click the Rectangle tool. (If necessary, click the arrow and select the Rectangle tool.)

16. Press and drag the tool to draw a rectangle that snaps to the guides around the title.

By default, the rectangle is pale yellow. Next, you will change its color and move it behind the title.

17. Make sure the rectangle is selected. On the Draw toolbar, click the Fill Color arrow. On the color palette, click the Sahara Sand color chip.

The rectangle color is changed to a light sandy orange.

18. Right-click the selected rectangle, and click Order > Send to Back.

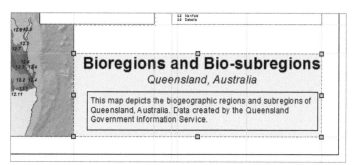

Finally, you will center the title within the rectangle. Because ArcMap aligns elements in relation to the last element selected, you will select the grouped text first.

19. Click the grouped element containing the title, subtitle, and description. Press and hold Shift and click the bottom of the colored rectangle, outside the grouped text elements selection box.

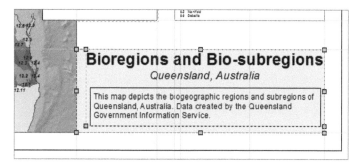

Both the grouped title and the rectangle should be selected. The rectangle's selection color is cyan, indicating it is the last element selected. The grouped-text selection color is changed to green.

20. Right-click either selected element, and click Align > Align Vertical Center.

21. Right-click either selected element, and click Align > Align Center.

1
2
3
4
5
6
7
8
9
10
11
12
13
14
15
16
17
18
19
20

1

2

3

4

5

6

7

8

9

10

11

12

13

14

15

16

17

18

19

20

This centers the title vertically and horizontally within the rectangle.

22. **Right-click either of the selected elements once more, and click Group. Then click an empty part of the virtual page to clear the grouped-element selection.**

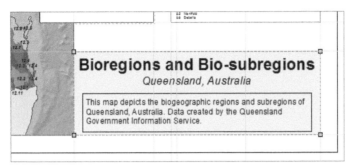

23. **Zoom back to the extent of the full page by clicking the Zoom Whole Page button.**

24. **To save your work, rename your map document, and save to your MyData folder for chapter 10.**

25. **If you are continuing to the next exercise, leave ArcMap open. Otherwise, exit the application, and click No if prompted to save your changes.**

In exercise 10c, you will add a north arrow, scale bar, and legend to your page layout.

Exercise 10c
Add standard map elements

Map elements such as the north arrow, scale bar, and legend provide useful information to readers about your map. The north arrow provides an indication of true north (the direction toward the geographic North Pole) instead of magnetic north (the direction of a compass needle pointing toward the magnetic North Pole) for orientation. Displaying scale or distance with a scale bar is important and can be indicated using graphical and numeric forms. A legend lists the symbols used on a map and what they symbolize. ArcMap has many styles for north arrows, scale bars, and legends and allows you to customize them.

These elements are dynamically associated with the data frame that is active when they are inserted. A scale bar changes when the display scale of the data frame changes. A north arrow

rotates with the data frame. A legend is updated when layers in the data frame are deleted or resymbolized. Although these elements can be moved anywhere on a map layout, it is good practice to keep them near or within the data frame to which they belong.

The people who view your presentation and your map may or may not be familiar with the bioregions and bio-subregions of Australia. Adding a legend is necessary to aid the interpretation of a map. In exercise 10a, you positioned a bio-subregions index in the layout. This can be considered a legend, too, because the values in it correspond to the values on the map. In this exercise, you will begin by creating a north arrow and scale bar and finish by creating a legend for the bioregions layer.

1. In ArcMap, open the ex10c map document from your Chapter10 folder. Maximize the window as needed, and on the Layout toolbar, click the Zoom Whole Page button.

The map is in layout view and looks as it did at the end of exercise 10b. The Overview data frame is active.

2. On the menu bar, click Insert > North Arrow. On the North Arrow Selector dialog box, click ESRI North 2, and click OK.

The north arrow is added and placed near the center of the page.

TIP

Use the Zoom In tool on the Layout toolbar to zoom in on the page if you have difficulty seeing the north arrow. Using the Zoom In tool on the Standard toolbar will change the map scale of a data frame. Don't forget to click the Select Elements button on the Tools toolbar after zooming in so you can select features again.

3. Center the mouse pointer over the north arrow until the pointer becomes a four-headed arrow. Using the pointer, drag the north arrow to the lower-left corner of the Overview data frame. (Use the graphic as a guide. If necessary, turn off Snap to Guides for more control of placement.)

4. Double-click the north arrow. On the North Arrow Properties dialog box, click the color chip, and click Gray 40%. In the Size box, type 40. Then click OK. When you are finished, click an empty part of the virtual page to clear the north arrow selection. (Click the Zoom Whole Page button if you zoomed in.)

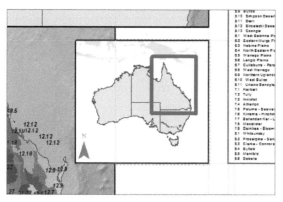

Next, you will add a scale bar to the Bio-subregions data frame. Scale bars tell you how to convert measurements on the map into the real distances they represent on the ground.

5. **In the layout, click the Bio-subregions data frame (the largest frame) to activate it. Then, on the menu bar, click Insert > Scale Bar.**

TIP

You can also activate a data frame by right-clicking its name in the table of contents and clicking Activate. When a data frame is activated, its name is shown in boldface in the table of contents.

6. **On the Scale Bar Selector dialog box, scroll down the list, and click Scale Line 1 Metric. Then click OK.**

The scale bar is added and centered in the layout. Before moving it into position, you will change some of its properties.

7. **Right-click the scale bar, and click Properties. (You can also double-click map elements to open their properties dialog box.) On the Scale Line Properties dialog box, click the Scale and Units tab.**

The scale bar currently has two divisions and four subdivisions. (The scale line division has subdivisions for more precise measuring.)

8. **On the middle of the dialog box, click the When resizing arrow, and click Adjust width.**

The resizing option will change the width of the scale bar if the display scale of the data frame is changed.

ADJUSTING SCALE BARS

A scale bar changes as you zoom in or out of a data frame. If you adjust the scale bar by division value, the distance represented by a scale bar division is variable. As you zoom in, it represents a smaller distance; as you zoom out, it represents a larger distance. If you adjust the scale bar by the number of divisions, the distance represented by a division stays the same, but there are fewer divisions as you zoom in and more as you zoom out. If you adjust the scale bar by width, the distance represented by a division and the number of divisions stay the same, but the entire scale bar grows wider as you zoom in and narrower as you zoom out. **Note:** If you adjust a scale bar by width, you cannot widen it by dragging it; its width changes only with the display scale.

9. In the Scale frame of the dialog box, in the Division value box, replace the current value with **250**. Click the Number of divisions down arrow, and change the value to **1**. Leave the Number of subdivisions value as 4.

10. In the Units frame, in the Label box, change the label Kilometers to **km**.

Next, you will make a slight change to the scale mark heights and text of the scale bar.

11. On the dialog box, click the Numbers and Marks tab. In the Marks frame, change Division Height to **5 pt** and Subdivision Height to **4 pt**.

1
2
3
4
5
6
7
8
9
10
11
12
13
14
15
16
17
18
19
20

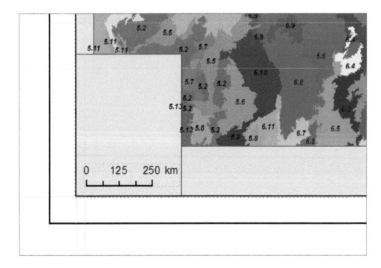

12. Click the Format tab. In the Text frame, change the size to **9**, and click **OK** to apply the format changes.

In the layout, the scale bar is updated to reflect the new properties you set.

13. Press and drag the scale bar to the lower-left corner of the Bio-subregions data frame. (Zoom in on the layout to position the scale bar more accurately. Remember to click the Select Elements tool after zooming.)

Next, you will insert an additional scale bar for the Bioregions data frame. This time, to speed up the process, you will copy the scale bar you just created and paste it when the Bioregions data frame is active. This is possible because the scale bar is dynamically linked to an active data frame. The values and width will adjust accordingly.

14. With the scale bar selected, on the menu bar, click **Edit > Copy**. Activate the Bioregions data frame, and click **Edit > Paste**.

A second scale bar is added to the layout and placed near the Bio-subregions scale bar. It doesn't look quite right because the Bioregions map scale (1:24,000,000) is different from the Bio-subregions map scale (1:13,000,000).

15. Double-click the newly created scale bar designated for the Bioregions data frame.

Notice that it has the same properties as the other scale bar. Change this one to a more appropriate width for its map scale.

16. On the Scale Line Properties dialog box, click the Scale and Units tab. In the Scale frame, in the Division value box, replace the current value with **500**. Change the Number of subdivisions value to **2**, and click **OK**.

1
2
3
4
5
6
7
8
9
10
11
12
13
14
15
16
17
18
19
20

Scale Line Properties

Scale and Units | Numbers and Marks | Format | Frame | Size and Position

Scale

Division value: 500 km

Number of divisions: 1

Number of subdivisions: 2

☐ Show one division before zero

When resizing…

Adjust width

Units

Division Units:

Kilometers

Label Position:

after labels

Label: km Symbol…

Gap: 2.7 pt

OK Cancel Apply

17. Zoom to the whole page if necessary, and drag the scale bar to the lower-left corner of the Bioregions data frame.

Next, you will create a legend for the Bioregions data frame.

18. With the Bioregions data frame active, on the menu bar, click Insert > Legend to open the Legend Wizard.

By default, the legend includes all the layers on the map, and the number of legend columns is set to one (1). Because the only layer you want to show in the legend is Bioregions, remove the other layers from the Legend Items list. Also, you will increase the number of columns in the legend.

19. In the Legend Items list, click Provinces, and click the left-facing arrow (<) button. Remove the Background and Ocean layers as well.

These layers are removed from the list, and only the Bioregions layer remains.

20. Change the Set The Number Of Columns In Your Legend value to 2, and click Next.

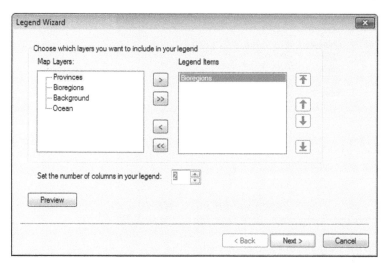

On the second panel of the wizard, you choose a title and font properties for the legend.

21. Replace the default legend title with **Bioregions**. Click Next.

On the third panel of the wizard, there are options to add a border. You will skip this option.

22. Click Next.

Clicking Preview at any time displays the legend in its current form on the map and allows you to exit the wizard or return and make further changes.

On the fourth panel of the wizard, there are options to change the symbol patch, but you will accept the default settings for now. You will change the symbol patch size later.

1 2 3 4 5 6 7 8 9 10 11 12 13 14 15 16 17 18 19 20

23. Click Next.

On the fifth panel of the wizard, you can change the spacing between elements, which you will do next.

24. Change the spacing between Title and Legend Items to **1** and the Patches (vertically) value to **1**, and click Finish.

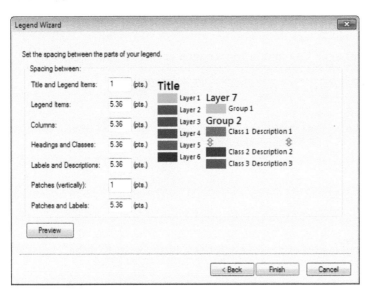

The legend is added to the page layout and centered.

Finally, you will adjust a few more settings and move the legend.

25. Double-click the legend.

This opens the Legend Properties dialog box. On the dialog box, you can further tweak and fine-tune how the legend looks.

26. On the General tab, in the Title frame, click the Symbol button. In the Symbol Selector, change the font size to **8 pt**, and click OK.

27. On the Legend Properties dialog box, click the Items tab. In the lower-left corner, click the Style button to open the Legend Item Selector dialog box, and click the Properties button.

28. On the General tab, in the Appearance frame, uncheck both the Show Layer Name and Show Heading check boxes. Check the Override default patch size check box, and change Width to **8 pt** and Height to **6 pt**. Next, click Label Symbol. In the Symbol Selector, change the font size to **5 pt**, and click OK.

29. Click OK to close the Legend Item dialog box, and then OK again to close the Legend Item Selector. Finally, click OK to close the Legend Properties dialog box.

The legend is updated and reflects all the changes you made.

30. Drag the legend above the Bio-subregions index. (Use the guides and snap to them if you want.) When you are finished, click an empty part of the virtual page to clear the legend selection.

31. To save your work, rename your map document, and save to your MyData folder for chapter 10.

32. If you are continuing to the next exercise, leave ArcMap open. Otherwise, exit the application, and click No if prompted to save your changes.

In exercise 10d, you will add a few more map elements and prepare to print your map. Here is a look at the work you have done so far.

Exercise 10d
Add final touches and set print options

Before sending a map to print, you should check your page setup options and preview the map in ArcMap. It is very good practice to check the overall composition of your layout, including the alignment of your data frames and map elements and the spelling of your titles and labels. Do an objective assessment: If your intended audience would benefit from more information (or in some cases, less), now is the time to make those final improvements.

The map is almost ready. In this exercise, you will add a neatline, information about the map projection, a title for the bio-subregions index, and a picture graphic. Then, you will preview the map and send it to a printer. Throughout the exercise, zoom in on the layout as much as you need to.

1. **In ArcMap, open the ex10d map document from your Chapter10 folder.**

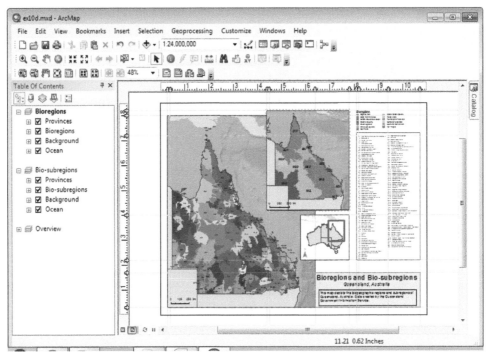

The map is in layout view. It looks as it did at the end of exercise 10c. Maximize the ArcMap application window as needed, and click the Zoom Whole Page button on the Layout toolbar.

2. **Right-click anywhere outside the map display, and click Guides > Snap to Guides to turn off snapping. (Your snapping may already be turned off, depending on how you worked with the layout in the previous exercise.)**

3. **On the menu bar, click Insert > Text.**

A text box appears in the middle of the layout with the word "Text" highlighted.

4. **In the text box, type Geographic Coordinate System: Geocentric Datum of Australia 1994, and press Enter. The text is selected. On the Draw toolbar, change the font size to 8.**

5. Drag the text to the lower center of the Bio-subregions data frame, and position it to the right of the scale bar. Clear the text selection.

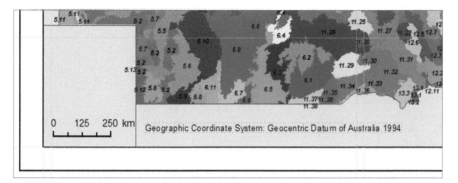

Next, you will add a title for the Bio-subregions index, and position it at the top of the index rectangle.

6. Insert a new text box. In the text box, type **Bio-subregions**, and press Enter. On the Draw toolbar, change the font size to **8**, and click the Bold button.

7. Drag the text to the right of the Bio-subregions index. (Don't worry if it is off the page— you will reposition the index slightly to make room for the title.)

8. Use the Zoom and Pan tools on the Layout toolbar to zoom to the Bio-subregions index. Make sure you can see some white space below the index rectangle.

9. On the Tools toolbar, click the Select Elements tool. Click the Bio-subregions index to select it. Nudge the index rectangle down twice.

10. Right-click the selected index, and click Ungroup. (The rectangle and the text are grouped together.) With the rectangle still selected, press and drag the upper-center selection handle so that it aligns with the horizontal guide just below the Bioregions legend.

11. Drag the Bio-subregions title to the space between the upper border of the rectangle and the top of the index text, and align it with the left column. Change the drawing order so the text is visible.

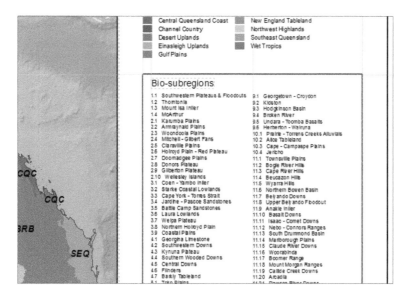

Next, you will add a small picture of your agency logo to the layout.

12. On the Layout toolbar, click the Zoom Whole Page button.

13. On the menu bar, click Insert > Picture. On the Open dialog box, browse to
..\GTKArcGIS\Chapter10\Data. Click agency-logo.gif, and click Open.

14. Press and drag a corner selection handle to resize the graphic until it is approximately 0.5 inch wide. (Alternatively, right-click the image, and change the width on the Size and Position tab in the picture properties.) Drag the image to the upper-left corner of the Bio-subregions data frame.

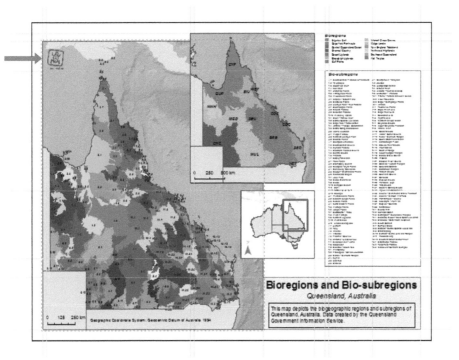

You can insert other kinds of picture file formats, including JPEG, TIFF, EMF, and BMP.

The map is almost finished. The last element you will add is a neatline border, a bounding line that frames the other elements.

15. Clear the logo selection.

16. On the menu bar, click Insert > Neatline.

17. On the Neatline dialog box, click the option to place the neatline around all elements. Click the Border arrow, and click 1.5 Point. Click the Background arrow, and click None. Then click OK.

1
2
3
4
5
6
7
8
9
10
11
12
13
14
15
16
17
18
19
20

18. Clear the neatline selection.

Before you print a map, you should make sure your map page size and printer setup specifications match. Otherwise, your map may be cut off or misaligned.

19. On the menu bar, click **File > Page and Print Setup.**

In exercise 10a, you changed the map page orientation from Portrait to Landscape. Next, you will change your paper orientation, too.

20. On the Page and Print Setup dialog box, in the Paper frame, click the Landscape option. If necessary, click the Name arrow, and click the printer you want to use. Accept the default paper size (Letter). Click OK.

1
2
3
4
5
6
7
8
9
10
11
12
13
14
15
16
17
18
19
20

You are ready to preview your map. The preview shows how the map will look on the printed page, so you can correct any mistakes in advance.

21. On the menu bar, click **File > Print Preview.**

At this point, you should mainly be concerned with the alignment of the elements on the page. Your alignment will depend on the type of printer you are connected to and its configuration. If the map elements overlap the page edge, this problem can be corrected.

Don't worry that the quality of the logo graphic is poor in the preview—it will be fine when it is printed. (**Note:** Printing at this step is optional.)

22. On the top bar of the application, click Print to open the Print dialog box.

23. If the map looked good in the print preview, click OK to print it.

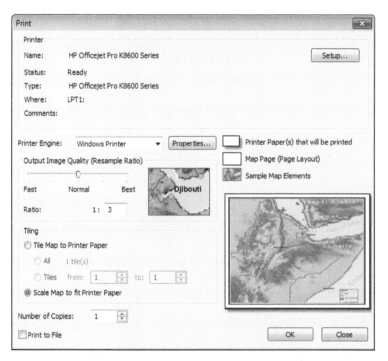

24. If the map went over the edge of the page, in the Tiling frame, click the Scale Map To Fit Printer Paper option. Click OK to print it. Click Close to return to the map.

The sample map in the lower-right corner of the Print dialog box shows the changes you make interactively. Printer settings vary depending on the default printer and its paper settings. There are many printer-specific options that may help you print your map.

Printing is not the only option for sharing your maps. Next, you will explore other formats for sharing your map.

25. On the menu bar, click File > Export Map.

You can export your map to many types of file formats, each with its own unique options. A common format that can be easily opened and shared is the Adobe Acrobat PDF file type, which you will use next.

26. Browse to your MyData folder for chapter 10. Click PDF (*.pdf) in the Save as type list. The file name is automatically entered as **ex10d.pdf.** To see file format options, at the bottom of the dialog box, click Options to expand it. Change the resolution to **360** dpi.

In the Options section, on the General tab, you can change both the resolution and the output image quality. Depending on the quality required, a resolution of 360 dpi (dots per inch) is sufficient for typical print jobs. Increasing the resolution will yield a slightly better image, but it will also increase the file size.

27. Under Options, click the Advanced tab.

```
┌─ ▽ Options ──────────────────────────────────────────────────┐
│                                                              │
│  General │ Format │ Pages │ Security │ Advanced │            │
│                                                              │
│  Layers and Attributes:    Export PDF Layers Only      ▼     │
│  ☑ Export Map Georeference Information                       │
│                                                              │
│ └──────────────────────────────────────────────────────────┘│
│  ☐ Clip Output to Graphics Extent                            │
└──────────────────────────────────────────────────────────────┘
```

The Layers and Attributes option shows what will be exported to the PDF file. Currently, it is set to export layers only, although ArcMap can also export all data attributes. Notice that the Export Map Georeference Information option is selected, which allows the georeference and coordinate system information to be included with the exported PDF file. Acrobat Reader contains Analysis tools that make the georeference information useful.

28. Click Save.

As the map is being exported, its progress is displayed at the bottom of the application window, on the ArcMap status bar.

29. When the export is complete, use File Explorer to browse to your MyData folder for chapter 10 and open ex10d.pdf.

TIP

Adobe Acrobat contains tools to navigate your layout. Find and open the Layers navigation panel. It displays the layers of the map similar to how they are in ArcMap. You can expand the layers and use the controls to hide and show layers. Depending on which Adobe Acrobat product you open the layout in, additional Analysis tools are available to find locations, measure distances, and add location markers.

You are now finished with the maps and page layout.

ON YOUR OWN

As you learned how to do in exercise 5c, share the map package to your ArcGIS Online account.

30. To save your work, return to ArcMap, rename your map document, and save to your MyData folder for chapter 10.

1 2 3 4 5 6 7 8 9 10 11 12 13 14 15 16 17 18 19 20

31. Close the Draw toolbar. If you are continuing to the next chapter, leave ArcMap open. Otherwise, exit the application, and click No if prompted to save your changes.

Summary

This chapter has given you a general overview of presenting data in ArcMap. You learned what to consider before placing and positioning elements in a page layout, including composition, data frame size, colors, titles, and text elements. With more practice, you should feel comfortable stylizing features, working with multiple data frames, and creating legends. You also learned other useful information such as how to create guides and how to snap data frames and map elements to them; that north arrows are oriented to true north; and elements can be arranged in different stacking order, aligned, and nudged into position. You should now realize that ArcMap provides many tools for presenting GIS data effectively, and you should feel comfortable creating a map on your own.

Terms

map extent
north arrow
scale bar
legend
neatline

CHAPTER 11
Building geodatabases

Create a geodatabase
Create feature classes

Learning objectives

- Create a file geodatabase
- Map x,y data
- Convert shapefiles to geodatabase feature classes
- Create an empty geodatabase feature class, and set its properties
- Add attribute fields to a new geodatabase feature class
- Establish a coded domain for attribute values, and select a default value

Spatial data (or "geodata") comes in a variety of formats that can be managed and mapped using ArcGIS Desktop. Spatial data formats include raster data, coverages (a format that is becoming less common), shapefiles, and geodatabase data. Tabular data that records x,y locations or attributes is also a type of spatial data that can be integrated into ArcGIS, as you will see in an upcoming exercise. Many data formats, including shapefile and geodatabase, organize spatial data into feature classes. As you learned in chapter 3, a feature class is a group of points, lines, or polygons that represent geographic objects of the same kind, such as countries or rivers.

A shapefile is a single, stand-alone feature class. Cities.shp, for example, might be a point feature class representing cities. (Shapefiles are actually composed of several files with different extensions, although you will see only one item listed in the Catalog window.) A geodatabase, by contrast, is a "container" in which sets of feature classes can be stored. For example, World.gdb might include a polygon feature class of world countries, a line feature class of world rivers, a point feature class of world cities, and more.

Most of the vector GIS datasets in use today are shapefiles or geodatabase feature classes. Although there are differences among spatial data formats, there is also a great deal of mutual compatibility, and you can usually convert one format to another. When you add spatial data to

ArcMap, you work with it as a layer, and you usually aren't interested in the format of the source file.

You should know, however, that the geodatabase, as the most sophisticated spatial data format, has certain advantages. One of them is the ability to store multiple datasets, including rasters, which makes the geodatabase a built-in tool for organizing data. Related feature classes can be grouped into a feature dataset—for example, sewer mains (lines) and manhole locations (points) could be grouped together in a Sewer feature dataset. When shapefiles are imported into a geodatabase, the size on disk is dramatically reduced, so the geodatabase is an ideal model for data sharing. Another advantage of using a geodatabase is the ability to create an attribute domain, which establishes and enforces valid values or ranges of values for an attribute field and minimizes data entry mistakes by prohibiting invalid values. For example, if "Open" and "Closed" were the domain values for a water valve status field, it would be impossible to enter any values in the attribute table other than Open or Closed.

Finally, the geodatabase has the ability to store behavior rules. Because of this, a geodatabase can support specialized types of data such as annotation feature classes, cartographic representations, geodatabase topologies, and networks, which all have specific behaviors. These special data types are increasingly important to GIS, although they are beyond the scope of this book. Once you have mastered the basics of ArcGIS, you can access resources to help you harness the extended capabilities of the geodatabase. For more information, go to ArcGIS Desktop Help.

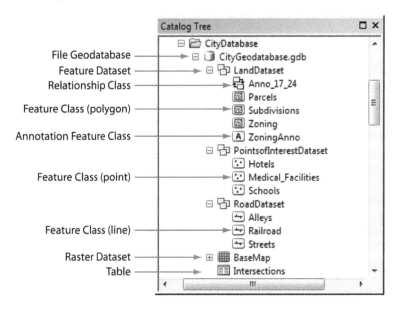

Geodatabases are sophisticated storage containers for feature classes, feature datasets, relationship classes, and rasters.

Geodatabases come in three types: file, personal, and enterprise (or multiuser). In this chapter, you will create and work with a file geodatabase. A file geodatabase is designed to store datasets in a folder of files on your computer. It can store large datasets (up to one terabyte) and can be accessed by several users at the same time, although feature classes can be edited by only one person at a time.

The original geodatabase introduced by Esri is the personal geodatabase (.mdb file extension). It uses the Microsoft Access data file structure and supports databases that are 2 GB or less. However, the effective database size is smaller, somewhere between 250 and 500 MB. Anything larger causes the database performance to slow down. For this reason, file geodatabases (.gdb) are recommended over personal geodatabases, particularly because datasets tend to grow larger over time.

Enterprise geodatabases are scalable multiuser geodatabases that can accommodate large work groups or enterprises. Using technology called versioning, enterprise geodatabases permit simultaneous editing of feature classes by different users. These geodatabases have no size limit or maximum number of users. Data is stored in a relational database management system (RDBMS), such as Microsoft SQL Server Express, SQL Server, Oracle, Informix, IBM DB2, or PostgreSQL.

Data

In this chapter, you will use the following datasets:

- health_areas.shp—a shapefile containing polygons that show boundaries of Manhattan districts created to aid city agencies in administering public services *(Source: NYC Department of City Planning)*
- MN_neighborhoods—a shapefile containing polygons that represent neighborhood boundaries in the Manhattan borough of New York City; modified for training purposes *(Source: NYC Department of City Planning)*
- MN_streets.shp—a shapefile containing street lines for the Manhattan borough of New York City *(Source: NYC Department of City Planning)*
- subwayroutes.shp—a shapefile containing lines that represent subway routes throughout New York City *(Source: CUNY Mapping Service at the Center for Urban Research)*
- subwaystops.dbf—a dBase file containing x,y coordinates and other attributes of New York City subway stations *(Source: CUNY Mapping Service at the Center for Urban Research)*

You will also create your own data—a feature class for playgrounds in Central Park. (You will populate the empty feature class with features in chapter 12.)

1 2 3 4 5 6 7 8 9 10 11 12 13 14 15 16 17 18 19 20

Exercise 11a
Create a geodatabase

You create geodatabases using the Catalog window in ArcMap (or you can use the ArcCatalog application). You can also create geodatabase feature classes, either from scratch or by importing a different data format into the geodatabase. In this exercise, you will create a file geodatabase and import shapefile data into it.

For this exercise, you work in the city planning department for New York City. You are part of a team that is deciding whether the city should convert its vast collections of shapefile data to the geodatabase format. One question you need to answer is how difficult or time-consuming the process would be. To find out, you will create a file geodatabase and import four shapefiles (streets, health areas, neighborhood boundaries, and subway routes) and a text file that contains coordinate points indicating the positions of subway stops. Because you are only investigating possibilities at this point, you have limited your study area to the borough of Manhattan. (This was accomplished by clipping the data layers beforehand. You will practice clipping data in chapter 18.)

1. **Start ArcMap and open the Blank Map template. Open the Catalog window. In the Catalog window, browse to Folder Connections and double-click C:\EsriPress\ GTKArcGIS. (You should already have a folder connection established. If not, create one now.)**

 TIP

 > You will use the Catalog window exclusively in the exercises in this chapter. You may find it easiest to undock the window, expand it, and reduce the size of the ArcMap window.

2. **Double-click Chapter11. Right-click MyData and click New > File Geodatabase.**

 A file geodatabase is created in your MyData folder.

3. **Rename the geodatabase CityData.gdb.**

Next, you will import four shapefiles into the new CityData geodatabase.

4. In the Catalog window, right-click CityData.gdb, and click Import > Feature Class (multiple).

5. Move the Feature Class to Geodatabase (multiple) dialog box away from the Catalog window. In the Catalog window, expand Ex11aData, and click and drag health_areas. shp to the Input Features box of the tool.

The path to the feature class is displayed in the Input Features box. In the Output Geodatabase box, Catalog correctly assumes you want to import the data into the CityData geodatabase. (You could change the output location if desired.)

6. Drag the remaining three shapefiles to the Input Features box, and click OK.

1
2
3
4
5
6
7
8
9
10
11
12
13
14
15
16
17
18
19
20

When the operation is complete, a pop-up notification appears in the lower-right corner of your screen. The shapefiles (two polygons, two lines) are converted to geodatabase feature classes in the CityData geodatabase.

7. **Expand your MyData folder, and double-click CityData.gdb to see the new feature classes.**

Next, you will preview the data to make sure it looks OK.

8. **In the Catalog tree, right-click MN_neigborhoods, and click Item Description. In the Item Description window, click the Preview tab.**

9. **In the Catalog window, click the other feature classes in CityData.gdb.**

The Item Description window now previews each feature class.

You still have a .dbf (dBase) file to import into your geodatabase. It is common to find data in a nonspatial format such as this. In this case, Manhattan subway stop coordinates are stored in a spreadsheet, along with other attributes. You will map the coordinates to create points that can be seen on the map.

10. **Close the Item Description window. In the Catalog window, browse to the Ex11aData folder, right-click subwaystops.dbf, and click Create Feature Class > From XY Table.**

This opens the Create Feature Class From XY Table dialog box. The table already has an X field and a Y field designated, but they are not correct.

11. **Change the X Field option to POINT_X. Change the Y Field option to POINT_Y.**

TIP

Note the table also includes latitude and longitude measurements, but for the points to be displayed properly, the longitude and latitude values need to be translated, respectively, into x- and y-coordinates, as they have been here (using the Add XY Coordinates tool). For a refresher about geographic coordinate systems, read the chapter 6 introduction.

Next, you will specify the spatial reference.

12. On the dialog box, click the Coordinate System of Input Coordinates button. On the Spatial Reference Properties dialog box, click the Add Coordinate System drop-down arrow, and choose Import.

13. Import the coordinate system used by one of the other shapefiles in CityData.gdb.

Is the imported coordinate system projected, or is it a geographic coordinate system?

14. Click OK. Back on the Create Feature Class From XY Table dialog box, specify the output as **subwaystops** in CityData.gdb. Be sure the file type is set to File and Personal Geodatabase feature classes, and then click Save.

15. Click OK to create the feature class, and preview the subway stops.

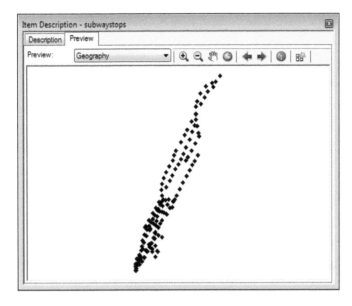

ON YOUR OWN

Create a thumbnail description for subway stops.

TIP

When you create spatial data from a nonspatial source, such as tabular XY data or GPS points, it is good practice to create metadata, which will appear on the Description tab in the Item Description window. To create metadata, click the Edit button on the Description tab. Metadata helps make your data searchable and offers valuable information such as where the data was obtained and what changes have been made, if any.

16. If you are continuing to the next exercise, leave ArcMap open. Otherwise, exit the application, and click No if prompted to save your changes.

Exercise 11b
Create feature classes

When you converted shapefile data in exercise 11a, you accepted the spatial properties of the existing data. When you mapped x,y points from tabular data, you had to define the spatial reference. In the same way, when you create an empty feature class for features that have yet to be collected, you have to define the spatial reference as well as the feature class geometry–point, line, or polygon.

You now know that you can easily convert the city's existing spatial data to a geodatabase format. You also need to know how to create data in the geodatabase. Even though the city has GIS data for many of its managed assets, there are still some datasets that need to be created. In this exercise, you will make a polygon feature class for city playgrounds, although you won't put any features in it until chapter 12. You will add attribute fields for inspection dates and the borough of New York in which each playground resides. For this latter attribute, you will create an attribute domain, which prevents error by limiting values to valid borough names.

1. If you have completed exercise 11a, skip to step 2. Otherwise, start ArcMap and open the Catalog window. Expand your GTKArcGIS folder. Double-click Chapter11, and double-click Ex11bData. Right-click CityData.gdb and click Copy. Right-click MyData and click Paste.

2. Open a blank map in ArcMap and open the Catalog window. In the Catalog tree, expand your ..\GTKArcGIS\Chapter11\MyData folder. Right-click the CityData file geodatabase and click New > Feature Class.

1
2
3
4
5
6
7
8
9
10
11
12
13
14
15
16
17
18
19
20

This opens the first panel of the New Feature Class wizard.

3. In the Name box, type **playgrounds.**

By default, the Type property is set to Polygon Features, so you don't need to make a change.

4. Click Next.

If you completed exercise 11a, you will see the correct coordinate system under Favorites.

5. Under Favorites, click **NAD_1983_StatePlane_New_York_Long_Island_FIPS_3104_Feet. Then skip ahead to step 8.**

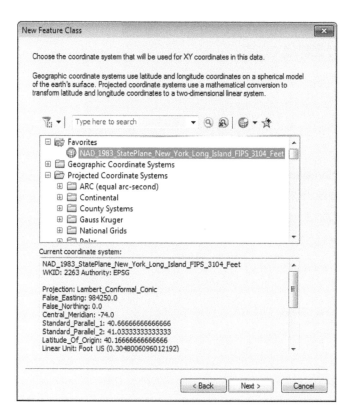

6. Alternatively, click the Add Coordinate System button arrow and click Import.

7. Browse to Chapter11\MyData, double-click CityData.gdb, choose any feature class, and click Add.

The current coordinate system information is updated.

8. Click Next.

The XY tolerance defines the minimum distance between two coordinates before they are considered equal. You will accept the default value.

9. Click Next.

A database storage configuration allows you to fine-tune how data is stored in a file geodatabase. You will accept the default option.

10. Click Next.

The fifth panel of the wizard displays the field names, data types, and field properties that belong to the new feature class. OBJECTID and SHAPE are required fields for geodatabase feature classes and are automatically added by the software. The OBJECTID field stores a unique ID number for each feature in the class. The SHAPE field stores each feature's shape and location.

Depending on the type of feature class created, ArcGIS also creates and maintains measurement fields. A SHAPE_Length field is created for line feature classes. SHAPE_Length and SHAPE_Area fields are created for polygon feature classes. (The SHAPE_Length attribute for polygons stores perimeter lengths.) These measurement fields appear only after the feature class has been created.

11. In the Field Name column, click SHAPE. The field properties for the SHAPE field are displayed.

Notice that the geometry type is set to Polygon.

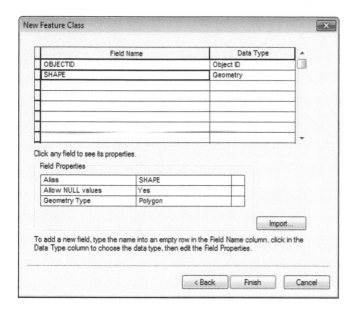

12. In the Field Name column, type **Inspect_Date** in the first empty row.

TIP

Field names cannot contain spaces or special characters, such as hyphens, brackets, parentheses, or symbols such as $, #, %, or &. But underscores are okay.

13. In the Data Type column, click the empty cell next to Inspect_Date and click Date.

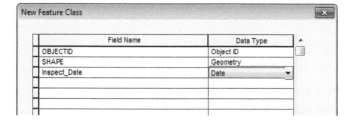

TIP

For more information about field data types, consult ArcGIS Desktop Help.

1
2
3
4
5
6
7
8
9
10
11
12
13
14
15
16
17
18
19
20

SUBTYPES

Another way to categorize geodatabase features is to use subtypes. With subtypes, you can group features into separate classes without having to create distinct feature classes. For example, you may want to classify a Roads feature class with the subtypes Main, Highway, and Utility. To create subtypes, you must have a short integer or long integer field in the feature class. The integer is a code that is associated with a description on the Subtypes tab on the Feature Class Properties dialog box. Attribute domains may be established for each subtype—for example, you may want to establish a domain range for the speed limit attribute, which will vary by road type.

14. In the Field Name column, in the next empty row, type **Name**. For Data Type, choose Text.

15. In the next empty row, type **Borough**. For Data Type, choose Text.

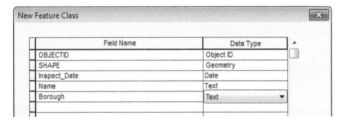

16. Click Finish. In the Catalog tree, notice that a new feature class, playgrounds, has been added to CityData.gdb, making a total of six feature classes.

Now that you have added the two attribute fields, you will next create a domain and apply it to the Borough field. You create domains in the geodatabase properties.

17. In the Catalog window, right-click CityData.gdb (in your MyData folder) and click Properties to open the Database Properties dialog box. Click the Domains tab.

18. In the Domain Name column, in the first cell, type **NYC_borough**. Click the first cell in the Description column, and type **NYC borough**.

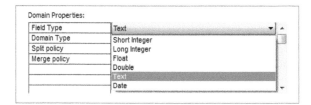

In the Domain Properties list, the default Field Type value is Long Integer and the Domain Type value is Range. The field type must match the data type of the feature class to which it is being applied. Because you made Borough a text field in step 15, you must do the same here and change the field type to text.

19. Under Domain Properties, click Long Integer and change it to Text.

The domain type is changed automatically from Range to Coded Values.

A Coded Values domain is used when the attribute values you want to enforce are best defined in a list. (The values may still be numeric because numbers are commonly used as codes for descriptions.) A Range domain is used when the values you want to enforce are quantities with upper and lower limits.

20. In the Coded Values list on the lower portion of the dialog box, click the first empty cell in the Code column. Type **M.** Press the Tab key or click the first cell in the Description column, and type **Manhattan.**

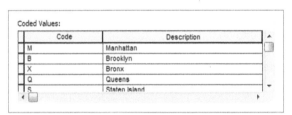

21. In the next four rows, type the following codes and descriptions:

- **B Brooklyn**
- **X Bronx**
- **Q Queens**
- **S Staten Island**

Coded Values:

Code	Description
M	Manhattan
B	Brooklyn
X	Bronx
Q	Queens
S	Staten Island

After a coded value domain has been applied to a feature class, you will be picking from a list of these descriptions when you enter attribute values. The coded values are stored in the geodatabase, but you will see the descriptions in the attribute table only.

22. Click OK to apply the domain to the playgrounds feature class.

Now that you have created the attribute domain, you will next apply it to the Borough field of the playgrounds feature class.

23. In the Catalog window, right-click the playgrounds feature class (in MyData\CityData. gdb) and click Properties. On the Feature Class Properties dialog box, click the Fields tab.

24. In the Field Name column, click Borough. In the Field Properties frame, click the empty cell next to Domain and click NYC_borough.

You have now created an attribute domain and applied it to the Borough field.

Finally, you will assign the field a default value. You are focusing on Manhattan as your study area, so you will set "Manhattan" as the default. Whenever a playground is added to the feature class, it will automatically get this value in the Borough field. (But you can easily change the value as needed.)

25. In the Field Properties frame, click the empty cell next to Default Value and type **M**, the code for Manhattan. Then click OK.

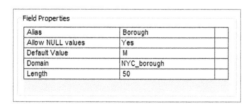

You can also set a default value for a field without creating a domain. If a domain is applied, however, the default value must be a valid domain value.

26. Restore the Catalog window to the position you prefer, and exit the application. Click No if prompted to save your changes.

Summary

In this chapter, you learned how simple it is to create a geodatabase container, as well as to convert other data types to geodatabase feature classes. Both tasks can be completed with the click of a button. Creating a feature class from x,y tabular data takes only one additional step–selecting a coordinate system. Setting up a new feature class from scratch takes a few more steps, because you must set properties, determine the spatial reference information, create attribute fields, and establish attribute domains, if desired.

Terms

feature dataset
attribute domain

1
2
3
4
5
6
7
8
9
10
11
12
13
14
15
16
17
18
19
20

CHAPTER 12
Creating features

Draw features
Use more construction tools

Learning objectives
- Open an edit session in ArcMap
- Create geodatabase features using various construction tools
- Use commands that facilitate the feature creation process
- Set the snapping environment in an edit session
- Change the default attribute value of a new feature

Much of the vector spatial data used in a GIS comes digitized from paper maps and aerial or satellite photographs. In the past, digitizing data often involved placing a map or a photograph on a digitizing tablet (a drawing table connected to a computer) and tracing features with a puck, which is a device similar to a mouse. In a variation called "heads-up digitizing," more common today, features are drawn directly on the computer screen with a mouse by tracing an aerial photograph, a scanned map, or other spatial data. Spatial data points can also be collected from GPS-enabled devices, which can then be uploaded to (or automatically update) a GIS database.

Italy's Lago di Bolsena digitized "heads up" from a satellite image. Map and data courtesy of i-cubed, information integration and imaging LLC, distributed through i-cubed's DataDoors Archive Management at www.datadoors.net.

Features can also be digitized without tracing. ArcGIS has several tools for creating circles, rectangles, curves, and other shapes of exact dimensions.

Points are the simplest features to digitize—all it takes is a single click. Lines are more complicated, because they start, end, and may change direction. Polygons, from the point of view of digitizing, are just special cases of lines—they are lines that return to their origin.

Points that define the shape of a line or polygon are called **vertices** (singular **vertex**). Points that mark the ends of a line segment are called **endpoints**. The segment between vertices is called an **edge**.

| Digitized line | Endpoints | Vertices | Edges |

Features that can be digitized in ArcGIS.

When you look at line or polygon features on a map, you don't see them as edges, vertices, and endpoints, although you do when you create or edit features. When you see a feature in this

mode, you are looking at an **edit sketch**. A new feature exists only as an edit sketch until you save it. After you save it, it becomes a feature.

In ArcMap, all digitizing–whether it is to create new features or modify existing ones–is done in the course of an **edit session**. An edit session begins when you click the Start Editing command from the Editor menu on the Editor toolbar and ends when you click the Stop Editing command. After you start an edit session, you create features by using feature templates. Feature templates define all the information required to create a feature: the layer where the feature will be stored, the attributes the feature is created with, and the default tool used to create that feature.

Editor menu Create features

Editor toolbar.

The Create Features window is the central place for creating and managing feature templates. The window has three main components: a toolbar to manage your templates and their properties, a list of templates used to create new features, and a set of construction tools used to define the shape of the features. You will learn how to use this window and its options in this chapter.

When you digitize features, you will sometimes make mistakes. You can undo mistakes with the Undo button 🔄 on the Standard toolbar.

Create Features window.

Data

In this chapter, you will use the New York City data in the CityData geodatabase you created in chapter 11. If you didn't complete chapter 11, that's OK. We supply a copy for you. Additionally, the data for this chapter includes an imagery basemap, served from ArcGIS Online.

- World_Imagery–basemap *(Source: ArcGIS Online; Esri, DigitalGlobe, Earthstar Geographics, CNES/Airbus DS, GeoEye, USDA FSA, USGS, Getmapping, Aerogrid, IGN, IGP, swisstopo, and the GIS User Community)*

Note: You need to be connected to the internet to use the ArcGIS Online basemap layer.

Exercise 12a
Draw features

In this exercise, you will update the New York City geodatabase you created in chapter 11 by digitizing playgrounds in Central Park. You will also create a new subway line that is currently under construction below 2nd Avenue in Manhattan. For feature integrity, you will use snapping to connect the subway line to existing subway stops. You will use feature templates and ArcGIS editing tools to create the new features.

1. **Start ArcMap. Open the ex12a map document from your C:\EsriPress\GTKArcGIS\ Chapter12 folder.**

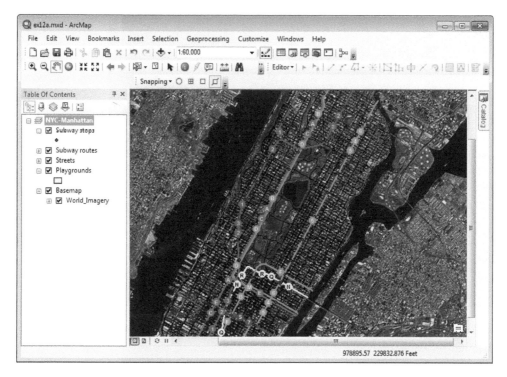

TIP

The imagery basemap is a web map served from ArcGIS Online. If you don't have a stable internet connection, or if you find your map takes too long to load or navigate, you can turn off the basemap and use the book's graphics for guidance.

The map is centered on the Manhattan borough of New York City. In addition to the World Imagery basemap, layers for subway stops, subway routes, streets, and playgrounds are listed in the table of contents. (The neighborhoods and health areas feature classes you may recall from chapter 11 aren't used again until chapter 13.) The playgrounds feature class you created in exercise 11b has no features yet. Before starting an edit session, you will zoom to the area where you are going to digitize the first playground.

2. On the menu bar, click Bookmarks > Playgrounds1.

1

2

3

4

5

6

7

8

9

10

11

12

13

14

15

16

17

18

19

20

The display zooms to an area of Central Park where two playground areas are visible. You will digitize these two areas.

First, you will add the Editor toolbar, and then start an edit session.

3. On the Standard toolbar, click the Editor Toolbar button 🖉.

This opens the Editor toolbar. The toolbar appears dimmed, meaning the tools are unavailable.

4. Move the Editor toolbar away from the map display or dock it. On the Editor toolbar, click Editor > Start Editing.

Some of the editing tools are now activated.

5. To open the Create Features window, click the Create Features button 🗐 on the Editor toolbar. Move or dock the window as you prefer.

By default, no feature template is selected.

6. In the Create Features window, click the Playgrounds template. In the Construction Tools panel, click Polygon, if it is not already selected.

Notice the symbology has already been set for playgrounds in the map document, even though there are currently no features.

Take a look at the larger of the two playgrounds, which you will digitize first. Notice the shape has two straight lines and two curved lines.

7. **Move the mouse pointer over the map display. The pointer changes to a crosshair. Move the pointer to the middle of the street-side edge of the large playground, and click to create the first vertex, as shown in the graphic.**

1

2

3

4

5

6

7

8

9

10

11

12

13

14

15

16

17

18

19

20

Notice that the Feature Construction toolbar appears. It is slightly transparent until the mouse points to it, and it may move depending on the area of the map you are editing. By default, the Straight Segment tool is active.

TIP

The Feature Construction toolbar provides quick access to editing tools. If it doesn't appear, from the Editor menu, click Options. In the Editing Options window, on the General tab, select the Show Feature Construction Toolbar check box. Alternatively, you can find the tools on the Editor toolbar.

8. Move your pointer to the northernmost edge of the playground, just before the curved end, and click to place another vertex.

The edit sketch of the line moves with the pointer (an effect known as "rubberbanding").

9. On the Feature Construction (or the Editor) toolbar, click the End Point Arc Segment tool. Click once to place a vertex on the eastern edge of the curved end. Move your mouse northward to adjust the curvature of the line. Click to create the curved line.

TIP

Remember, to be sure you are using the correct tool, move your mouse pointer over a tool button to see the ToolTip.

10. Activate the Straight Segment tool . Click to place another vertex just before the southern curved end of the playground.

TIP

If you make a mistake, click the Undo button on the Standard toolbar and repeat any steps necessary.

11. Switch to the End Point Arc Segment tool, and create another curved line to match the image layer. Click to finish the curved line.

12. The final segment of the polygon is anticipated by ArcMap. Right-click and click Finish Sketch to complete the polygon.

The playground polygon inherits the symbology that was set in the map document for the Playgrounds layer. The finished sketch is selected, evident by the bright blue outline and the X in the center.

Next, you will draw the smaller playground using an edit command, Square and Finish.

13. **Make sure the Straight Segment tool is active. Digitize the second, smaller playground. D raw two segments of the square, and then move your pointer slightly away from the most recently placed vertex. Right-click and choose Square and Finish.**

14. **With the second, smaller playground still selected, on the Editor toolbar, click the Attributes button** [icon].

The Attributes window opens.

The Attributes window can show one or more features at a time. Use the Edit tool on the Editor toolbar to select features to see their attributes. Alternatively, open the attribute table of the Playgrounds layer in the table of contents to view all the records without selecting them.

The length of segments and area of polygon have been calculated in the SHAPE_Length and SHAPE_Area fields in feet. (Length and area attributes maintained by the software accurately measure features in the geodatabase, but they do not have legal authority. An official planning database would include an "as-built" length attribute containing the field results of actual surveys.)

Note: Your SHAPE values may vary from those shown.

The Inspect_Date and Name fields are null because no attribute values have been entered for them. The Borough attribute values have been set to Manhattan for both playgrounds, in accordance with the default value for Borough you specified in exercise 11b. The default Borough attribute is correct, but you will manually enter the other attributes.

15. **For the Inspect_Date field, enter today's date. For the Name field, type Adventure Playground B.**

ON YOUR OWN

Select the larger playground (the first one you digitized) and update its attributes with an inspection date and the name Adventure Playground A.

You have finished digitizing for now and can save your work.

16. **Close the table. On the Editor toolbar, click Editor > Stop Editing. Click Yes to save your edits, and turn off the Editor toolbar if you want.**

TIP

To close a docked toolbar, right-click in an empty area near the toolbar and turn off the toolbar from the shortcut menu.

There is no need to save your map document. Your edits are saved to the feature class in the CityData geodatabase. Any map document that contains the Playgrounds layer will include the new features.

17. **If you are continuing to the next exercise, leave ArcMap open. Otherwise, exit the application, and click No when prompted to save your changes.**

1
2
3
4
5
6
7
8
9
10
11
12
13
14
15
16
17
18
19
20

Exercise 12b
Use more construction tools

Your job as GIS specialist in the city planning department requires you to maintain not only current datasets, but also GIS records for new proposed and under-construction features. In this exercise, you will digitize the footprint of a proposed playground in Central Park, as well as a new subway line, which is currently under construction with plans to open in two years. The basic procedure–starting an edit session and choosing a feature template–is the same as in exercise 12a, but you will use different tools, and this time you will take advantage of a technique called snapping.

New features are often connected to existing features. For example, the boundaries of new land parcels may be adjacent to boundaries that have already been digitized. New streets intersect or extend to existing ones. New power lines run to and from existing poles. Making sure that features are connected would be difficult if this depended entirely on eyesight.

ArcGIS uses snapping to automatically connect features placed within a certain distance of each other. The rules that specify which features, and which feature parts, snap to others make up the snapping environment. The distance at which snapping occurs is called the snapping tolerance.

1. In ArcMap, open the ex12b map document from your Chapter12 folder.

This map is identical to ex12a.mxd.

2. **Zoom to the Playgrounds1 bookmark. If you completed exercise 12a, you will see the two playground footprints you created.**

 Now you will create a third playground footprint without imagery to guide you because you are digitizing a proposed playground, not an existing one.

3. **Zoom to the Playgrounds2 bookmark.**

4. **Display the Editor Toolbar and start an edit session.**

5. **Open the Create Features window. Choose the Playgrounds template and the Polygon construction tool.**

1
2
3
4
5
6
7
8
9
10
11
12
13
14
15
16
17
18
19
20

You will use your notes to draw a rough outline of the proposed playground. (This is only meant for information-sharing purposes internally—the actual footprint would be provided by the builder and would be surveyed once built.) The playground is a triangle, so you will create three vertices. You will place the first vertex at a specific x,y location as recorded in your notes. You will place the second at a specific length from the first, parallel to a street centerline, and the third at a specific angle and length measurement.

6. **Place your mouse pointer in the approximate position shown in the graphic, and then right-click and choose Absolute X,Y. For X type 993928.234, and for Y type 222462.291, and then press Enter.**

7. **Move your mouse pointer to hover over the street centerline as shown in the graphic. (The ToolTip for Streets: Edge or Streets: Vertex may appear.) Right-click the street feature, and then click Parallel.**

TIP

If you left-click by mistake and add a vertex, click the Undo button on the Standard toolbar.

TIP

You can press the Escape key (Esc) on your keyboard to close the shortcut menu without clicking a command. Pressing Esc will also cancel the effect of a menu command.

The sketch line is now restricted to one that is parallel to the street. Next you will explore some other editing commands: Length and Direction/Length.

8. Right-click and choose Length. In the Length box, type **125**, and then press Enter.

1
2
3
4
5
6
7
8
9
10
11
12
13
14
15
16
17
18
19
20

9. To place the third vertex, right-click and choose Direction/Length. Set the direction angle in the left box to **210** and the length in the right box to **225**, and then press Enter.

TIP

If your Direction pop-up box looks different from the one shown in step 9, your setting for Direction Type may be incorrect. To change the setting, on the Editor toolbar, click Editor > Options. On the Editing Options dialog box, click the Units tab. Change Direction Type to Polar. Set Direction Units to Decimal Degrees, and then click OK.

Angular Units

Used by editing functions in which directions are specified.

Direction Type: Polar

Direction Units: Decimal Degrees

Display angles using: 4 decimal places.

10. Press F2 to complete the sketch.

Next, you will digitize a new subway line currently under construction. The first phase of the new line is actually an extension of the Q line, which will run along 2nd Avenue from East 60th Street to 96th Street. First, you will set the snapping environment.

11. Zoom to the Q Line bookmark, and then turn on labels for Streets. Turn off the basemap for easier feature visibility.

12. Open the Snapping toolbar. (On the Editor menu, click Editor > Snapping > Snapping Toolbar.)

SNAPPING SETTINGS

The Snapping toolbar contains buttons for the main snapping settings, although additional tools are available on the Snapping menu. As you digitize a new feature, the cursor snaps to existing features according to the buttons that are clicked. (Clicking either the Vertex Snapping or the Edge Snapping button for a point layer, for example, makes the cursor snap to point features.) The cursor snaps when it comes within a specified distance of a feature. To set this snapping tolerance, from the Editor menu, click Snapping > Options.

The Snap To Sketch command on the Snapping menu sets the snapping rules for an edit sketch. It is useful, for example, when you are creating a feature that intersects or connects to itself, such as a boundary line that forms a closed loop. For more information, consult ArcGIS Desktop Help.

13. On the Snapping toolbar, turn off all buttons except Edge Snapping. (Click to activate or deactivate.)

14. On the Snapping toolbar, click the drop-down arrow. Notice the default setting is to use Snapping, and Intersection Snapping is turned on. (If it's not, turn it on now.) These settings will apply to all ArcMap sessions until you change them.

15. In the Create Features window, scroll through the subway routes and choose the template for the Q Broadway line. In the Construction Tools panel, the Line tool is automatically selected.

16. Move your pointer close to the yellow subway line (which serves routes N, Q, and R). Notice how the cursor snaps to its edge. (You will see a SnapTip for either Subway routes: Edge, or Streets: Edge when the subway and street centerlines are coincident.)

17. Move your pointer close to 2nd Avenue. Notice the same snapping behavior to the edge of the street line.

18. Move your pointer to the intersection of the yellow subway line and 2nd Avenue. When you get to the Intersection SnapTip, click to begin the line.

19. Move your pointer northeast along 2nd Avenue. Press C on your keyboard (the Pan shortcut key) to pan northeast while remaining in sketch mode. Make sure you remain snapped to the edge of 2nd Avenue. When you reach the intersection of 2nd Avenue and 96th Street, double-click to end the line.

*Is it possible to turn SnapTips off? (**Hint:** Explore the Snapping options.)*

ON YOUR OWN

Zoom back to the Q Line bookmark. Change the feature template and construction tool to Subway stops/Point. Add a subway stop at the intersection of the new Q segment and the existing N Q R line.

20. On the Editor toolbar, click Editor > Stop Editing. Click Yes to save your edits.

ON YOUR OWN

Start an edit session and experiment with the edit tools you didn't use in this chapter. Don't forget about the Undo button. You don't need to save your edits—just practice and figure out how some of the tools work.

You don't need to save your map document. Your edits were made directly to the feature class in the geodatabase. When you saved them, the geodatabase was updated.

21. **Close the Editor toolbar and the Create Features window. If you are continuing to the next chapter, leave ArcMap open. Otherwise, exit the application, and click No when prompted to save your changes.**

Summary

Although you didn't get to use all the construction tools ArcMap has to offer, this chapter taught you the basics of feature creation: You learned how to start an edit session, choose options in the Create Features window, use various edit commands to facilitate feature creation, and use various sketch tools that were logical choices for the features you were creating. Thankfully, it is not necessary to create all new features freehand, although that is an option. With these basic skills, you should feel more comfortable trying new tools the next time you have to create geographic features.

Terms

digitizing
vertex (vertices)
endpoints
edge
edit sketch
edit session
snapping

CHAPTER 13
Editing features

Delete and modify features
Split and merge features
Edit feature attribute values

Learning objectives

- Delete features in an edit session
- Modify a feature's geometry
- Create a topology to enforce geometry rules
- Split a single feature into two features
- Merge multiple features into one
- Edit attribute values
- Add a new attribute field and calculate its values

Creating features is one part of maintaining a geodatabase. Another part, equally important, is making changes to existing features.

You can edit features in several ways. You can delete, move, split, merge, resize, reshape, and buffer them. In addition to the tools on the Editor toolbar, there are also commands on the Editor menu for changing features.

When you edit features, you may need to update their attributes as well. In a geodatabase, the existing length and area attributes are updated automatically by ArcMap, but you will need to manually update other attributes. You may need to edit attribute values even when features don't change. For example, when a land parcel is sold, the owner name needs to be changed. You may also add new attributes to a table.

With ArcGIS Desktop, you can edit shapefile features and those found in file and personal geodatabases, as well as multiuser geodatabase features, relationship classes, and other behaviors. To edit raster layers, you need to use the ArcGIS Spatial Analyst extension. You will use Spatial Analyst in chapter 20, although editing rasters is beyond the scope of this book.

Data

In this chapter, you will work with the same geodatabase you created in chapter 11, overlaid on the imagery basemap you saw in chapter 12, although you won't work with all the feature classes.

Exercise 13a
Delete and modify features

To modify a feature's geometry, you add, delete, or move vertices on the feature's edit sketch. If a feature shares an edge or vertex with another feature, you might want to ensure that no gap or overlap results. You can make features with coincident geometry behave as if they are connected by creating a map topology within your edit session.

For this exercise, you will continue your work for the city planning department. You have already tested the ease of switching to the geodatabase data model, and you have updated the geodatabase with new playground features and an under-construction subway line extension. Now you need to perform some quality control. With such vast amounts of spatial data, there are bound to be errors that need to be corrected and updated on a regular basis. Today's to-do list entails the following: (1) delete two proposed playgrounds, (2) align an incorrectly digitized road, (3) change the boundaries of a neighborhood polygon using map topology, (4) split a health area polygon into two, (5) merge two health area polygons into one, and (6) edit health area feature attributes.

1. **Start ArcMap. Open the ex13a map document from your C:\EsriPress\GTKArcGIS\ Chapter13 folder.**

TIP

The basemap is helpful but not required for the exercises in this chapter. So if you don't have a stable internet connection, or you find the map service is slowing ArcMap's performance, feel free to turn it off.

2. Zoom to the bookmarks Playgrounds 1 and Playgrounds 2. These playgrounds are like the ones you created in chapter 12.

Deleting an unwanted feature is easy. Make sure you are in an edit session, select one or more features, and click the Delete tool.

3. Zoom to the Playgrounds 3 bookmark.

4. Display the Editor toolbar and start an edit session. Make sure the Edit tool ▶ is active.

5. Click one of the playgrounds to select it and, holding down the Shift key, click the other playground.

Both playgrounds are selected.

The small blue *x* in the center of the selected parcels is called a selection anchor. Every selection in an edit session, whether of one feature or many, has one. (Selection anchors are used to rotate and scale features, which you won't do in this exercise.)

6. Click the Delete button ☒ on the Standard toolbar.

The playgrounds are deleted.

 TIP

You could also press Delete on your keyboard.

7. **Save your edits (on the Editor toolbar, click Editor > Save Edits).**

 It is also easy to modify a feature's geometry—a fairly common GIS task for error correction or to make updates when features are changed (when a road is rerouted or parcel lines are redrawn, for example).

8. **Zoom to the Road error bookmark, and turn on the Streets layer.**

Using the imagery basemap as a guide, you can see that the road in the center of the view needs to be moved.

ON YOUR OWN

If you find the streets difficult to see, change their symbol color to one of your choice.

9. Make sure you are still in an edit session. With the Edit tool active, double-click anywhere on the road segment to select it and display its edit sketch.

On the edit sketch, the vertices are green squares, and the end vertex is a red square. ArcMap has automatically activated the Edit Vertices toolbar. You can also activate the Edit Vertices toolbar by clicking the Edit Vertices button, which is now highlighted on the Editor toolbar.

10. Place your pointer over the first vertex to be moved. The pointer becomes a four-headed arrow.

11. Use the pointer to drag the vertex down to the center of the road in the basemap, and then release the mouse button.

The new line shows the new shape of the feature.

TIP

The highlighted cyan line remains in place as a reference. It will disappear when you clear the selection or select another feature.

12. Continue dragging vertices to match the road as seen in the basemap.

13. On the Edit Vertices toolbar, click the Finish Sketch tool ☒. (Alternatively, press F2, or double-click the line.)

14. Clear the selection, and save your edits.

1
2
3
4
5
6
7
8
9
10
11
12
13
14
15
16
17
18
19
20

Finally, you have been asked to correct a neighborhood boundary in a feature class that is shared on the city's public web map.

15. Zoom to the Yorkville bookmark. Turn on the Neighborhoods layer, and turn off all other layers.

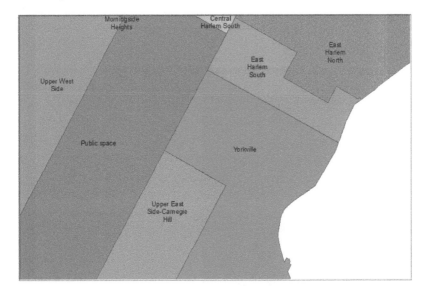

The Yorkville neighborhood extends too far uptown. The border it shares with the East Harlem South neighborhood should be moved down to 96th Street (the upper limit of the Upper East Side/Carnegie Hill neighborhood). You could use the Edit tools to move each boundary line separately. Because you want the neighborhoods to remain adjacent, this would mean setting the snapping environment and then moving one polygon's vertices, moving the second polygon's vertices, and snapping the latter's vertices to those of the first. This wouldn't be too hard this time, but if you had to do it several

times, or if you had four or five polygons sharing an edge, the process might start to get tedious.

A better approach is to create a map topology. Map topology helps you become aware of features' coincidence to other features so that common vertices and edges can be moved together.

In what situations can you imagine map topology would be useful?

16. Turn on the Topology toolbar. (Right-click an empty area at the top of ArcMap to see a menu of toolbar options. You may need to scroll down to see Topology.)

17. Click the Select Topology button .

On the Select Topology dialog box, you can choose the layers for the map topology. You can also set a cluster tolerance, which is the distance at which ArcMap considers nearby edges or vertices to be coincident, or shared.

TIP

To set the cluster tolerance, expand the Options menu.

18. On the Select Topology dialog box, select the box next to Neighborhoods, and then click OK.

19. On the Topology toolbar, click the Topology Edit Tool ![icon].

20. Double-click the edge between Yorkville and East Harlem South. (ArcMap will first build a topology cache, so you may have to repeat the double-click.) It is important that you double-click the edge to reveal the line segment vertices.

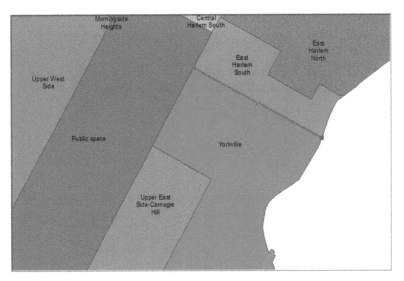

The selection is indicated by a bright purple color. Double-clicking activated the Modify Edge tool, which in turn activated the Edit Vertices toolbar.

21. On the Topology toolbar, click the Shared Features tool ![icon].

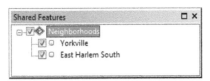

The Shared Features window shows you which features share a selected point or edge; in this case, Yorkville and East Harlem South share the selected edge.

Note: You may see the object IDs rather than the neighborhood names.

TIP

Click each feature in the Shared Features window to see it flash on the map. When using editing tools, it is very helpful to use shortcut navigation so you can move around the map without deactivating the desired editing tool:

Zoom in: Z key, or scroll down using the mouse's scroll wheel if you have one, or hold the Ctrl key and zoom using the mouse's middle button.

Pan: C key, or drag the map using the mouse's middle button.

Zoom out: X key, or scroll up with the scroll wheel.

22. Hover your pointer over the left-most green vertex. Click and drag the vertex down to the northeast corner of Upper East Side/Carnegie Hill. Snap to the intersection and release to indicate the new vertex.

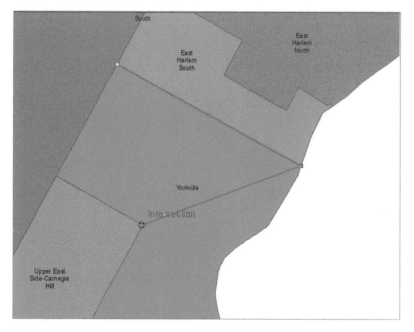

23. Drag the red vertex so it forms a straight line and snaps to the eastern edge of the Neighborhoods feature class.

1 2 3 4 5 6 7 8 9 10 11 12 13 14 15 16 17 18 19 20

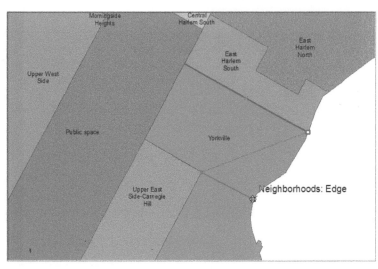

24. Finish the sketch (press F2), and click away from an edge to clear the selection.

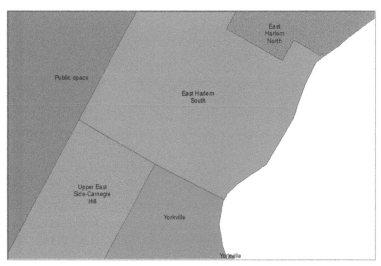

25. Save your edits, and close the Topology toolbar and the Shared Features window.

ON YOUR OWN

Examine the sketch tools available on the Editor toolbar and the Advanced Editing toolbar. Read the expanded ToolTips, and feel free to experiment with feature editing tools. You don't have to save your edits. Just end the session and your features will revert to what they were.

The errors have been corrected so the Manhattan layers are available for use internally and for external websites. You don't need to save your map document because your edits were made directly to the feature class in the geodatabase.

MAP TOPOLOGY AND GEODATABASE TOPOLOGY

Map topology can be created for shapefiles and geodatabase feature classes. It lasts only for the duration of an edit session and allows you to edit features (within the same layer) as if they were connected to each other.

ArcGIS Desktop also offers the ability to create a more sophisticated topology for geodatabase feature classes, geodatabase topology. Geodatabase topology lets you enforce more than 30 spatial rules within and between feature classes, helping to guarantee the integrity of your data. For example, you can specify that polygons may not overlap or have gaps, that lines must be connected to other lines, or that points in one feature class must lie within polygons of another feature class. Geodatabase topology is created within a geodatabase feature dataset; it is independent of a particular map document or edit session.

26. **If you are continuing to the next exercise, leave ArcMap open. Otherwise, exit the application, and click No when prompted to save your changes.**

Exercise 13b
Split and merge features

To split a polygon or a line (create two features from one), you digitize a line across the feature where you want to split it. You can split several features at once.

Select polygons to split, and draw a split line.

Each selected polygon is split in two.

To merge polygons or lines (create one feature from two or more), you select them and click the Merge command on the Editor menu. The merged features do not have to be contiguous. For example, you could merge a large polygon representing a country and smaller polygons representing its offshore islands. The country and islands would then exist as a multipart feature with a single record in the attribute table.

Select polygons to merge.

The three polygons are merged into one.

When features in geodatabase feature classes are split or merged, ArcMap updates their Shape_ Length and Shape_Area attributes. ArcMap does not automatically update features in other data formats, such as shapefiles. For all feature classes, ArcMap updates the new feature's identifier attribute, which keeps track of the number of features. Changes to other attributes, including user-defined IDs, such as parcel numbers or legal measurements of length and area, must be made manually.

For this exercise, the planning department is adjusting the boundaries of its designated health areas—districts created to aid city agencies in administering public services. The changes are based on population analysis of each area and a desire to boost efficiency, so no designated health area serves either too many or too few residents. One area will be split in two, and two

smaller areas will be merged into one. Again, for the purpose of simplicity all datasets have been limited to the borough of Manhattan.

1. **In ArcMap, open the ex13b map document from your Chapter13 folder.**

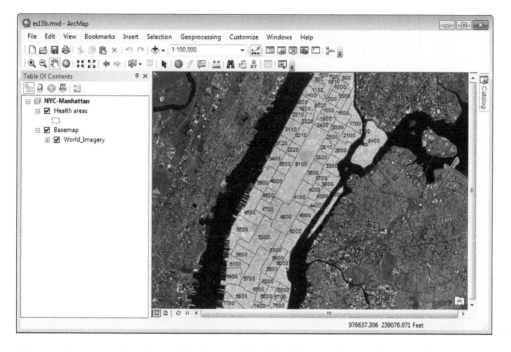

The map is centered on Lower Manhattan. The Health areas layer is displayed at 30 percent transparency so you can see the basemap image through the polygons. Labels are turned on to indicate the health areas, which are designated by numbers.

2. **Go to the 5700 bookmark.**

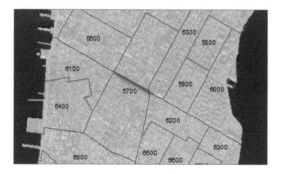

The red line (which appears in the book but not on your screen) shows where the parcel is to be split. When the parcel is split, ArcMap will update the attribute table

with the correct area of each new parcel. To see this for yourself, you will check the area measurement of health district 5700 before you split it.

3. On the Tools toolbar, click the Identify tool and click health area 5700.

The Shape_Area field shows a value of 17,983,789 square feet, before the decimal place. Obviously, square feet is not a useful measurement for such a large area. In the next exercise, you will add a new field showing acres.

4. Close the Identify window.

5. Turn on the Editor toolbar. Start an edit session and open the Create Features window, if it is not already displayed.

6. In the Create Features window, click Health areas (the only editable layer) and the Polygon tool. Close the window if you want.

TIP

Even in a map with only one editable layer, it is necessary to choose the correct feature template in the Create Features window to activate specific edit tools.

7. On the Editor toolbar, click the Edit tool ▸, if it is not already active.

8. Click area 5700 to select it. You will draw the split line between the adjacent polygon edges indicated by the two arrows in the graphic.

Because you are drawing the split line where adjacent polygon edges intersect the selected polygon, you shouldn't need to open the snapping toolbar because Intersection Snapping is turned on by default for every ArcMap session unless you turn it off.

9. On the Editor toolbar, click the Cut Polygons tool ⊕.

10. On the display, the pointer changes to a crosshair. Move the pointer to the position shown in the graphic. When the cursor snaps to the intersection, click the mouse button.

TIP

If you don't see an Intersection SnapTip, on the Snapping toolbar click the drop-down arrow and activate Intersection Snapping.

11. Move your pointer over the second intersection. When the cursor snaps to it, double-click to end the split line. (If you moved your pointer away from the end point, press F2 to finish the sketch.)

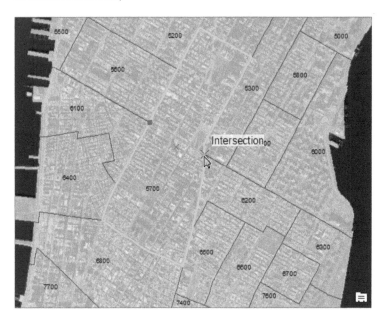

The parcel is split in two. Each parcel is labeled with the same HealthAreaID value, 5700. In exercise 13c, you will update the HealthAreaID attribute for the smaller parcel.

12. Clear the selection, and then save your edits.

13. On the Tools toolbar, click the Identify tool. Click each of the two new parcels, and examine the Shape_Area values.

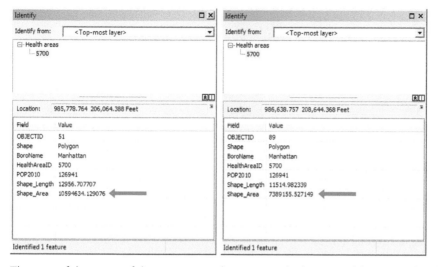

The sum of the areas of the two new polygons equals the area of the original polygon. The HealthAreaID number and the Pop2010 value, both user-defined attributes, do not change. However, the new polygon is assigned a new OBJECTID number.

Your next task is to merge two smaller health area polygons into one.

14. Close the Identify window. If necessary, pan slightly to the east, so health areas 6700 and 6300 are in view.

15. Activate the Edit tool, and then click anywhere on area 6700 to select it. Press and hold the Shift key and click area 6300 to add it to the selection.

16. On the Editor toolbar, click Editor > Merge. On the Merge dialog box, click 6700 (Health areas), and then click OK.

The two parcels are now one. The merged parcel retains the user-defined attributes (such as HealthAreaID), but the software-generated attributes (Shape_Length and Shape_Area) have been combined from the input polygons.

17. Open the attribute table for the Health areas layer. Click the Show selected records button at the bottom of the table to display the selected records.

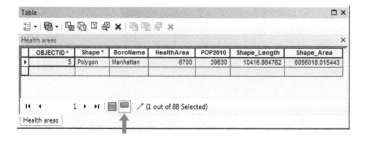

The record that remains is for the feature you selected on the Merge dialog box.

18. **Close the table. On the Editor toolbar, click Editor > Stop Editing. Click Yes to save your edits.**

 You don't need to save your map document, because your edits were made directly to the Parcels feature class in the geodatabase.

19. **If you are continuing to the next exercise, leave ArcMap open. Otherwise, exit the application, and click No when prompted to save your changes.**

Exercise 13c
Edit feature attribute values

In ArcMap, you can change attribute values one at a time by typing new values in the table during an edit session. You can also change values for a selected set of records or for all records in a table using the Field Calculator. The Field Calculator assigns values to a field according to an expression that may simply be a value, such as 101 or "Residential," or an expression that uses math operators. The Field Calculator can be used in or out of an edit session.

In addition to editing attribute values, you can add and delete records or add new fields to a table and define properties for those fields. Records are added and deleted during an edit session. Fields are added and deleted outside of an edit session.

Having split health area 5700 in exercise 13b, you need to assign a new HealthAreaID number to one of the new polygons. After that, you will add an attribute to the feature class. Currently the health area sizes are provided in square feet, but an acreage value would be more practical. Currently the feature class doesn't have an acres attribute. You will add one and then have ArcMap calculate values for it, converting square feet to acres.

1
2
3
4
5
6
7
8
9
10
11
12
13
14
15
16
17
18
19
20

1. In ArcMap, open the ex13c map document from your Chapter13 folder.

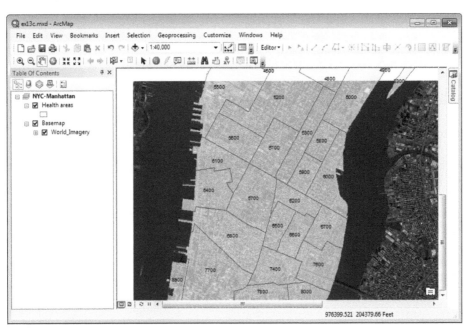

The map is centered on the health area polygons you split in exercise 13b. Both have an ID of 5700. You will assign a new ID number to the smaller parcel.

2. Display the Editor toolbar, if it is not already available, and start an edit session.

3. With the Edit tool activated, click the smaller health area 5700 (the northernmost one) to select it.

4. Open the Health areas attribute table. Click the Show selected records button at the bottom of the table.

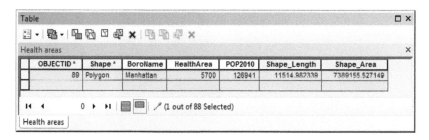

Field names on gray backgrounds in the table indicate fields that are maintained by ArcMap. They cannot be edited. Names on white backgrounds indicate fields that you can edit. (Note that with some display schemes, the color difference may be hard to see.)

5. Click in the cell under the HealthAreaID field. Highlight and delete the existing value of 5700. Type **5750**, and press Enter.

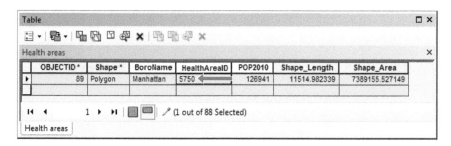

The value 5750 does not duplicate an existing health area ID.

6. On the Editor toolbar, click Editor > Stop Editing. Click Yes to save your edits. Then move the table away from the map display.

When the edit session is finished, no records in the table or features on the map are selected. On the map, the smaller parcel is labeled with its new health area ID.

Ideally, you would also update the population field for the new features. The best way to do that would be to perform a spatial join with census tracts (because health areas are composed of multiple census tracts) and summarize the population field. You will learn about spatial joins in chapter 17.

Next, you will add a field to the table to store the health areas in acres.

7. At the bottom of the Health areas attribute table, click the Show All Records button.

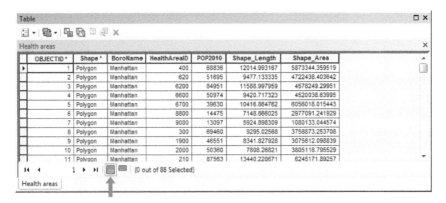

Currently, the Shape_Area field contains the square footage of each health area. After adding the new field, you will write an expression in the Field Calculator to convert square feet to acres.

What expression do you anticipate creating to perform this conversion?

8. At the top of the table, click the Table Options arrow and click Add Field.

TIP

If the Add Field command is not available, you may not have concluded the edit session in step 6.

The Add Field dialog box is where you name the new field, specify its data type, and set its field properties. (This is equivalent to what you did in chapter 11 when you added fields to the playgrounds feature class in Catalog.)

ADDING FIELDS FROM ArcCATALOG

New fields can be added from ArcCatalog or ArcMap, as long as you have write access to the data.

9. On the Add Field dialog box, in the Name box, type **Acres**. Click the Type arrow and click Float. (Float is a numeric field that allows decimal points.) Click OK.

The new field is added to the table. Its values are set to Null.

Next, you will calculate values for the Acres field, which you can do whether you are in an edit session or not. Field calculations made in an edit session take longer but have an Undo option. Calculations made outside an edit session are faster but cannot be undone. (To replace an incorrect calculation, you can make a new one or delete the field and start over.) For the sake of speed, you will make your calculation without starting an edit session.

10. In the attribute table, right-click the Acres field name and click Field Calculator.

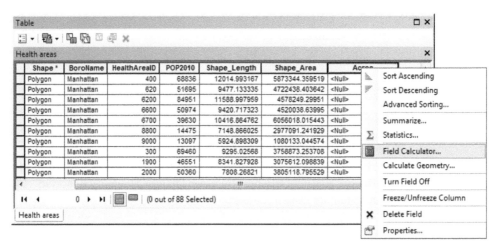

11. If you get a warning about calculating outside an edit session, click Yes to dismiss it. (You may also select the check box to no longer show this warning.)

The Field Calculator appears.

Next, you will create an expression to convert feet to acres. One acre equals 43,560 square feet, so dividing the values in the Shape_Area field by 43,560 will give each parcel's acreage.

12. In the Fields box of the Field Calculator, double-click Shape_Area to add it to the expression box below. Click the division (/) button. Press the Spacebar, and type **43560**. Click OK.

The values are calculated and written to the Acres field of the attribute table.

TIP

If you had any features selected, the field expression would be calculated for the selected records only.

Shape *	BoroName	HealthAreaID	POP2010	Shape_Length	Shape_Area	Acres
Polygon	Manhattan	400	68836	12014.993167	5873344.359519	134.8334
Polygon	Manhattan	620	51695	9477.133335	4722438.403642	108.4123
Polygon	Manhattan	6200	84951	11588.997959	4578249.29951	105.1021
Polygon	Manhattan	6600	50974	9420.717323	4520038.63995	103.7658
Polygon	Manhattan	6700	39630	10416.864762	6056018.015443	139.027
Polygon	Manhattan	8800	14475	7148.866025	2977091.241929	68.34461
Polygon	Manhattan	9000	13097	5924.898309	1080133.044574	24.79644
Polygon	Manhattan	300	69460	9295.02568	3758873.253708	86.29185
Polygon	Manhattan	1900	46551	8341.827928	3075612.098839	70.60634
Polygon	Manhattan	2000	50360	7808.26821	3805118.795529	87.35351

(0 out of 88 Selected)

Health areas

When you edit a feature in a geodatabase feature class, ArcMap automatically updates its Shape_Length and Shape_Area values. ArcMap does not, however, manage fields

1

2

3

4

5

6

7

8

9

10

11

12

13

14

15

16

17

18

19

20

that you have added yourself. Therefore, when you make changes to parcels which affect their size, you will have to recalculate the Acres field.

You don't need to save your map document. Your edits were made directly to the feature class attributes in the geodatabase.

13. **Close the attribute table. Close the Editor toolbar. If you are continuing to the next chapter, leave ArcMap open. Otherwise, exit the application, and click No when prompted to save your changes.**

Summary

Editing GIS data is an important skill. Features change, new information is received, and in some cases, data needs to be simplified. In this chapter, you learned basic editing skills that should provide a foundation for learning the more advanced editing techniques available in ArcMap.

Terms

map topology
geodatabase topology

CHAPTER 14
Geocoding addresses

Create an address locator
Match addresses
Rematch addresses

Learning objectives

- Create and manage address locators
- Learn how addresses are eligible for geocoding
- Manually match unmatched addresses

In chapter 12, you learned that a lot of spatial data was originally digitized from paper maps. You can also create spatial data–specifically, point features–from information that describes or names a location. The most common kind of information that is used for this is an address. The process of creating map features from addresses, place-names, and similar information is called geocoding.

The value of geocoding is that it lets you map locations from data that is readily available. While working on projects, you often can't buy an atlas with a ready-made map of your area of interest. With a list of your points of interest, however, you can create this map yourself.

Geocoding addresses requires that you have an address table–a list of addresses stored as a database table or a text file. You also need a set of reference data, such as streets, on which the addresses can be located.

FID	Shape	Address	City	State	ZIP
0	Point	737 Redman Ave	Haddonfield	NJ	0803
1	Point	45 Treaty Elm Lane	Haddonfield	NJ	0803
2	Point	130 Chestnut St	Haddonfield	NJ	0803
3	Point	412 Kings Hwy E	Haddonfield	NJ	0803
4	Point	601 Grove St	Haddonfield	NJ	0803

Address table. Data from Data and Maps for ArcGIS 2010, courtesy of Tele Atlas, US Census, Esri (Pop2010 field).

Reference data. Data from Data and Maps for ArcGIS 2010, courtesy of Tele Atlas, US Census, Esri (Pop2010 field).

ArcGIS uses address information in the attribute table of the reference data to figure out where address points are located. The more detailed the reference data, the more accurately addresses can be located.

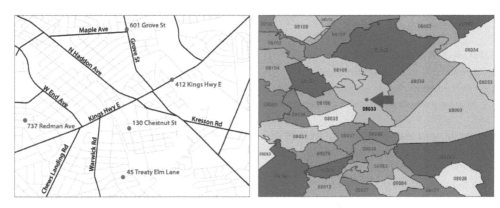

The same table of addresses is geocoded using two sets of reference data. *Left:* On the street map of Haddonfield, New Jersey, address points are placed in the correct location on the correct street. *Right:* On the zip code map of the United States (zoomed to western New Jersey), the same address points are placed in the center of the correct zip code. Because all five points have the same location, they look like one point. Data from Data and Maps for ArcGIS 2010, courtesy of Tele Atlas, US Census, Esri (Pop2010 field).

Street-level reference data is commercially available. Its attributes include street name, street type (avenue, boulevard, and the like), and directional prefixes and suffixes needed to avoid ambiguity in address locations. Each street feature is divided into segments that have starting and ending addresses, such as those seen on neighborhood street signs. This makes it possible to estimate the position of an address along the length of a street segment. There may be separate address ranges for each side of the street, allowing an address to be placed on the correct side of the street.

2846 Central Avenue

2700 2796 2800 2892

2701 2797 2801 2893

Central Avenue

To geocode the address 2846 Central Avenue, ArcGIS examines the reference data and finds the street segment of Central Avenue with the correct address range. Because 2846 is midway between the range of 2800 to 2892 for Central Avenue, the address point is placed in the middle of the street segment, on the even-numbered side of the street. Data from Data and Maps for ArcGIS 2010, courtesy of Tele Atlas, US Census, Esri (Pop2010 field).

In addition to an address table and reference data, geocoding requires an address locator. An address locator is a file that contains the reference data (the relevant attributes of the address table, such as street name and zip code) and various geocoding rules and tolerances (which are defined by an address locator style). Many styles are available, and successful geocoding is dependent on choosing the right style to most appropriately match the reference data and address table. In ArcGIS, address locators are made in ArcCatalog and then used in ArcMap.

The output of the geocoding process is either a shapefile or a geodatabase feature class of points. The geocoded data has all the attributes of the address table, some of the attributes of the reference data, and optionally, some new attributes, such as the x,y coordinates of each point. For more information on what fields are included in the output feature class for geocoding results, consult ArcGIS Desktop Help.

Data

The exercises in this chapter use the following datasets:

- High School Catchment.shp—several zip codes used as a background layer *(Source: School District of Philadelphia, Office of Capital Programs)*
- Philadelphia Streets.shp—Philadelphia street and interstate data *(Source: Streets Department/Office of Innovation and Technology, City of Philadelphia)*
- Schools.csv—a table of school addresses *(Sources: Philadelphia City Planning Commission in cooperation with the School District of Philadelphia, the Archdiocese of Philadelphia, and the Mayor's Office)*

Exercise 14a
Create an address locator

Geocoding starts with creating an address locator. Address locators come in different styles, depending on the attributes of the reference data. For example, the style called "Single Field" is used with reference data that contains only a single attribute of geographic information. In a reference map of the United States, for instance, this would probably be a state name field. Using this type of locator, the data to be geocoded could be located in the correct state but could not be located more precisely.

The address locator style ZIP 5 Digit is used with reference data that has a zip code attribute. A table of US addresses could be located within the correct ZIP Codes using this type of locator.

The US Address-Dual Ranges style is used with street reference data that contains a street name attribute and starting and ending address ranges for each side of the street. A table of US addresses could thus be geocoded to their approximate positions along the length of a street and placed on the correct side of the street as well.

Each address locator style requires that certain attributes be present in the address table. For more information about address locator styles, consult ArcGIS Desktop Help.

For this exercise, you work for a school board in Philadelphia and are in charge of creating a map that shows high schools and their respective catchment areas, to aid in planning and to provide information to future students about which school they will attend. The catchment areas have already been provided; your job is to geocode the schools using their addresses. Being inexperienced with geocoding, you will work with a provided subset of city schools.

1. Start ArcMap and open the ex14a map document from your C:\EsriPress\GTKArcGIS\ Chapter14 folder.

The map shows a portion of downtown Philadelphia, Pennsylvania, with layers of streets and school catchment areas. The School Catchment Area layer displays at 35 percent transparency, allowing you to see the basemap. The ArcGIS Online basemap is turned on. (An internet connection is required to view the basemap. If you can't connect to the internet, make sure the Philadelphia Streets layer is visible. The basemap can be turned off at any time.) Use the bookmarks and navigation controls to view the map extent or details more closely.

2. At the top of the table of contents, click the List By Source button.

In the List By Source view, the table of contents shows nonspatial tables, such as dBase files, that belong to the map document. In this case, there is a table of schools that contains all the high school addresses and other attributes that belong to the school catchment areas. This is the address table you will geocode in exercise 14b.

TIP

It may be helpful to use spreadsheet software, such as Microsoft Excel, to prepare data for geocoding. Excel is useful in formatting your tables so that ArcMap will recognize the data.

The address locator style you choose depends on the attributes in your reference layer (the local streets layer) and in your address table. Next, you will open the two tables to look at this information.

3. **In the table of contents, open the Schools table.**

Table — Schools.csv

FACIL_ID	NAME	ADDRESS	GRADE_LEVE	GRADE_ORG	ZIPCODE
1020	West Philadelphia HS	4901 Chestnut St	High School	09-12	19139
2000	South Philadelphia	2101 South Broad St	High School	09-12	19148
8030	Washington, George	10153 Bustleton Ave	High School	10-12	19116
6030	Roxborough	6498 Ridge Ave	High School	09-12	19128
4330	Lamberton, Robert E.	7501 Woodbine Ave	Elem/Mid/High	K-12	19151
8020	Northeast High	1601 Cottman Ave	High School	09-12	19111
2130	Vaux, Roberts	2300 Masterst	High School	07-09	19121
1080	University City	3601 Filbert St	High School	09-12	19104
2160	Furness, Horace	1900 S 03rd St	High School	09-12	19148
6060	King, Martin Luther	6100 Stenton Ave	High School	09-12	19138
2010	Franklin, Benjamin	550 North Broad St	High School	09-12	19130

1 ▶ ▶I (0 out of 24 Selected)

The attributes include school name and ID, street address, five-digit zip code, enrollment, and grade level. For geocoding, the relevant attributes are ADDRESS and ZIPCODE.

4. **Close the Schools table. In the table of contents, open the table for the Philadelphia streets layer.**

Scroll horizontally to see more attributes. The four address range attributes are L_F_ADD (left from address), L_T_ADD (left to address), R_F_ADD (right from address), and R_T_ADD (right to address). These are the starting and ending street numbers (to and from) on each side (left or right) of a street segment.

There are several other address attributes, as follows:

- PRE_DIR is a direction that precedes a street name, such as the "W" in 499 W Grange Ave. SUF_DIR is a direction that follows a street name, such as the "E" in 364 Roosevelt Blvd "E."
- ZIP_LEFT and ZIP_RIGHT are the ZIP Codes for the left and right sides of a street segment, respectively.
- ST_NAME and ST_TYPE are the familiar attributes of a street address.

The attributes in the streets table most closely match the US Address-Dual Ranges address locator style that you will be creating next.

5. **Close the table. Open the Catalog window in ArcMap.**

6. **In the Catalog tree, browse to ..\GTKArcGIS\Chapter14\MyData. Right-click MyData and click New > Address Locator.**

TIP

You may have noticed that there is a command called Composite Address Locator. A composite address locator consists of two or more individual address locators that allow addresses to be matched against either locator to find the best match. If an address does not match one of the locators, it has a chance of matching the other. For more information, consult ArcGIS Desktop Help.

The Create Address Locator dialog box appears.

7. **Click the Browse button next to the Address Locator Style box. Choose US Address – Dual Ranges, and click OK.**

The Dual Ranges style is used for address ranges that have data for both sides of a street segment.

Notice that an error message appears in the upper portion of the Create Address Locator dialog box. Scroll down and click the warning icon next to the Field Map label. The error message, which reads "Field(s) required by this style are missing," simply means that field names need to be matched to alias names from the reference data.

8. Click the Reference Data box drop-down arrow and choose Philadelphia streets. (Alternatively, browse to the ..\Chapter14\Ex14aData folder, click the Philadelphia Streets shapefile, and click Add.)

The reference data is added to the list and has the role of Primary Table. (You may have to expand the window to see the full path.) Below it, the Field Map section is populated with the fields required to use the Dual Ranges locator style.

9. Scroll down to view the Field Map list.

The Field Map list shows the types of address information contained in the reference data that the address locator style can use. Required information for geocoding is signified by an asterisk (*) in front of the field name. The values in the corresponding lists under Alias Name show the reference data attributes that contain this information. In each case, ArcMap has already selected the correct field (alias) name. (If it hadn't, you could use the lists to choose the corresponding fields.)

How does using more fields in the Field Map affect geocoding?

10. In the lower portion of the Create Address Locator dialog box, click the Browse button next to the Output Address Locator box. Save it as **Philadelphia Schools** in the **..\GTKArcGIS\Chapter14\MyData** folder.

11. Click OK to create the address locator.

Be patient—geocoding can take time. When the Create Address Locator process is complete, a pop-up notification appears in the notification area, at the far right of the taskbar. Click the notification link if you want to view the results.

ArcMap creates the new address locator in the output folder specified.

12. If you are continuing to the next exercise, leave ArcMap open. Otherwise, exit the application, and click No if prompted to save your changes.

In exercise 14b, you will add the Philadelphia Schools address locator to ArcMap and use it to match addresses.

Exercise 14b
Match addresses

Now that you have created an address locator in exercise 14a, you can compare the table of addresses to the reference data attributes and look for matches.

The first step in address matching is to standardize the addresses in the address table—that is, divide them into their component parts. This is done by ArcGIS in the address locator. For example, the address "500 Rankin Street NE" has four parts: the street number (500), the street name (Rankin), the street type (Street), and a suffix (NE). The reference data is already standardized—each address part is listed as a separate attribute.

ArcGIS takes each standardized address from the address table and looks for features in the reference data that have matching parts. For each address, it generates a list of probable or possible matching locations, called candidates, in the reference data. Each candidate gets a score that rates its likelihood of being the correct location. For ArcGIS to make a match and create

1 2 3 4 5 6 7 8 9 10 11 12 13 14 15 16 17 18 19 20

a point feature, a candidate's score must reach a certain level. You can choose both the score that defines a location as a candidate and the score that defines a candidate as a match. By default, the minimum candidate score is 75, on a scale of 1 to 100, and the minimum match score is 85.

When ArcGIS has made as many matches as it can, it creates a new dataset that has a point feature for each address, whether it is matched or not. Features that have unmatched addresses are not assigned spatial coordinates and are therefore not displayed. At that point, you can adjust the geocoding options to try to find matches for unmatched addresses.

Most of the time, geocoding is done from a table of addresses, but as long as you have created an address locator, you can geocode by typing address information in the Find dialog box. Geocoding in this manner does not create new features, but it will flash or place a graphic on a matching location if there is one. This process can also add the matching location to the map. This is useful in situations where you need to locate a single address or find a few addresses quickly.

For this exercise, before geocoding your Schools table, you will test the address locator by locating an address in downtown Philadelphia.

1. **In ArcMap, open the ex14b map document from your Chapter14 folder.**

The map shows the school catchment areas of Philadelphia and streets. The basemap is currently visible and provides more detail around the school catchment areas, but it can be turned off. The Schools table is near the bottom of the table of contents. (If you don't see it, click the List By Source button at the top of the table of contents.)

2. **On the Customize menu, click Toolbars > Geocoding.**

The Geocoding toolbar appears.

Next, you will add the address locator to the map document.

3. **On the Geocoding toolbar, click the Select Address Locator arrow and click <Manage Address Locators>.**

The Address Locator Manager dialog box appears.

4. **On the Address Locator Manager dialog box, click Add. Browse to your ..\GTKArcGIS\ Chapter14\MyData folder. Click the Philadelphia Schools address locator you created in exercise 14a and click Add. (If you don't have the address locator, return to exercise 14a and create it.)**

The Philadelphia Schools address locator and a description of the style appear on the Address Locator Manager dialog box. An asterisk denotes the active address locator.

5. Close the Address Locator Manager dialog box.

Now that the Philadelphia Schools address locator has been added to the map document, you can find addresses on the map, which you will do next.

6. On the Tools toolbar, click the Find button ![M]. On the Find dialog box, click the Locations tab. The Philadelphia Schools address locator is displayed in the Choose A Locator list. If it isn't, click the list arrow and click Philadelphia Schools.

7. In the Full Address box, type **2101 South Broad St Philadelphia**, the address of South Philadelphia High School, and then click Find.

Two candidate street segments appear in the lower section of the dialog box. Their scores are 100 and 92.43.

ON YOUR OWN
Open the Schools table and quickly look up a few more addresses to see where they are on the map. Return to the Find dialog box when you are done.

8. Scroll across the dialog box and look at the candidates' address information.

The address range for odd street numbers, AddNumFrom and AddNumTo, is correct because it includes the street number 2101. The street name, street type, and suffix direction are also correct. The candidate with a score of 92.43 is actually 2101 North Broad St, so that's why it doesn't match and has a lower candidate score. The candidate with a score of 100 is the correct address of 2101 South Broad St.

9. **Move the Find dialog box so you can see the main ArcMap window. Click the first candidate to highlight it. Notice that a green point flashes at the location of the South Philadelphia High School. Right-click this candidate and click Add Point. Close the Find dialog box.**

 ArcMap places a colored dot on the map at the location of the candidate.

 You temporarily geocoded the South Philadelphia High School to verify that the address locator was working. This point is actually a graphic (much like a point that can be drawn using the Draw toolbar) and not a feature associated with a shapefile or a feature class. Next, you will delete the graphic.

10. **On the Tools toolbar, click the Select Elements tool ▶. Click the graphic to select it (cyan selection handles appear). Press Delete.**

 Next, you will geocode the Schools table.

11. In the table of contents, make sure the Schools.csv table is highlighted. On the Geocoding toolbar, click the Geocode Addresses tool .

TIP

You can also right-click the Schools table in the table of contents and click Geocode Addresses.

On the Choose An Address Locator To Use dialog box, the Philadelphia Schools address locator is highlighted.

TIP

An Esri world geocoding service is also available. This is a geocoding service based on ArcGIS Online geocoding technology and allows you to find addresses using single-line input on the Geocoding toolbar. You can also use it to find an address with a spatial offset (a specified distance from a line) or by latitude and longitude coordinates. **Note:** Using the Esri world geocoding service uses ArcGIS Online credits. Visit the Esri ArcGIS Online information page to learn more at www.esri.com/software /arcgis/arcgisonline.

12. Click OK to open the Geocode Addresses dialog box.

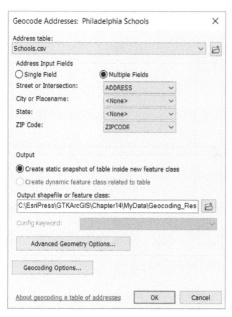

The City or Placename and the State lists are designated as None because the Schools table does not contain this information.

Geocoding creates a point feature for every address. The points can be saved as a shapefile or as a geodatabase feature class. Your Philadelphia data is in a shapefile format, so you will save the address points as a shapefile as well.

13. **Click the Browse button next to the Output shapefile or feature class box.**

14. **On the Saving Data dialog box, browse to ..\GTKArcGIS\Chapter14\MyData and name the output Geocoded_Schools.shp. Make sure the Save as type list is set to Shapefile, and click Save.**

The Geocode Addresses dialog box is updated with the new path.

15. **Click OK to begin the geocoding process.**

1 2 3 4 5 6 7 8 9 10 11 12 13 14 15 16 17 18 19 20

ArcMap geocodes the addresses and creates the output shapefile. When the process is complete, a Geocoding Addresses results window appears.

Of the 24 addresses in the Schools table, 21 are matched, and 3 addresses are unmatched. You will match the unmatched addresses in exercise 14c. For now, you will look at the attributes of the geocoded shapefile.

16. Close the Geocoding Addresses results window.

The Geocoding Result: Geocoded_Schools layer is listed at the top of the table of contents. Meanwhile, 21 point features are displayed on the map, representing the locations of the matched addresses.

17. Open the attribute table for the Geocoding Result: Geocoded_Schools layer.

Table

Geocoding Result: Geocoded_Schools

FID	Shape	Status	Score	Match_type	Match_addr	Addr_type	A
0	Point	U	0	A			
1	Point	M	100	A	10153 BUSTLETON AVE, Philadelphia, PA, 19116	StreetAddress	10
2	Point	M	100	A	3601 FILBERT ST, Philadelphia, PA, 19104	StreetAddress	36
3	Point	M	100	A	1601 COTTMAN AVE, Philadelphia, PA, 19111	StreetAddress	16
4	Point	U	0	A			
5	Point	M	100	A	5000 OXFORD AVE, Philadelphia, PA, 19124	StreetAddress	50
6	Point	M	100	A	5500 LANGDON ST, Philadelphia, PA, 19124	StreetAddress	55
7	Point	M	100	A	550 N BROAD ST, Philadelphia, PA, 19130	StreetAddress	55
8	Point	M	100	A	3133 RIDGE AVE, Philadelphia, PA, 19132	StreetAddress	31
9	Point	M	100	A	5898 LANCASTER AVE, Philadelphia, PA, 19131	StreetAddress	58
10	Point	U	0	A			
11	Point	M	100	A	3201 RYAN AVE, Philadelphia, PA, 19136	StreetAddress	32

1 ▶ ▶I (0 out of 24 Selected)

Geocoding Result: Geocoded_Schools

The table contains 24 records, one for each address in the Schools table. The Status attribute tells you whether each address was matched (M), unmatched (U), or tied (T) with another address for the highest score. (In this case, none of the addresses scored a tied match.) Point geometry is created for records with status M only. The Score attribute contains the match score, and the Side attribute tells you whether an address is on the right side or left side of the street.

The next two attributes that are used, ARC_Street and ARC_ZIP, contain the information from the ADDRESS and ZIP fields, respectively, in the Schools table. The remaining attributes are the same as those in the Schools table.

Although the ARC_Street and ARC_ZIP attributes contain the same values as ADDRESS and ZIP, they serve a different purpose. When you rematch unmatched addresses, you can edit the values in ARC_Street and ARC_ZIP to make matches. You would do this, for example, if you found mistakes in the Schools table. The ADDRESS and ZIP attributes, on the other hand, preserve the original information from the Schools table and should not be edited.

18. Close the table. You have already saved the geocoded shapefile. There is no need to save your map document.

19. If you are continuing to the next exercise, leave ArcMap open. Otherwise, exit the application, and click No when prompted to save your changes.

Exercise 14c
Rematch addresses

When you geocode at the street address level, it is common to have addresses that aren't matched. Occasionally, this is the result of reference data that has errors or is incomplete. For instance, an address may belong to a subdivision that is newer than the reference data. More often, there is a mistake, such as a spelling error, in the address table. In some cases, the problem is with the way ArcGIS has standardized an ambiguous address.

Unmatched addresses can be rematched automatically (ArcGIS does all of them at once) or interactively (you do them one at a time). If you choose to rematch automatically, you need to adjust the ArcMap geocoding matching options, particularly the minimum match score and the spelling sensitivity, or your results won't change. If you rematch interactively, you have additional options. You can edit the way ArcGIS has standardized an address, and you can also match a candidate with a score below the minimum match score.

So far, the address locator has found matches for 21 addresses in the Schools table, leaving three addresses unmatched. In this exercise, you will match the remaining three. First, you will modify the address locator settings to generate more candidates. Then, you will interactively match each address.

1. **In ArcMap, open the ex14c map document from your Chapter14 folder.**

The map layers include the familiar Philadelphia streets and school catchment areas of downtown. The online basemap layer is visible as well. If not displayed, click the List By Source button at the top of the table of contents to see all the layers and tables.

Next, you will add the layer of geocoded addresses you created in exercise 14b. If you are continuing from the previous exercise, skip the following step.

2. Click the Add Data button to add Geocoded_Schools.shp from the ..\GTKArcGIS\ Chapter14\MyData folder. (If you don't have Geocoded_Schools.shp, return to exercise 14b and create it.) If a Geographic Coordinate Systems Warning dialog box appears, click Close.

> *Which coordinate systems are present in the map document?* **Hint:** *Look at data frame properties.*

The Geocoded_Schools layer is added to the table of contents. It is symbolized with the default ArcMap point symbol in a random color.

ON YOUR OWN
If you want, change the symbol to make it easier to see. Try changing the point symbol to Circle 2, Leaf Green, and 7 pt size.

Next, you will rematch the unmatched addresses.

3. **In the table of contents, make sure the Geocoded_Schools layer is highlighted. Open the Geocoding toolbar, if it is not already open, and click the Review/Rematch Addresses button 🔳.**

The Interactive Rematch dialog box appears. (If needed, resize the dialog box to make it larger.)

In the Statistics panel in the upper-right corner, you get the results from exercise 14b (21 matched and 3 unmatched). The Show results box at the upper left allows you to specify which addresses you want to rematch.

4. **Click the Show results arrow and click Unmatched Addresses. Horizontally scroll to the right to see the actual addresses of the unmatched records (the ADDRESS field).**

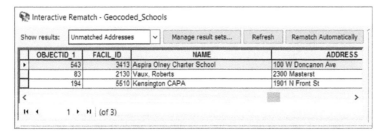

5. **Right-click the ADDRESS heading and choose Sort Descending.**

The table displays the three unmatched records with scores of 0. The first one is highlighted in cyan. Below the table and to the left, in the Address panel, the Street or

Intersection box and the ZIP Code box contain the current address information for the highlighted record. Edits to this information will be written to the ARC_Street and ARC_ZIP fields of the Geocoded Schools table.

The large box in the lower midsection will display match candidates for a highlighted record. There are no match candidates for "2300 Masterst." In the Address panel, notice that the address 2300 Masterst does not have a street suffix. It looks like the address may also have a spelling error that the geocoding engine is not detecting.

6. **Select the record with the address "2300 Masterst." In the Address panel, change the address from 2300 MASTERST to 2300 MASTER ST. (Insert a space in between the R and ST to separate the letters, and then press Enter.)**

Notice that two candidates appear and the top match score is 100. The second candidate (with a score of 79) closely resembles the first candidate; however, it is not the correct numerical address. The first candidate is the correct one for matching the address.

TIP

In the lower-left corner of the Interactive Rematch dialog box there is a Zoom to Candidates button. When geocoding, use this feature to zoom to the candidates listed. Move the dialog box so you can see the map.

7. **Make sure the first candidate is highlighted and click Match.**

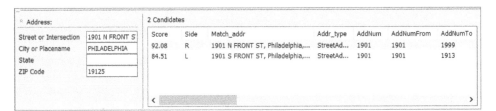

The candidate is matched. In the address table, the record's Status value is changed from U to M, the Score value is changed from 0 to 100, and its match address value is added. (You may need to scroll left or right to see these changes.) Twenty-two addresses are now reported as matched. Next, you will match the other two addresses.

8. **Select the second record in the table.**

There are no match candidates for the 1901 N Front St record. It is possible that not enough required information is available for this address.

9. **In the Address panel, type Philadelphia into the City or Placename box, and then press Enter.**

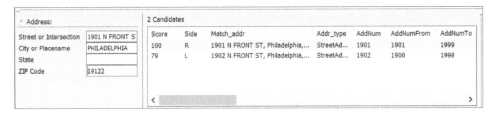

It is possible that the wrong zip code was entered in the Schools table. By adding the city or place-name into the Address panel, you are adding more information to be even more specific. Two candidates appear in the large box where the top candidate has a score of the 92.08. So next, you will view the zip code of the Primary table (Philadelphia streets) to see if it is different from that in the Schools table.

10. **In the Candidates table, horizontally scroll to the right and notice that the zip code for the top candidate is 19122, which is different from the ZIP Code of the unmatched candidate of 19125. (Expand the ZIPCODE column if you can't fully see its values.)**

11. **In the Address panel, highlight the ZIP Code value 19125 and type 19122. Then press Enter.**

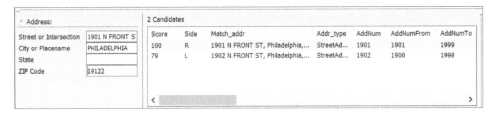

The top candidate now has a perfect score of 100. (You may need to horizontally scroll to the left to see the candidate score.)

12. **Make sure the first candidate is highlighted and click Match.**

The candidate is matched. Its ARC_ZIP value is changed from 19125 to 19122. (You may need to horizontally scroll the address table to see these changes.) In the Statistics panel, 23 addresses are now reported as matched.

Continuing with the last unmatched address, there are no match candidates for the 100 W Doncanon Ave record. It is possible that the spelling of the address may be incorrect.

13. **In the table, select the record with the address "100 W Doncanon Ave" and click the Geocoding Options button. Change the Spelling sensitivity to 50 and click OK.**

Two candidates appear in the large box where the top candidate has a score of 95.96. By the looks of it, the school address was completely misspelled. By changing the spelling sensitivity, the geocoding engine was able to find an address that was similar. Also notice that the other candidate is similar but has the wrong direction prefix; therefore, the candidate with the highest score is correct.

GEOCODING OPTIONS

Geocoding Options can be accessed from the Address Locator Manager, Interactive Rematch, or Geocoding Addresses dialog boxes. These options allow you to fine-tune the interactive-matching experience. You can adjust spelling sensitivity, minimum candidate score, and minimum match score to control how closely an address must match a candidate to score a match. For more information, consult ArcGIS Desktop Help.

14. Make sure the first candidate with a score of 95.96 is selected and, under the Candidates table, click Match.

The candidate is matched. Its match address value is added as 100 W Duncannon Ave. (You may need to scroll left or right to see these changes.) In the Statistics panel, 24 addresses are now reported as matched. You have matched all the addresses in your Schools table.

15. Close the Interactive Rematch dialog box.

On the map, there are now 24 geocoded school points.

16. In the table of contents, right-click the Geocoded_Schools layer and click Label Features.

The points are labeled with the names of the schools.

17. Open the attribute table for the Geocoded_Schools layer.

All the records have the value M in the Status field. Edits that you made during rematching ("MASTERST" to "MASTER ST," for example) are reflected in the ARC_Street and ARC_ZIP fields.

18. Close the table. If necessary, return the table of contents to the List By Drawing Order view and close the Geocoding toolbar.

ON YOUR OWN

As you learned how to do in exercise 5c, share the map package to your ArcGIS Online account.

19. If you want to save your work, rename your map document and save to your MyData folder for chapter 14.

20. If you are continuing to the next chapter, leave ArcMap open. Otherwise, exit the application, and click No if prompted to save your changes.

1
2
3
4
5
6
7
8
9
10
11
12
13
14
15
16
17
18
19
20

Now that you know how to geocode addresses, you can create digital maps for reports and projects to present to the school board.

Summary

In this chapter, you learned how to create an address locator in ArcCatalog by choosing an address locator style and using street data as the primary reference. Then in ArcMap, you were able to use the locator to geocode your table of addresses and create a new point layer. You also used geocoding options to search for unmatched addresses and rematch them. Geocoding is a common GIS task that has many practical applications. Although it is true that the quality of derived data is only as good as its source, geocoding options allow you to manually correct those inevitable data entry errors.

Terms

geocoding
address locator

CHAPTER 15
Querying data

Select and find features
Use attribute queries
Create reports

Learning objectives

- Identify and search for features
- Become familiar with various querying methods
- Open feature hyperlinks
- Modify the selectable layers on a map
- Select features by using Structured Query Language (SQL) queries based on attribute values
- Generate useful reports

A great strength of GIS is that it doesn't just show you where things are, it also tells you a lot about them. If you are curious about a road leading to the next town—for example, US Interstate 70— you may be able to find out such things as its direction, length, speed limit, and the number of lanes it has. If that road leads to a place such as Topeka, Kansas, you may want to know the city's population (127,473, based on the census of 2010), the median household income ($40,826), the median home value ($95,600), and many other things, in addition to the city's geographic attributes.

A lot of information about features can be conveyed by the way they are symbolized and labeled. But for most features, it isn't possible to display everything that is known about them on a single map.

In ArcMap, there are several ways to retrieve unseen information about features. As you may have already done, you can click features with the Identify tool to display their attributes. This is called identifying features. You can click features on the map with the Select Elements tool to highlight them and look at their records in the layer attribute table. This is called selecting features interactively. You can write a query that automatically selects features that meet specific qualitative and quantitative criteria combined (e.g., three-bedroom houses with swimming

pools). This is called selecting features by attribute. Or you can provide ArcMap with a piece of information, such as a name, and see which feature it belongs to. This is called finding features. You can also select features by location, which you will learn more about in chapter 16.

Attributes are not the only kind of information that can belong to a feature. Pictures, text documents, and web pages can also be associated with features through hyperlinks. Clicking a hyperlinked feature opens a file on disk, points your web browser to a URL, or runs a macro (a sequence of automated commands).

Data

This chapter uses the following datasets:

- cabbagetown.shp—polygons of the postal code areas that make up the Cabbagetown neighborhood in Toronto, Ontario, Canada *(Source: Courtesy of Generation5, a Motricity Business © 2011 Motricity Canada Inc.)*
- toronto_background.shp—polygons of the postal code areas that make up a region of Toronto, Ontario, Canada *(Source: Courtesy of Generation5, a Motricity Business © 2011 Motricity Canada Inc.)*
- toronto_streets.shp—lines of select streets around the Cabbagetown neighborhood in Toronto, Ontario, Canada *(Source: City of Toronto)*

Exercise 15a
Select and find features

The fastest way to get information about a single feature is to identify it. If you want to compare information about several features, the best way to do it is to select the features on the map and look at their records in the layer attribute table. When you have a piece of information—a place-name or address, for instance—but don't know which feature it belongs to, you can use that information to search the map for the feature.

A hyperlink links features to things that can't be stored as attribute values. A point feature representing a government office, for example, could be linked to the office website.

For this exercise, you are a marketing analyst in Toronto, Canada. Many of your clients are major retail chains located in the area that also operate online e-commerce websites. You have access to marketing data about neighborhoods throughout the city, and you have decided to use

ArcGIS to help you identify current market profiles. A market profile attempts to characterize a neighborhood based on its demographics and socioeconomic composition.

Currently, you are working with a large retail chain that sells everything from household products to banking services. Your client has asked you to look at online sales for potential customers of goods because it wants to concentrate its marketing and advertising efforts on consumers who meet a certain criteria (which you will see in the exercise). The target area is in a neighborhood on the east side of downtown Toronto called Cabbagetown, which got its name from an influx of Irish immigrants who populated the area in the 1840s and were known to grow cabbage on their properties. Today, the neighborhood is known for its Victorian row homes, beautiful gardens, and mixture of residents of various income levels. You are going to explore the composition of the neighborhood first before you analyze its marketing statistics.

1. **Start ArcMap. Open the ex15a map document from your C:\EsriPress\GTKArcGIS\ Chapter15 folder.**

 The map shows target postal code areas (in light purple) in the Cabbagetown neighborhood, surrounding postal code areas (in light brown), and local roads. You may need to widen your table of contents to see layer names in their entirety (drag the table of contents divider).

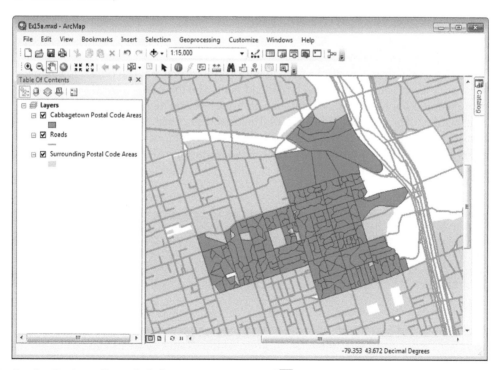

2. **On the Tools toolbar, click the HTML Pop-up tool 💬. The mouse pointer becomes a hand and pointer finger.**

1
2
3
4
5
6
7
8
9
10
11
12
13
14
15
16
17
18
19
20

3. On the map, click a postal code feature, such as the one in the graphic.

The HTML pop-up window displays the feature's attribute values. The bold title at the top tells you which postal code feature is being identified (in this case, M4X1R1). Included is some demographic and socioeconomic data, such as age range (e.g., 30_to_34) and total number of households (Total_Hous). The Canadian postal code format is similar to the US ZIP+4 format in that it uses a base code plus additional digits to identify a geographic segment such as a city block. (M4X represents a larger area, and 1R1 identifies a subunit of this area.)

TIP

The HTML pop-up window can be customized as a table of the visible fields, as a URL on the internet, or as a formatted page based on a spreadsheet template. To learn more about HTML pop-ups, consult ArcGIS Desktop Help.

4. Click a few more postal code areas. A new HTML pop-up window appears each time you click a feature. Arrange the HTML pop-up windows side by side so you can see each one.

Using the HTML Pop-up tool is a quick way to view feature attributes. As you have seen in previous chapters, using the Identify tool is also a fast way to get information about features. Neither method, however, is a convenient way to compare the attributes of several features. To do this, you will select features on the map and look at their records in the layer attribute table.

5. Close the HTML pop-up windows. In the table of contents, click the List By Selection button.

By default, all layers are selectable. You want to select features from the Cabbagetown Postal Code Areas layer only, so you will adjust for that next.

6. For both the Roads and Surrounding Postal Code Area layers, click the toggle key to make them not selectable.

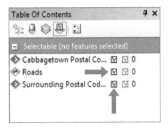

Both the Roads and Surrounding Postal Code Area layers are now in the Not Selectable category.

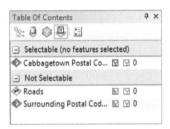

7. On the menu bar, click Selection > Selection Options.

On the Selection Options dialog box, the selection tolerance is set to 3 pixels. Because you are working with small features that have many features adjacent to them, you will reduce the tolerance. This ensures that when you click a location, you won't select more than one feature at a time.

8. In the Selection tolerance box, replace the value of 3 with 0, and then click OK.

TIP

You can change the default selection color using the Selection Options dialog box. Or you can override the default for individual layers by changing the selection color on the Selection tab of the layer properties.

9. **On the Selection menu, click Interactive Selection Method > Add to Current Selection.**

 The Add to Current Selection option allows you to click features without the need to press and hold Shift. When the Create New Selection option is chosen, you must press and hold Shift to select multiple features.

10. **On the Tools toolbar, click the Select Features tool 🔲.**

11. **On the map, click the postal code areas shown in the graphic, one at a time. Each area you click is added to the selection set. The table of contents shows how many postal codes are selected and the name of each postal code.**

TIP

If you select a postal code area by mistake, on the Selection menu, click Interactive Selection Method > Remove From Current Selection, and then click the selected postal code. Then switch back to Add to Current Selection to select the remaining postal code areas.

The eight selected postal code areas are outlined in cyan (or a different color if you changed the selection color). In the attribute table of the layer, the records corresponding to the selected postal codes will also be highlighted.

TIP

You can change the selection color for attribute tables by modifying your ArcMap options. (On the Customize menu, click ArcMap Options, and then click the Tables tab.)

12. In the table of contents, right-click Cabbagetown Postal Code Areas and click Open Attribute Table.

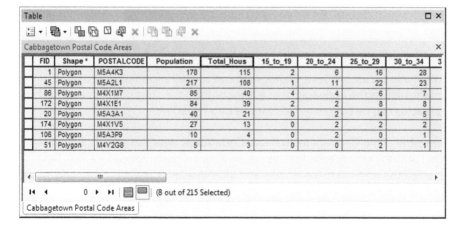

The records for selected postal code areas are highlighted but not grouped together.

13. At the bottom of the attribute table, click the Show selected records button to show the selected postal code records only.

14. Scroll to the right in the table to see its attributes. You can now compare the attributes of the selected features for such values as average household income, total population, and age ranges. Right-click the Total_Hous field name. On the shortcut menu, click Sort Descending.

Table

Cabbagetown Postal Code Areas

FID	Shape *	POSTALCODE	Population	Total_Hous	15_to_19	20_to_24	25_to_29	30_to_34	3
1	Polygon	M5A4K3	178	115	2	6	16	28	
45	Polygon	M5A2L1	217	108	1	11	22	23	
86	Polygon	M4X1M7	85	40	4	4	6	7	
172	Polygon	M4X1E1	84	39	2	2	8	8	
20	Polygon	M5A3A1	40	21	0	2	4	5	
174	Polygon	M4X1V5	27	13	0	2	2	2	
106	Polygon	M5A3P9	10	4	0	2	0	1	
51	Polygon	M4Y2G8	5	3	0	0	2	1	

I◄ ◄ 0 ► ►I (8 out of 215 Selected)

Cabbagetown Postal Code Areas

Now the records are sorted in order, from the most households per postal code to the fewest. The first record shows a value of 115 (that is, 115 households for postal code M5A4K3).

15. **Click the gray tab to the left of the first record (postal code M5A4K3).**

The record is highlighted in yellow (because it has already been selected in cyan), as is the corresponding map feature. You may need to move the attribute table to see the highlighted feature.

16. **Click the gray tabs for the other records to familiarize yourself with the locations and attributes of each postal code. When you are done, close the attribute table. The yellow highlighting on the map features disappears. On the menu bar, click Selection > Clear Selected Features.**

Attribute information is useful, but a photograph can convey information that table attributes can't–such as the appearance of a neighborhood. Pictures, documents, web pages, and macros can be connected to features by a hyperlink.

The attribute table of the Cabbagetown Postal Code Areas layer has a field called IMAGE that contains paths to photographs that represent what neighborhoods look like. Next, you will hyperlink the features to the photos and then view them.

17. In the table of contents, switch back to the List By Drawing Order view and open the properties of the Cabbagetown Postal Code Areas layer. On the Layer Properties dialog box, click the Display tab. In the Hyperlinks frame, select the Support Hyperlinks Using Field check box. Below the check box, click the field list arrow and click IMAGE. Below the list, make sure the Document option is selected, and then click OK.

Next, you are ready to see photos of the area to get a better sense of what kinds of buildings and residential homes make up the Cabbagetown neighborhood.

18. On the Tools toolbar, click the Hyperlink tool.

On the map, hyperlinked features are outlined in blue. Pausing the mouse pointer over a feature that is hyperlinked displays the path location to the document (in this case, a photograph).

19. Click any hyperlinked feature to display a photograph of the property. The photograph appears in your default image-browsing software. After you look at a few more photographs, close the application that opened it.

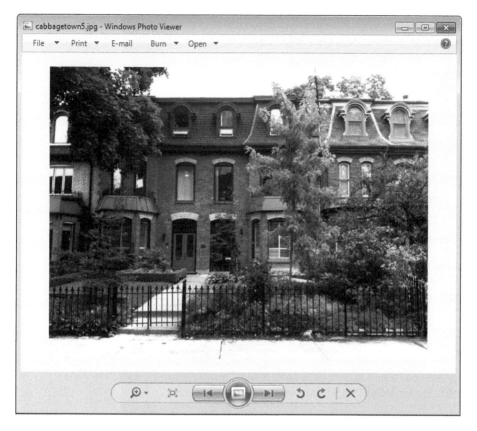

Where are the photographs linked from? Where else can they be linked from?

CREATING HYPERLINKS

There are two ways to create hyperlinks. One is to add document paths or URL addresses to a field in the layer attribute table. This is efficient if you are hyperlinking many features. The other way is to click a feature using the Identify tool, and then right-click the feature in the Identify dialog box and click Add Hyperlink. This is easier if you are setting just a few hyperlinks. It is also the only way that lets you hyperlink a feature to more than one object. For more information, consult ArcGIS Desktop Help.

Prior to your work in ArcGIS, your coworker suggested that you investigate the average household income of a postal code M4X1C5. So next, you will use the Find tool to locate the area on the map and display its attributes.

1
2
3
4
5
6
7
8
9
10
11
12
13
14
15
16
17
18
19
20

20. On the Tools toolbar, click the Find tool to open the Find dialog box. On the Features tab, in the Find box, type **M4X1C5.**

To speed things up, you can specify a layer, and even a field, to search in.

21. Click the In arrow and click Cabbagetown Postal Code Areas. Under Search, click the In field option, and then click the arrow below and click POSTALCODE. Then click Find.

ArcMap searches the POSTALCODE field of the Cabbagetown Postal Code Areas attribute table for M4X1C5 and displays the matching record.

Now that you have found the feature, you can locate it on the map and get its attributes.

22. Move the Find dialog box away from the map but so you can still see it. In the lower portion of the dialog box, right-click M4X1C5. On the shortcut menu, click Identify.

The postal code flashes briefly in the middle of the map, and the Identify window appears. (If the Identify window covers the postal code, move it and repeat the preceding step.)

23. In the Identify window, scroll down to view the average household income.

This postal code area has a high household income, $143,870. Similarly, if you identify the immediate surrounding areas, you will see that they also have high household incomes.

24. Scan the other attribute values. When you are finished, close the Identify window and the Find window.

25. If you want to save your work, rename your map document and save to your MyData folder for chapter 15.

26. If you are continuing to the next exercise, leave ArcMap open. Otherwise, exit the application and click No if prompted to save your changes.

1
2
3
4
5
6
7
8
9
10
11
12
13
14
15
16
17
18
19
20

The area of Cabbagetown you just looked at may meet your client's target criteria for customers. You will look at your client's specific target criteria in exercise 15b.

Exercise 15b
Use attribute queries

Your large retail chain client has asked you to target potential customers in the Cabbagetown neighborhood who meet specific criteria. The client will target households in these postal code areas for customized marketing campaigns focused on three different products sold online (groceries, home and garden products, and vacation packages). By targeting these postal codes, you will be able to provide your client with information about the neighborhood so that advertisement efforts can be focused on customer needs.

Interactive selection works when you can see what you are looking for on the map–the largest postal code, for instance. To select features according to criteria you can't see on the map (such as households that have an average income of at least $100,000), you will need to write a query.

A valid query results in a selected set of features that meet the specified conditions. The simplest query expression consists of an attribute (such as income), a value (such as $100,000), and a relationship between them (such as "greater than or equal to"). Complex queries can be created by connecting simple queries with terms such as "and" and "or." Relationship terms and connecting terms are called operators.

Queries are not written in ordinary English but in Structured Query Language (SQL). This is not as daunting as it might sound. To build a query expression, all you need to do is open a dialog box that contains a query builder interface (such as Select By Attributes), double-click an attribute, and then click an operator and a value.

The marketing information for your retail client shows online sales of groceries, home and garden supplies, and vacations. These values show the percentage of households (not the dollar amounts they spent). For example, the GROCERY attribute shows the percentage of households that used the internet to purchase groceries within the past year. The related attributes GRCYL250 and GRCYG250 show the percentage of households that spent less than $250 or more than $250, respectively, on groceries purchased online within the same time period.

Rather than target advertisements toward the entire neighborhood, your client wants to target specific postal codes. For groceries, your client wants to win over new customers and has determined that it wants to advertise to postal code areas where less than 5 percent of

households purchase groceries online. For home and garden products, your client wants to advertise to postal code areas with an average household income of $75,000 to $150,000 and where households make online home and garden purchases of more than $250 (regardless of the percentage of population that spends this amount). Finally, for vacations, your client wants to target postal code areas that have an average income of $175,000 or more and where more than 35 percent of households spend more than $1,000 on online vacation purchases.

1. **In ArcMap, open the ex15b map document from your Chapter15 folder.**

 The map of neighborhood postal codes is displayed.

2. **On the menu bar, click Selection > Select By Attributes. On the Select By Attributes dialog box, set the Layer option to Cabbagetown Postal Code Areas.**

 The fields from the postal codes attribute table appear in the Fields list in the upper portion of the dialog box. When a field is highlighted, you can display its attribute values in the Unique Values box, in the center of the dialog box. The buttons to the left of the Unique Values box are used to choose operators, such as greater than or equal to (>=) or less than or equal to (<=).

1 2 3 4 5 6 7 8 9 10 11 12 13 14 15 16 17 18 19 20

The first target you want to find is postal code areas with less than five percent of households purchasing groceries online.

3. In the Fields list, scroll to "GROCERY" and double-click it. Click the less than (<) button, press the spacebar, and then type **0.05**. Your query expression is displayed in the expression box in the lower portion of the dialog box.

When you apply the query, ArcMap searches the attribute table for records that have a value in the GROCERY field that is less than 0.05. The corresponding features are selected on the map.

4. Click OK.

On the map, postal code areas that meet the desired criteria are selected based on your query.

If you had just looked at the map at face value, you never would have known that these postal code areas meet the selected criteria. On the status bar, notice that 123 features are selected. (If you don't see it, switch to the List By Selection view to see how many are selected, and then switch back to the List By Drawing Order view.) Since this is one of the targets your client wants, you will next create a layer from this selection.

5. Right-click the Cabbagetown Postal Code Areas layer and click Selection > Create Layer From Selected Features.

A new layer is created, called Cabbagetown Postal Code Areas selection. It is symbolized by a random fill color.

6. Rename the new layer Grocery target.

1
2
3
4
5
6
7
8
9
10
11
12
13
14
15
16
17
18
19
20

7. On the menu bar, click Selection > Clear Selected Features.

You now have a new selection layer that you can provide to your client which shows postal code areas with less than 5 percent of households purchasing groceries online. Your client will now be able to increase advertisements to postal codes in this segment that generally do not make many online purchases.

Next, you will work on the home and garden criteria your client wants. You will target postal code areas in which households have an average income of $75,000 to $150,000 *and* in which at least one household spends more than $250 online each year for home and garden products.

8. Turn off the Grocery target layer. On the menu bar, click Selection > Select By Attributes. Make sure Cabbagetown Postal Code Areas is chosen in the Layer list. Below the expression box, click the Clear button to clear the previous query.

9. In the Fields box, scroll to "AvgHHIncom" and double-click it. Click the greater than or equal to (>=) button, press the Spacebar, and type 75000. Click the And button and double-click "AvgHHIncom" once more. Then click the less than or equal to (<=) button, press the Spacebar, and type 150000. Your query is displayed in the expression box. Click Apply.

TIP

When building query expressions, incorrect syntax, such as unnecessary use of commas or periods, can cause errors. The Verify button can be used to check whether the query syntax is correct. For more information on building a query expression, consult ArcGIS Desktop Help.

When you apply the query, the corresponding features (93) that have an average household income of $75,000 to $150,000 are selected on the map. Next, you will make the query slightly more complex to target the postal code areas that have households that spend more than $250 on home and garden products online (represented by the HMGRG250 attribute field).

10. On the Select By Attributes dialog box, click the And button. In the Fields box, double-click "HMGRG250", click the greater than (>) button, press the Spacebar, and then type 0. Your query should match the graphic. Click OK.

ArcMap searches the HMGRG250 attribute field for values greater than zero (0), which means that all households that have spent more than $250 are chosen from the selection. On the map, fewer postal code areas (79) are selected because fewer of them satisfy both conditions.

As with the previous grocery selection, you will create a new layer for your client.

11. Right-click the **Cabbagetown Postal Code Areas** layer and click **Selection > Create Layer From Selected Features.**

12. Rename the new layer **Home and Garden target.**

13. Switch to the List By Selection view. In the Selectable category, click the Click To Clear Layer Selection button for the **Cabbagetown Postal Code Areas** layer. After that, don't forget to switch back to the List By Drawing Order view.

Finally, you will create one more query to target households that have an average income of $175,000 or more *and* that have a high proportion (35 percent or above) of online vacation purchases that exceed $1,000 (represented by the V_G1000 attribute field).

14. Turn off the Home and Garden target layer. On the menu bar, click **Selection > Select By Attributes.** Make sure **Cabbagetown Postal Code Areas** is chosen in the Layer list. Click **Clear** to clear the previous query.

15. In the Fields box, scroll to "AvgHHIncom" and double-click it. Click the greater than or equal to (>=) button, press the Spacebar, and type **175000.** Click the And button and double-click "V_G1000". Then click the greater than or equal to (>=) button, press the Spacebar, and type **0.35.** Your query is displayed in the expression box. Click **OK.**

There are far fewer areas (17, to be exact) to target, considering the combined criteria of a high level of income along with a high proportion of online purchases of vacations costing more than $1,000.

ON YOUR OWN

As with the previous segment selections, create a new layer from the current selection. Rename the new layer Vacation target. When you are finished, clear your selection.

TIP

Remember, these layer files are elements of the map document, pointing to cabbagetown.shp. If you want to make this a permanent shapefile or feature class, you can right-click the layer in the table of contents and click Data > Export Data to create it. Another option is to create a layer package (.lpk) that includes both the layer properties (symbolization, labeling, table properties) and the dataset referenced by the layer. It can be easily shared with ArcGIS Online through ArcMap. For more information, consult ArcGIS Desktop Help.

16. If you want to save your work, rename your map document, and save to your MyData folder for chapter 15.

17. If you are continuing to the next exercise, leave ArcMap open. Otherwise, exit the application and click No if prompted to save your changes.

In exercise 15c, you will make a report of the query results that your clients can take with them.

Exercise 15c
Create reports

Reports let you organize, format, and print the information contained in an attribute table. ArcMap has a built-in report generator that allows you to create simple reports that can be added to your map layout.

In exercise 15b, you selected postal code areas in the Cabbagetown neighborhood that meet certain criteria defined by your client. The client will target households in these postal code areas for customized marketing campaigns focused on three different products sold online (groceries, home and garden products, and vacation packages). The grocery target areas will receive advertisements designed to entice new customers who have rarely if ever purchased groceries online. The home and garden target areas will receive advertisements designed to appeal to existing online customers of such products. And the vacation target areas will get information about family and luxury resort vacation packages.

For this exercise, you will create a report so your client can create a customized mailing list. The report you make will be based on the Vacation target layer, where average household income is $175,000 or more, and 35 percent or more of households spend more than $1,000 annually on online vacation purchases.

1. In ArcMap, open the ex15c map document from your Chapter15 folder.

The map displays the Vacation targets layer (with 17 postal code areas) at the top of the table of contents.

2. On the View menu, click Reports > Create Report. The Report Wizard dialog box appears.

1 2 3 4 5 6 7 8 9 10 11 12 13 14 15 16 17 18 19 20

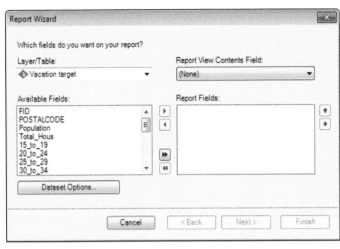

The Layer/Table list specifies the layer or table you will work with. The Available Fields list lets you choose the attributes to include in your report.

You are going to provide your client with information including the target postal codes' average household income, proportion of households that make online vacation travel purchases, and proportion of households that spend more than $1,000 annually on online vacation travel purchases.

3. Make sure the Layer/Table list shows **Vacation target**. In the Available Fields list, click **POSTALCODE**. Click the Add fields button (right-facing arrow) in the middle of the dialog box. The **POSTALCODE** field moves from Available Fields to Report Fields.

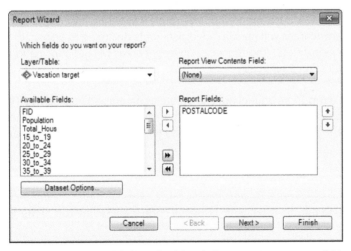

4. In the Available Fields list, double-click **AvgHHIncom** to add it to the Report Fields list. Do the same for the **VACATRAV** and **V_G1000** fields.

5. In the upper-right portion of the dialog box, click the Report Viewer Contents Field arrow and click POSTALCODE. This means this field will be displayed when you preview the report.

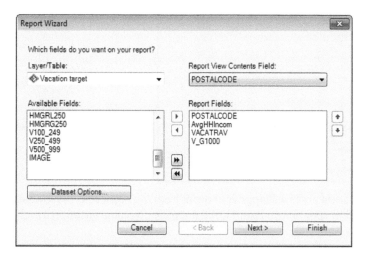

TIP

The Vacation target layer you created already has the postal code areas you want to report. If you didn't create this layer, you could still create a report that uses the selected postal code areas. To do this, click Dataset Options on this wizard panel and click the Selected Set option. This outputs only the selected features to the report.

6. On the Report Wizard dialog box, click Next, and then click Next again. You will not be adding any grouping levels.

7. In the next panel, in the Fields column, select V_G1000 from the drop-down list, and in the Sort column, select Descending from the drop-down list in the first row. Click Next.

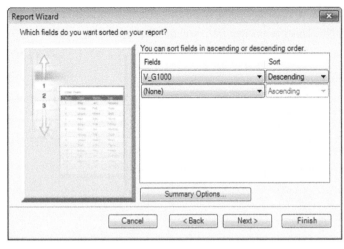

In the report, target postal code areas will appear in order from the highest to the lowest proportion of households that spent more than $1,000 annually on online vacation travel.

8. In the next panel, set the orientation to Landscape, and then click Next.

9. The Havelock (Default) style is selected. Scroll down the list and click Toronto, and then click Next. (These styles have more to do with colors than any real relation to the city.)

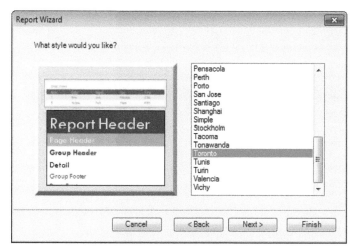

10. Replace the text in the title box with **Vacation Target Report.** Make sure the Preview The Report option is selected and click Finish.

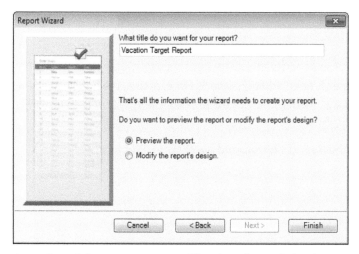

A preview of the report appears. The size of your dialog box may be different, but you probably can't see the whole report. The list at the top of the dialog box lets you change the preview scale. (It doesn't change the printed size of the report.)

If you like the way the report looks, you could click Print to send it to your printer.

Most reports, including this one, need additional work. For example, the headings need to be better aligned. In steps 11 through 16, you will give the report a professional look by setting a few of its properties.

11. In the upper-left corner of the Report Viewer window, click the Edit button.

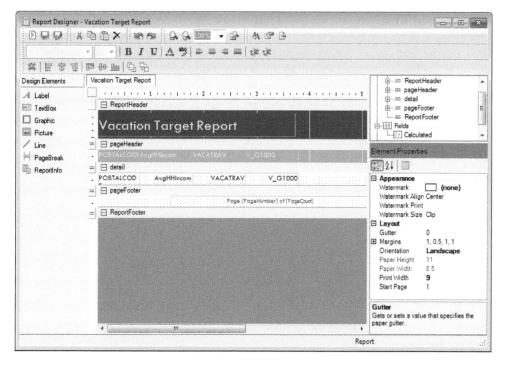

The window changes to the Report Designer window. The Report Designer is a template layout tool to help you edit the appearance of your reports. The upper-right box lists the report components you can customize. Below it are the element properties associated with the highlighted component and the current value of each property. When a report component is highlighted, the element is selected in the middle of the report template view. The template area represents how your report pages will look.

The headings, located in the pageHeader section of the template, need to be aligned properly.

12. **In the template area, click the AvgHHIncom pageHeader (not detail) component.**

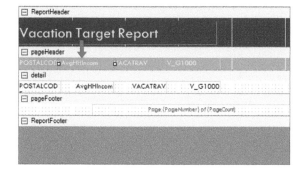

Items selected in the template area have their properties displayed under Element Properties. Scroll down the list to see the different categories: Appearance, Behavior, Data, Design, and Layout. You can use this list to confirm that the correct item in the template has been selected by viewing the Text section under Data.

13. In Element Properties, under the Appearance category, change the Alignment property to Right.

14. Repeat the same right alignment for the VACATRAV and V_G1000 pageHeader components. In the upper-left portion of the window, click the Run Report button ▷ to preview the report.

Vacation Target Report

POSTALCODE	AvgHHincom	VACATRAV	V_G1000
M4X1S3	191687	0.55	0.46
M4X1E7	197717	0.59	0.45
M4X1H7	181489	0.57	0.43
M4X1H7	181489	0.57	0.43
M4X1T6	182857	0.53	0.41
M4W3A2	189338	0.54	0.41
M4W3A2	189338	0.54	0.41
M5A2L9	179933	0.53	0.4
M4X1R7	201391	0.49	0.38
M4X1T7	243427	0.45	0.38
M5A2M1	195900	0.47	0.38
M4X1B3	205439	0.42	0.37
M5A2L7	238710	0.4	0.36
M4X1B4	224881	0.48	0.36
M4X1X7	199839	0.46	0.35
M4X1A3	282750	0.47	0.35
M4X1T8	234183	0.42	0.35

The AvgHHIncom, VACATRAV, and V_G1000 headings are now aligned properly to the column values.

ON YOUR OWN

Change some of the template element properties to see how the changes affect your report. You can edit the text of headings, change spacing between values, change text appearance, and more.

15. If you want to print the report, in the upper portion of the Report Viewer window, click the Print button. On the Print dialog box, click Print to print the report.

16. If you want to save the report, on the Report Viewer window, click the Save Report output to file button. On the Save Report Layout dialog box, name your report and save to your MyData folder for chapter 15.

17. Close the Report Viewer window. If prompted to save your changes, click No.

ADDING REPORTS TO LAYOUTS

Within a map document, you can add a report to a layout by clicking the Add Report To ArcMap Layout button 🔲 on the Report Viewer window. If you have saved a report on disk, you can add it to any map layout by navigating to View > Reports > Load Report. For more information, consult ArcGIS Desktop Help.

There is no need to save your map document. You haven't made any changes to it.

18. If you are continuing to the next chapter, leave ArcMap open. Otherwise, exit the application, and click No when prompted to save your changes.

1
2
3
4
5
6
7
8
9
10
11
12
13
14
15
16
17
18
19
20

Summary

In this chapter, you learned how to identify and search for features by using tools on the Tools toolbar. You also used the Select By Attributes tool to generate queries and target specific features that satisfy them. Query results and selections don't change the original dataset, so you can always clear your query and start a new one. Finally, you learned how to create a report from ArcMap so you can easily share your query findings with others.

Terms

hyperlink
query expression
operators

CHAPTER 16
Selecting features by location

Learning objectives
- Create a selected set of features according to location-based queries
- Create a map layer from selected features
- Select features by combining attribute and location queries

In chapter 15, you were introduced to the Select By Attributes tool, so you could select features based on specific criteria. In other chapters, you manually selected features by highlighting rows in an attribute table or by clicking features with the Select Elements tool. In this chapter, you will use the Select By Location tool to create a location query (also called a spatial query) that selects features according to their spatial relationship to other features, whether in another layer or within the same layer.

The four types of spatial relationships that can be used are distance, containment, intersection, and adjacency.

Left to right: Points within a given distance of the red point are selected, points contained by the red polygon are selected, lines that intersect the red line are selected, and polygons adjacent to the red polygon are selected.

1

2

3

4

5

6

7

8

9

10

11

12

13

14

15

16

17

18

19

20

ArcGIS defines 15 spatial relationships, each a variation of one of the four basic types. For example, there is containment (where the contained feature may touch the boundary of the containing feature) and complete containment (where the boundaries cannot touch).

Data

The exercises in this chapter use the following datasets:

- BayArea.shp–polygons of the counties in the Bay Area region of California *(Source: County of Sonoma GIS Central)*
- Cities.shp–points representing cities, towns, and unincorporated communities near the Russian River in Northern California *(Source: National Atlas of the United States)*
- Guerneville.jpg–orthophotograph of downtown Guerneville and adjacent area *(Source: County of Sonoma GIS Central)*
- GuernevilleParcels.shp–polygons showing parcel locations within the limits of the Guerneville community *(Source: County of Sonoma GIS Central and the county's Permit and Resource Management Department)*
- RussianRiver.shp–line feature representing the Russian River *(Source: County of Sonoma GIS Central)*
- RussianRiverFloodZone.shp–polygon layer identifying areas that could potentially be at risk of flooding from the Russian River and its tributaries *(Source: County of Sonoma GIS Central)*

Exercise 16a
Use location queries

To select features by location, you choose two layers and specify a spatial relationship. Features in the first layer are then spatially compared to features in the second layer and selected if they satisfy that relationship.

If the spatial relationship is one of distance, you also choose the distance value and measurement units you want to apply.

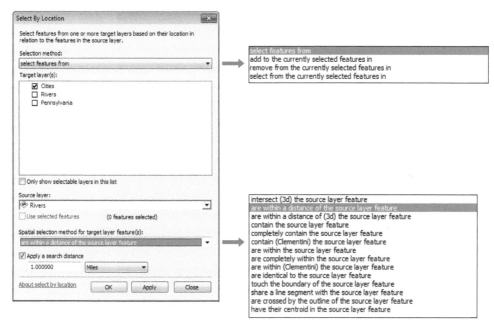

In this example, cities are selected if they lie within one mile of rivers. ArcMap measures the distances from cities to rivers and selects cities that satisfy that relationship.

The Pennsylvania cities selected on the map (*left*) and in the table (*right*) lie within one mile of a river. Data from Data and Maps for ArcGIS 2010, courtesy of ArcUSA, US Census Bureau, National Atlas of the United States, US Geological Survey, and Esri (Pop2010 field).

In this exercise, you will work with parcels in a small town bordering a major river in Sonoma County, California. You are an emergency response analyst who has been asked to identify parcels in the town of Guerneville that may be at risk of flooding. Once the at-risk parcels are

identified, they will be related to phone data to generate an emergency call list in accordance with the county's disaster plan.

1. **In ArcMap, open the ex16a map document from your C:\EsriPress\GTKArcGIS\ Chapter16 folder.**

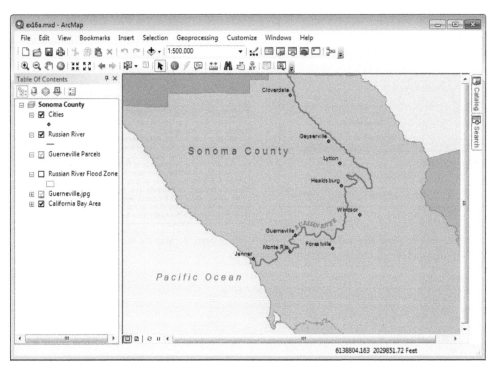

The map is centered on Sonoma County. You can see a blue line feature representing the Russian River, selected Sonoma County cities and towns represented with red points (only towns near the river are included in this dataset), and adjacent county polygons. Two layers, Guerneville Parcels and Guerneville.jpg, are turned off at this scale–they are visible only when you zoom in closer.

You can set a scale dependency for a layer's symbology on the Layer Properties dialog box, on the General tab.

ON YOUR OWN

Perhaps you want to have only the community of Guerneville displayed at this scale. Create a definition query to limit the display of cities so only Guerneville is shown. You learned how to do this in chapter 7. (**Hint:** Explore the layer properties.)

2. Go to the Guerneville 1 bookmark.

You are zoomed in on Guerneville, and the Guerneville Parcels layer is turned on.

Depending on the size of your ArcMap window, the dynamic label for the Russian River may be in a different spot.

3. Turn on the Russian River Flood Zone layer.

1
2
3
4
5
6
7
8
9
10
11
12
13
14
15
16
17
18
19
20

This layer shows areas at risk of flooding from the Russian River. This is a dataset generated by Sonoma County, based on Federal Emergency Management Agency (FEMA) 100- and 500-year flood zones along the Russian River. By selecting parcels that fall within this flood zone, the county can generate a call list to warn residents of possible flooding. You can get a general idea of which parcels fall within the flood zone by looking at the map, but next you will create a selected set of parcels so you can generate a parcel call list.

4. Zoom to the Guerneville2 bookmark. Turn off the Cities and Russian River Flood Zone layers.

The orthophotograph is turned on at this scale. You will keep your map display focused here, on the center of town. Remember, though, your selections may include parcels that are not in view. First, you will select all parcels that intersect the flood zone layer. This means that any parcel that has a boundary within or touching the boundary of the flood zone layer will be selected.

5. **On the menu bar, click Selection > Select By Location.**

6. **On the Select By Location dialog box, maintain the default Selection method option to Select Features From. In the Target layer(s) list, select Guerneville Parcels. For Source layer, click the arrow and click Russian River Flood Zone. For the spatial selection method for target layer feature(s), choose Intersect The Source Layer Feature. Click OK to make the selection.**

Select By Location

Select features from one or more target layers based on their location in relation to the features in the source layer.

Selection method:

```
select features from
```

Target layer(s):

- ☐ Cities
- ☐ Russian River
- ☑ Guerneville Parcels
- ☐ Russian River Flood Zone
- ☐ California Bay Area

☐ Only show selectable layers in this list

Source layer:

```
Russian River Flood Zone
```

☐ Use selected features (0 features selected)

Spatial selection method for target layer feature(s):

```
intersect the source layer feature
```

☐ Apply a search distance

```
400.000000     Feet
```

About select by location [OK] [Apply] [Close]

TIP

The target layer is the layer to which the selection is applied, based on its relationship to the source layer. Notice that you can choose any layers on the map for the source and target layers, regardless of whether they are turned on. The default selection method is Intersect The Source Layer Feature, but ArcMap recalls whichever selection method was chosen before.

Selected parcels are outlined in bright blue.

7. Turn on Russian River Flood Zone again so you can see that the parcels that intersect the flood zone are selected, even if it is just a corner or a boundary of the parcel that intersects the flood zone.

TIP

You can change the selection color by right-clicking a layer and modifying the options on the Selection tab of the Layer Properties dialog box.

8. Open the attribute table of Guerneville Parcels and dock it or place it so it does not obscure the map display.

 How many parcels are selected?

 Perhaps you are also interested in locating parcels that are within a certain distance of the flood zone.

9. Open the Select By Location tool dialog box. Again, select Guerneville Parcels as the Target layer and Russian River Flood Zone as the Source layer. For the spatial selection method, again select Intersect The Source Layer Feature.

10. Check the Apply A Search Distance check box and set the search distance to 400 feet.

1
2
3
4
5
6
7
8
9
10
11
12
13
14
15
16
17
18
19
20

Select By Location ☒

Select features from one or more target layers based on their location in relation to the features in the source layer.

Selection method:

| select features from | ▼ |

Target layer(s):

☐ Cities
☐ Russian River
☑ Guerneville Parcels
☐ Russian River Flood Zone
☐ California Bay Area

☐ Only show selectable layers in this list

Source layer:

| ◈ Russian River Flood Zone | ▼ |

☐ Use selected features (0 features selected)

Spatial selection method for target layer feature(s):

| intersect the source layer feature | ▼ |

☑ Apply a search distance

| 400 | Feet ▼ |

About select by location [OK] [Apply] [Close]

TIP

The search distance settings will be recalled the next time you use the tool, so be sure to uncheck the check box if you don't want to apply a search distance to the selection the next time you use the tool.

11. Click Apply and move the Select By Location dialog box out of the way.

Now how many parcels are selected?

Perhaps time is of the essence and you want to prioritize land owners at highest risk. You will refine your selection to only those parcels that are mostly within the flood zone. This will weed out the parcels that have only a small part of land in the danger zone. All residents of parcels within the flood zone will be contacted, but you want to determine how many parcels are likely to require emergency assistance. By calculating the number of parcels that have boundaries mostly within the flood zone, you can safely estimate and request emergency resources for this area.

12. On the Select By Location dialog box, maintain the previously chosen target and source. For the spatial selection method, click the arrow, scroll to the bottom of the list, and click Have Their Centroid In The Source Layer Feature.

13. Uncheck the Apply A Search Distance check box. Click Apply.

Notice how the selection is reduced.

Finally, how many parcels are selected?

ON YOUR OWN

Select target features that are within the source layer feature and examine the results. Explore other selection parameters if you want. When you are finished, close the Select By Location tool dialog box.

14. To save your work, rename your map document, and save to your MyData folder for chapter 16.

15. If you are continuing to the next exercise, leave ArcMap open. Otherwise, exit the application, and click No if prompted to save your changes.

1
2
3
4
5
6
7
8
9
10
11
12
13
14
15
16
17
18
19
20

Exercise 16b
Combine attribute and location queries

Location and attribute queries can be used together to solve a problem. For example, if you are doing emergency planning, you might want to select cities that have a population of 100,000 or greater (an attribute query) and that lie within 10 miles of a major fault line (a location query). In this exercise, you will select parcels that are zoned residential (an attribute query) and that intersect the flood zone (a location query).

1. **In ArcMap, open the ex16b map document from your Chapter16 folder.**

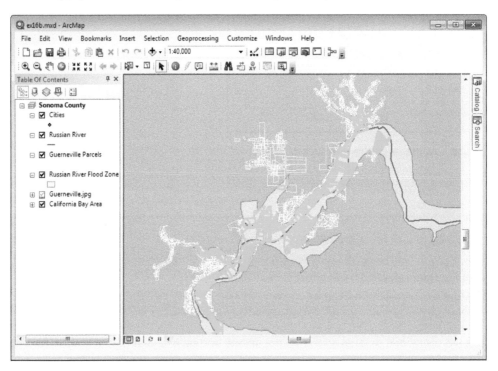

In this exercise, you want to further fine-tune the selection so it includes only parcels that are zoned residential. In this way, you will combine an attribute query and a location query. You completed the location query in exercise 16a, so now you will do the attribute query.

2. **On the menu bar, click Selection > Select By Attributes.**

3. On the Select By Attributes dialog box, for Layer, click the arrow and click Guerneville Parcels. For Method, change the default setting to Select From Current Selection.

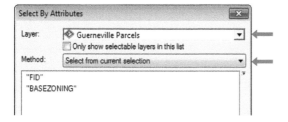

4. In the fields list, double-click "BASEZONING" to add it to the expression box. Click the equal sign (=) to add it to the expression.

5. Click Get Unique Values. The values for the BASEZONING field are listed.

6. In the list of values, double-click 'Residential' to add it to the expression. Click OK to make the selection.

1
2
3
4
5
6
7
8
9
10
11
12
13
14
15
16
17
18
19
20

The selection now indicates parcels that are within the flood zone *and* zoned as residences.

7. Open the attribute table of Guerneville Parcels.

How many parcels are currently selected?

Next, you will save the selection as a new layer on the map.

8. In the table of contents, right-click Guerneville Parcels and click Selection > Create Layer From Selected Features.

A new layer named Guerneville Parcels selection is added to the map.

TIP

To save the selected features permanently, you could export the selection to a new feature class (right-click the layer and click Data > Export Data).

9. Close the attribute table, clear the current selection, and turn off Guerneville Parcels.

10. In the table of contents, click the default polygon symbol for the new layer. In the Symbol Selector, use a hollow fill (no color) and a red outline, and click OK.

11. Zoom to the Guerneville 2 bookmark and turn off the flood zone layer.

As expected, most of the downtown area is not included in the selection. Thanks to your work, a flood watch call list for at-risk residences can be created.

12. To save your work, rename your map document and save to your MyData folder for chapter 16.

13. If you are continuing to the next chapter, leave ArcMap open. Otherwise, exit the application, and click No if prompted to save your changes.

Summary

In this chapter, you learned how to perform location queries using the Select By Location tool and how to combine location and attribute queries by applying the Select By Attributes tool to the results of a location query. These are common GIS tasks that will facilitate many projects and workflows. The results of a selection aren't permanent, so you can always clear your selection and rebuild the query until you get the correct results. You can save a selected set of features to a new layer in the map document.

Terms

location query
attribute query

CHAPTER 17
Joining and relating data

Learning objectives

- Examine tabular data and how it might be displayed spatially
- Join nonspatial data to spatial data to display new information on a map
- Select features on a map
- Export a selected set of features to a new feature class
- Determine how and why to use a join or relate function
- Relate the tables of two datasets
- Spatially join two layers
- Symbolize join results to display new information

Layer attribute tables contain the characteristics of geographic features, such as names (of cities, lakes, schools), codes (land use, zoning, feature type), and statistics (population, length, price). But a layer may not have all the attributes you want. If there is another data table that contains the desired attributes, you can join these attributes to the original attribute table, provided the two tables share a common field so the connection can be made—a Name field, for instance, or a field that contains a unique identification number.

Typically, an attribute join operation is used when you want to append nonspatial data (attributes that describe characteristics, but not geometry or spatial location) to a spatial dataset. In the following scenario, table 1 contains just a few basic spatial facts about world countries.

ObjectID *	Shape *	CNTRY_NAME	SQKM	SQMI
130	Polygon	Afghanistan	641358.44	247628.48
80	Polygon	Albania	28798	11118.91
69	Polygon	Algeria	2323510.25	897107.19
181	Polygon	American Samoa	84.2	32.51
70	Polygon	Andorra	336.24	129.82
162	Polygon	Angola	1252934.88	483758.22
200	Polygon	Anguilla	74.48	28.76
232	Polygon	Antarctica	12277658	4740404.5
35	Polygon	Antigua & Barbuda	255.13	98.51
5	Polygon	Argentina	2776912.75	1072166.13

An attribute table of a world countries layer.

Suppose you wanted to make a world map of average life expectancy. You would need some attributes such as those found in table 2, which has socioeconomic data.

CNTRY_NAME	POP2007	BIRTH RATE	DEATH RATE	LIFE EXPECT
Afghanistan	31889923	46.21	19.96	43.77
Albania	3600523	15.16	5.33	77.6
Algeria	33333216	17.11	4.62	73.52
American Samoa	57663	21.83	3.24	76.25
Andorra	71822	8.45	6.45	83.52
Angola	12263596	44.51	24.81	37.63
Anguilla	13677	13.97	5.34	77.46
Antarctica	-99	-99	-99	-99
Antigua & Barbuda	69481	16.62	5.31	72.42
Argentina	40301927	16.53	7.55	76.32

A nonspatial table containing population, birth rate (number of births per 1,000 population), and mortality data for world countries.

Unlike the previous table, this is a nonspatial table. It contains a lot of attributes but no Shape field. In other words, it stores descriptions but not spatial information. It is the sort of table you might find in a non-GIS database, or that you might even create yourself from facts based on your own research. By itself, this table can't be used to map data because its records have no relationship to features. You can append the attributes in the second table, however, to those of the first to make one big table that contains both the spatial information and the attributes from the nonspatial table. Once the new attributes are appended, the map can display new information.

ObjectID *	Shape *	CNTRY_NAME	SQKM	SQMI	POP2007	BIRTH RATE	DEATH RATE	LIFE EXPECT
175	Polygon	Afghanistan	641358.44	247628.48	31889923	46.21	19.96	43.77
120	Polygon	Albania	28798	11118.91	3600523	15.16	5.33	77.6
106	Polygon	Algeria	2323510.25	897107.19	33333216	17.11	4.62	73.52
9	Polygon	American Samoa	84.2	32.51	57663	21.83	3.24	76.25
107	Polygon	Andorra	336.24	129.82	71822	8.45	6.45	83.52
217	Polygon	Angola	1252934.88	483758.22	12263596	44.51	24.81	37.63
26	Polygon	Anguilla	74.48	28.76	13677	13.97	5.34	77.46
209	Polygon	Antarctica	12277658	4740404.5	-99	-99	-99	-99
27	Polygon	Antigua & Barbuda	255.13	98.51	69481	16.62	5.31	72.42
18	Polygon	Argentina	2776912.75	1072166.13	40301027	10.50	7.55	76.32

☐ ☑ World Life Expectancy at Birth
Male and Female 2007
☐ No data
■ 32 to 49 years
☐ 50 to 64 years
☐ 65 to 74 years
☐ Over 75 years

The output table (*top*) includes all the attributes from both input tables. Now countries can be symbolized (*left*) based on one of the joined attributes–in this case, average life expectancy (see legend, *right*). Data from Data and Maps for ArcGIS 2010, courtesy of ArcWorld Supplement.

The process depends on matching records between the tables. For records to be matched, the tables must have a common attribute–that is, a field of shared values. In the preceding example, the CNTRY_NAME field would be a common attribute, because both tables have a field that contains country names. (The field name does not have to be the same; it just has to be the same type, e.g., text fields are joined to text fields and numbers to numbers.) When the common attribute–also called a key–has been specified, records can be matched. For example, the Afghanistan record in the nonspatial table (along with all its attributes) can be matched to the Afghanistan record in the layer attribute table. The same goes for all the rest of the countries in the world. If a record in the **target table** (which contains the spatial information you want to maintain) doesn't have a match in the **join table** (which contains the attributes that will be added to the spatial table), it will either be left out of the output table or have null values where the joined attributes would be. You choose which result you want when you set up the operation in ArcMap.

For a join operation to work, there can be only one matching attribute value in each table. In the case of countries, in this example, there should never be more than one feature that has the same country name, so the join will work correctly.

Sometimes, rather than join two tables, you will want to relate the tables instead. A relate keeps the two tables separate but linked, so that record selections in one table lead to corresponding record selections in the other. Unlike a join operation, however, attributes from one table are not appended to the other.

Whether you join or relate tables depends on a relationship called cardinality between the tables. Cardinality refers to the numerical relationship (one-to-one, one-to-many, and so on) between a layer's attribute table and the table with the information you want to connect. You join tables when each record in the target table has no more than one matching record in the join table.

Common attribute

FID	Shape *	ObjectID	NAME	ST	POP2007
122	Point	836	Abilene	TX	115018
250	Point	2640	Akron	OH	211218
193	Point	1637	Albany (NY)	NY	94140
118	Point	813	Albuquerque	NM	503375
282	Point	3187	Alexandria	VA	139748
272	Point	3103	Allentown	PA	110279
119	Point	824	Amarillo	TX	183256
71	Point	559	Anaheim	CA	348374
1	Point	8	Anchorage	AK	283369
181	Point	1499	Ann Arbor	MI	118788
130	Point	878	Arlington (TX)	TX	373933
283	Point	3188	Arlington (VA)	VA	203123
256	Point	2754	Atlanta	GA	463421

U.S. Cities data
(target table)

iso3	Citycode	City	pm10
USA	8400173	Abilene	17.62378
USA	8400122	Akron	19.6925
USA	8400051	Albany (NY)	18.37535
USA	8400061	Albuquerque	19.65676
USA	8400188	Alexandria	22.52458
USA	8400065	Allentown	14.75936
USA	8400125	Amarillo	15.34999
USA	8400107	Anaheim	36.4622
USA	8400115	Anchorage	20.10121
USA	8400196	Ann Arbor	21.17848
USA	8400104	Arlington (TX)	23.16416
USA	8400138	Arlington (VA)	16.07828
USA	8400012	Atlanta	25.28407

City pollution data
(join table)

FID	Shape *	ObjectID	NAME	ST	POP2007	iso3	Citycode	City	pm10
119	Point	836	Abilene	TX	115018	USA	8400173	Abilene	17.62378
247	Point	2640	Akron	OH	211218	USA	8400122	Akron	19.6925
190	Point	1637	Albany (NY)	NY	94140	USA	8400051	Albany (NY)	18.37535
115	Point	813	Albuquerque	NM	503375	USA	8400061	Albuquerque	19.65676
278	Point	3187	Alexandria	VA	139748	USA	8400188	Alexandria	22.52458
268	Point	3103	Allentown	PA	110279	USA	8400065	Allentown	14.75936
116	Point	824	Amarillo	TX	183256	USA	8400125	Amarillo	15.34999
69	Point	559	Anaheim	CA	348374	USA	8400107	Anaheim	36.4622
1	Point	8	Anchorage	AK	283369	USA	8400115	Anchorage	20.10121
178	Point	1499	Ann Arbor	MI	118788	USA	8400196	Ann Arbor	21.17848
127	Point	878	Arlington (TX)	TX	373933	USA	8400104	Arlington (TX)	23.16416
279	Point	3188	Arlington (VA)	VA	203123	USA	8400138	Arlington (VA)	16.07828
252	Point	2754	Atlanta	GA	463421	USA	8400012	Atlanta	25.28407

Each record in the target table (top left) has just one match in the join table (top right), so the information from both is included in the output table (below). Data from Data and Maps for ArcGIS 2010, courtesy of US Census and The World Bank (pm10 field).

There is one record for each city in both the target table and the join table. Another way to express this relationship is that the table cardinality is one-to-one. (It is OK if one table has some cities that aren't found in the other–not every record needs to have a match, but each record can have *no more than* one match.) You can also join tables with a many-to-one cardinality–for example, if you have a layer of land-use polygons that defines land use using a three-letter code, you may want to join a nonspatial table of zoning code descriptions to it. This is a many-to-one cardinality (many land-use codes to one description for each). The resulting table would duplicate the description of each feature that shared the same zoning code.

You relate tables when each record in the target table may have more than one match in the relate table, as in the next example. When related tables are viewed, a selection highlighted in one will also select and highlight the matching features in the other.

The table cardinality is one-to-many. Each record in the target table (*left*) has many matches in the related table (*right*) (because there are many cities in a country).

If you simply joined the tables, you would lose information. For example, only one of the matching records for Afghanistan in the cities table could be joined to the target layer attribute table. In this case, it is better to relate the tables rather than join them so information won't be lost.

You can also relate tables that have a one-to-one cardinality (if you don't want to create a permanent join) or a many-to-many cardinality–for example, relating a buildings layer to a table of building tenants, in which some tenants might be chain stores located in multiple buildings.

For more information about cardinality rules for joining and relating tables, consult ArcGIS Desktop Help.

Sometimes, rather than joining data based on a common attribute, you will want to join data based on location or spatial relationship. This operation is called a spatial join. It allows you to define a spatial relationship between two layers and combine their attributes in a new output layer. Depending on the types of data you want to join, you can choose how the spatial relationship is calculated. For instance, you may choose to generate a Count field that summarizes

the number of features in the joined layer that intersect each feature in the target layer. Or you may generate a Distance measurement between features in two layers. For example, you could define a relationship between city points and earthquake points based on distance, generating an output that shows the nearest city to each earthquake.

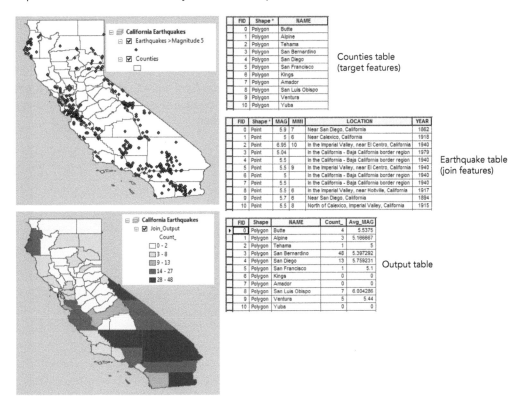

Top: A layer of California counties is spatially joined to a layer of earthquake points. *Bottom:* The output is a new feature class that includes a Count field for the total number of earthquakes in each county and an average of the earthquake attributes that fall within each county. Here you get the average magnitude, but you can choose different ways to calculate the attributes—a sum, for instance. The output layer is symbolized to show the number of earthquakes in each county. (Data includes earthquakes above magnitude 5, since 1769.) Data from Data and Maps for ArcGIS 2010, courtesy of National Atlas of the United States, US Geological Survey, ArcUSA, US Census, Esri (Pop2010 field), and ArcWorld.

Data

The exercises in this chapter use the following datasets, stored in a geodatabase named Wisconsin.gdb:

- USstates—polygons representing all 50 states, including demographic attributes (*Sources: Tele Atlas, US Census, Esri*)

- WisconsinCities–points representing cities (population greater than 10,000) in the state of Wisconsin *(Source: US Census)*
- WisconsinCounties–polygons representing Wisconsin county boundaries *(Sources: Tele Atlas, US Census, Esri)*
- WisconsinTracts–polygons representing Wisconsin census tracts, including demography *(Sources: Tele Atlas, US Census, Esri)*

The chapter data also includes the following:

- FoodDesert_Data.xls–a nonspatial table compiling research results from a food desert study *(Source: US Department of Agriculture's Economic Research Service)*

Exercise 17a
Join data by attribute

Once tables are joined, you can use the appended attributes of the nonspatial table to symbolize, label, query, and analyze the features in a layer.

A table join is preserved only within a map document–the tables remain separate on disk–and can be undone at any time. You can permanently append attributes by exporting a joined layer as a new feature class.

For this exercise, you are a freelance GIS analyst and mapmaker. The Economic Research Service (ERS), an agency of the US Department of Agriculture (USDA), has conducted a study to identify "food deserts," which for our purposes are defined as low-income neighborhoods with high concentrations of people who are far from a grocery store. ERS has created a tool to identify food-access statistics, called the Food Access Research Atlas (formerly called the Food Desert Locator). The Wisconsin state legislature has asked you to create a series of maps that will present the data effectively. The maps you create will be used by a task force that is charged with making nutritious food more readily available in food deserts.

The raw dataset is in tabular form. To visualize the data on a map, you need to use GIS to join the tabular data to geographic information. That way, you aren't required to peruse a list of data to find the information you need; rather, you can access a location, then quickly visualize the associated information. That's just what ERS did. To share important information that pertains to location, ERS used ArcGIS software to create an interactive web map to serve both the public and government agencies.

1
2
3
4
5
6
7
8
9
10
11
12
13
14
15
16
17
18
19
20

1
2
3
4
5
6
7
8
9
10
11
12
13
14
15
16
17
18
19
20

In this exercise, you will examine the Food Access Research Atlas. Then you will create the same type of map by joining a portion of the ERS raw data to a layer of Wisconsin census tracts.

1. To examine the results of the Food Access Research Atlas study, open a web browser, and go to http://www.ers.usda.gov/data-products/food-access-research-atlas. (Alternatively, in your web browser, search for "food access research atlas." The correct URL should be at the top of the search results.)

TIP

Due to the dynamic nature of the web, the URL in step 1 may change.

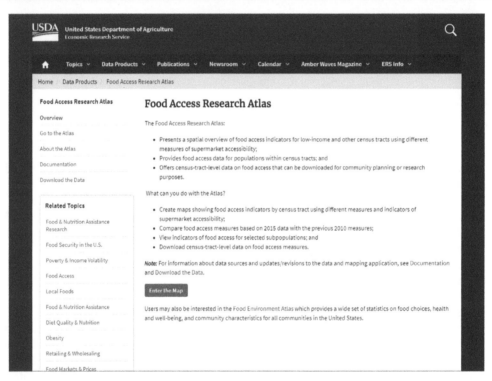

2. Under Overview, read about the atlas, and click Enter The Map. Use the navigation tools to pan and zoom around the map if you want. (If the site is unavailable or you don't have an internet connection, skip to step 4.)

Do you live in or near a food desert? Do you know anyone who is affected by limited access to groceries, especially fresh produce?

3. On the Locator, click any green square. Scroll through the pop-up box and notice the information that is provided. You will see this data later, in another form.

4. **On your browser, click the back button, then click the About the Atlas link. Read the sections titled "Overview" and "Measures of food access."**

 In this exercise, you will focus on the original food desert criteria–low income, with a significant number of residents living farther than 1 mile from a grocery store in urban environments, or more than 10 miles from a grocery store in rural environments.

 Next, you will learn how to create a similar map using an ERS dataset, limiting the study area to the state of Wisconsin.

5. **Close your web browser. Start ArcMap and open the ex17a map document from your C:\EsriPress\GTKArcGIS\Chapter17 folder.**

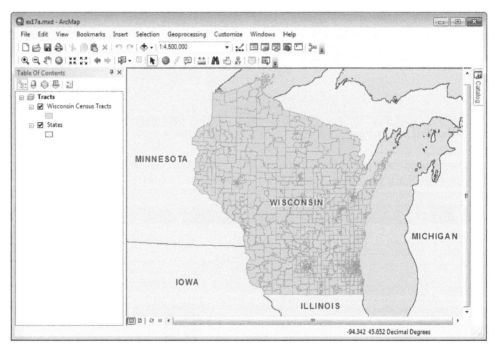

This map document contains US states and Wisconsin census tracts. Notice the blue lakes–they are not actual features but symbolized by means of a blue background color applied to the data frame.

6. **Open the attribute table of Wisconsin census tracts and examine the information stored there. Notice the FIPS_TRACT field.**

To join nonspatial data to spatial data, the two datasets must have a common field. Recall that the ERS data was collected for census tracts, each of which has a unique identification number (identified by the Federal Information Processing Standards, or FIPS).

Next, you will examine the ERS food desert dataset. The raw dataset is stored in a Microsoft Excel file.

7. **Close the attribute table.**

Steps 8 through 11 require the use of Microsoft Excel. As an alternative, you can preview the tables' sheets using the Item Description command in the ArcMap Catalog window.

8. **Start Excel and open FoodDesert_Data.xls from your ..\GTKArcGIS\Chapter17\Data folder.**

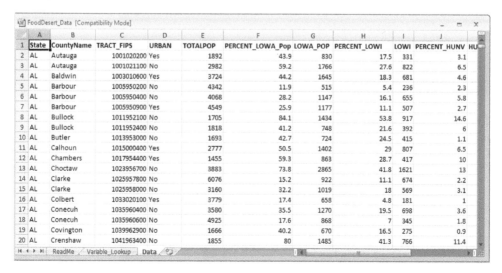

9. Scroll down through the rows of data. Notice that the spreadsheet includes data for all 48 continental states. (Hawaii and Alaska weren't included in the study.)

10. Scroll back to the top of the spreadsheet. Now scroll across and examine the column headings. To decipher the abbreviations, click the Variable_Lookup tab at the bottom of the spreadsheet and read the variable descriptions.

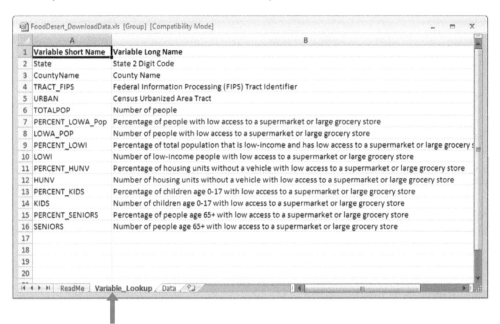

11. Minimize Excel. In ArcMap, on the Standard toolbar, click the Add Data button ⊕. Browse to your ..\Chapter17\Data folder, double-click FoodDesert_Data.xls, and click Data$. Then click Add.

TIP

Excel worksheets added to ArcMap have a $ symbol appended to their names.

The Data table is added to ArcMap. When a table is added to ArcMap, the view of the table of contents changes, and the button at the top is switched from List By Drawing Order to List By Source. That's because a table isn't drawn in ArcMap, but it is part of the map document.

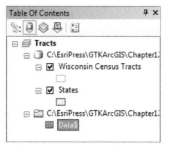

12. In the table of contents, right-click the Data$ table and click Open.

CountyName	TRACT_FIPS	URBAN	TOTALPOP	PERCENT_LOWA_Pop	LOWA_POP	PERCENT_LOWI	LOWI	P
Autauga	1001020200	Yes	1892	43.9	830	17.5	331	
Autauga	1001021100	No	2982	59.2	1766	27.6	822	
Baldwin	1003010600	Yes	3724	44.2	1645	18.3	681	
Barbour	1005950200	No	4342	11.9	515	5.4	236	
Barbour	1005950400	No	4068	28.2	1147	16.1	655	
Barbour	1005950900	Yes	4549	25.9	1177	11.1	507	
Bullock	1011952100	No	1705	84.1	1434	53.8	917	
Bullock	1011952400	No	1818	41.2	748	21.6	392	
Butler	1013953000	No	1693	42.7	724	24.5	415	
Calhoun	1015000400	Yes	2777	50.5	1402	29	807	
Chambers	1017954400	Yes	1455	59.3	863	28.7	417	
Choctaw	1023956700	No	3883	73.8	2865	41.8	1621	

1 ▸ ▸I (0 out of 6500 Selected)

Data$

The TRACT_FIPS attribute identifies census tracts and is the common attribute that will be used to join the tables. Notice the size of the table–there are 6,500 records. This table includes every census tract in the United States, but only the records that have a match in the Wisconsin census tracts attribute table will be brought over in the join operation. In other words, ArcMap finds all the food desert information that pertains to the state of Wisconsin and appends it to the Wisconsin tracts data. Then it may be displayed on the map.

13. Scroll down to find the data records for Wisconsin (WI in the State column). Notice that there are often several tracts per county. This will be important in exercise 17b.

14. Right-click TRACT_FIPS, and on the shortcut menu, click Freeze/Unfreeze Column. Scroll all the way to the right.

The TRACT_FIPS column is moved to the leftmost position in the table and does not move when you scroll.

TRACT_FIPS	LOWI	PERCENT_HUNV	HUNV	PERCENT_KIDS	KIDS	PERCENT_SENIORS	SENIORS
55081950800	447	3.9	53	7.7	316	3.5	144
55083990500	366	1.7	18	8.5	204	10.2	245
55083990600	290	1.5	17	7.9	225	5.7	161
55085970100	250	0.5	10	3.3	177	2.8	151
55087010100	368	6.6	75	2.9	94	4.5	148
55093960100	634	4.6	50	19.1	557	13.2	387
55093960800	325	1	12	11.3	412	6.2	224
55097960400	543	2.3	39	6	414	5.8	405
55099970600	511	2.1	22	12.7	329	9.3	240
55099970700	1021	6.2	69	25	688	15	413
55101000200	1404	7.6	174	16.8	1053	4.6	288
55101000300	1904	15.3	167	38.2	1389	5.3	192

(0 out of 6451 Selected)

Data$

What does the PERCENT_HUNV field measure? (If you have forgotten, revisit the Excel table's Variable_Lookup sheet.)

Both the FoodDesert_Data table and the Wisconsin census tracts layer attribute table contain one record for each census tract. The record relationship between the tables is therefore one-to-one: for each record in the Wisconsin census tracts table, there is no more than one matching record in the FoodDesert_Data table, thus making a good choice for a join operation.

15. **Close the table. In the ArcMap table of contents, right-click Wisconsin Census Tracts and click Joins and Relates > Join.**

TIP

You can also execute a join operation using the Add Join geoprocessing tool in ArcToolbox. For large join operations, the geoprocessing tool is recommended.

16. In the upper portion of the Join Data dialog box, in the first drop-down list, make sure Join Attributes From A Table is selected. In the first entry box, click the drop-down arrow and click FIPS_TRACT. This specifies FIPS_TRACT as the common attribute in the Wisconsin Census Tracts table.

17. In the second entry box, make sure the Data$ table is selected. In box 3, make sure the TRACT_FIPS field is selected. This specifies TRACT_FIPS as the common attribute in the Food Desert table.

Notice that ArcMap automatically populates the second and third entry boxes, but you should always check to be sure the correct fields are selected.

18. In the Join Options section, you want to keep all the records, so maintain the default setting.

Note that choosing to "keep all records" doesn't mean you are bringing all the food desert records over to the census tracts table; rather, it means all Wisconsin census tract records will remain, even if there's not a match in the food deserts table. In fact, in this case, there won't always be a match, because the food deserts table does not have a record for every census tract, but only for tracts that qualify as food deserts. In the resulting table, census tracts that do not qualify will have null values in the newly joined fields.

19. Click OK to run the join operation.

ArcMap joins the attributes of the FoodDesert_Data table to the Wisconsin Census Tracts table, matching the records by their FIPS tract numbers.

20. Open the attribute table of the Wisconsin Census Tracts layer. Scroll to the right until you get to the Shape_Area field.

To the right of Shape_Area are the fields from the FoodDesert_Data table, which are shown using field aliases. Field aliases are a more user-friendly description of the content of the field. Any alias can be changed in the field properties.

21. Turn off the field aliases by clicking the Table Options button 🔲▾, and click Show Field Aliases to clear the selection.

Now all field names show a prefix (in this case, Data$) that identifies the source table that the attribute comes from. This is a good way to distinguish joined fields from original fields if you are not sure.

22. Resize the Data$.TRACT_FIPS field to see its entire name.

23. Use the knowledge you gained in chapter 15 to perform a Select By Attributes query
using this SQL statement:

"Data$.TRACT_FIPS" IS NOT NULL

24. Apply the selection, and close the Select By Attributes dialog box. Show only the
selected records in the attribute table.

How many census tracts in the state of Wisconsin qualify as food deserts?

25. Move the table out of the way so you can see the map display. The selected records are also selected on the map.

The tracts outlined in cyan are food deserts. Because you have joined the food desert data to spatial data, you can present this information graphically on a map.

26. Close the table. Tracts will remain selected on the map.

Joins exist only within a map document—they aren't permanent. However, at this point, you can create a feature class of food desert polygons that includes their census attributes as well as their ERS attributes. To do this, you must export the selected features to a new, permanent feature class, so you will do that next.

27. In the table of contents, right-click Wisconsin Census Tracts and click Data > Export Data. Notice that the first option is set to export only the selected features of the current layer. Maintain the first two default options.

28. For the output location, browse to your MyData folder for chapter 17. Double-click MyWisconsin.gdb, and for Name, type **FoodDeserts.** Click the Save As type arrow and click File and Personal Geodatabase feature classes. Click Save.

29. **Click OK to export the data. When the operation is complete, click Yes to add the new layer to the map.**

30. **Clear the selected features. Change the table of contents to the Drawing Order view, and change the name of the new layer to Wisconsin Food Deserts.**

ON YOUR OWN

Symbolize the new layer using a 10% Ordered Stipple pattern, Gray 40% outline, and 35% transparency.

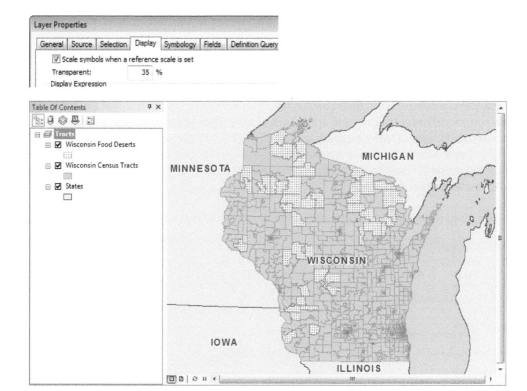

31. To save your work, rename your map document and save to your MyData folder for chapter 17.

32. If you are continuing to the next exercise, leave ArcMap open. Otherwise, exit the application, and click No if prompted to save your changes.

Exercise 17b
Relate data

Tables should be associated by a relate instead of a join when records in the layer attribute table (the target) may have many matches in the relate table (a one-to-many or a many-to-many relationship). When tables are related, you can highlight records in either table and see the matching records highlighted in the other. Whereas joins are performed to create layers for symbolizing, labeling, querying, and analysis, relates are used for data exploration. Although

1
2
3
4
5
6
7
8
9
10
11
12
13
14
15
16
17
18
19
20

the result is not a new dataset, it is valuable nonetheless in providing an organized view of the relationships between existing datasets.

The food desert data was calculated for census tracts in exercise 17a, although county governments in cooperation with federal programs are likely to tackle the problem. In this exercise, you will overlay food desert census tracts on Wisconsin counties to get a quick visual of how many food desert tracts there are in each county. Then you will relate the food desert table to a county layer to make it easy to pull up all the food desert details associated with each county. Because some counties contain multiple census tracts that qualify as food deserts, the relationship is one-to-many (one county, many census tracts). In this case, a join isn't helpful–you need to relate the tables instead.

1. **In ArcMap, open the ex17b map document from your Chapter17 folder.**

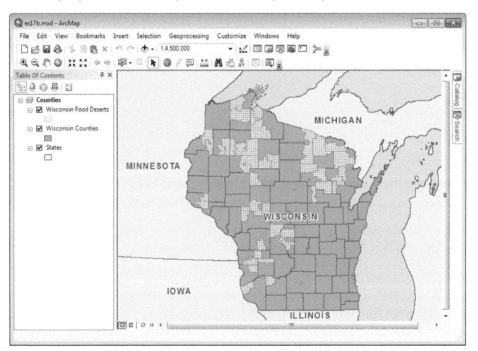

This map looks much like the map you worked with in exercise 17a, but instead of census tracts, you have polygons representing Wisconsin counties. Before you relate the tables, you will first examine the relationship between counties and food deserts.

2. **Turn the Wisconsin Food Deserts layer off and on to get an idea of where the food desert tracts are in relation to county boundaries.**

You can see that some counties have no food deserts, and others have multiple food deserts. (And there are many small food desert tracts that are difficult to see at this scale.) But even with a carefully chosen symbology scheme, it can be hard to tell which

food desert tracts fall within which counties. Next, you will use the Select Features tool to get a better understanding of the data before relating tables.

3. **Change the table of contents view to List By Selection.**

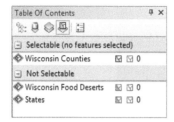

The map document was saved with a selection scheme that allows only the counties to be selectable.

4. **On the Tools toolbar, click the Select Features tool** 🔲. **(Just click the button, not the arrow next to it.)**

The default tool option for Select Features is to select by rectangle. This means you can either draw a rectangle to select features that intersect the rectangle or simply click to select a feature.

5. **On the map, click any county that looks, at this scale, as if it has multiple food desert census tracts.**

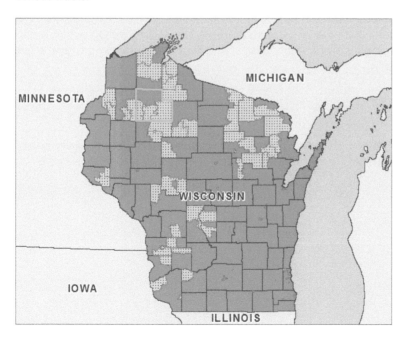

6. Click around the map to select other counties, one at a time. You may also try drawing a small rectangle to select a county. If you accidentally select several counties at once, just click outside the state of Wisconsin to clear the selection and start over.

TIP

There are several ways to clear selected features:
- Click outside the extent of the selectable layers.
- On the menu bar, click Selection > Clear Selected Features.
- On the Tools toolbar, click the Clear Selected Features button.
- In the table of contents (List by Selection view), click the hollow selection icon next to the layer in which features are selected.

ON YOUR OWN

Make Wisconsin Food Deserts selectable by clicking the first icon to the right of the layer in the table of contents. With the Select Features (Select by Rectangle) tool active, draw small rectangles to select individual counties and the food desert tracts within them for a quick visual analysis.

Notice in the table of contents that the NAME attribute is displayed for Wisconsin Counties, and the FIPS_TRACT attribute is displayed for Wisconsin Food Deserts. You can change the Display field on the Display tab of the Layer Properties dialog box.

A more accurate method of selecting food deserts within a specific county would be to run an attribute query using the Select By Attributes command, which you learned in chapter 15.

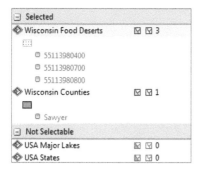

7. Clear any selections and switch the table of contents to List by Drawing Order view.

8. On the Tools toolbar, click the Select Elements tool ▶ to activate it.

9. Open the Layer Properties for Wisconsin Counties and click the Display tab. Verify that the Display Expression Field option is set to NAME. Check the Show MapTips Using The Display Expression check box. Click OK.

10. Now pause your mouse pointer over any county on the map. Notice that a MapTip appears, quickly showing a county's name.

Using MapTips is another quick way to get an idea of which counties contain food deserts.

Can you name a county whose entire extent qualifies as a food desert?

Where population is higher, there may be many, smaller census tracts. To see some food deserts that aren't easily visible at the full extent, you need to zoom in, which you will do next.

11. Go to the Southeast bookmark.

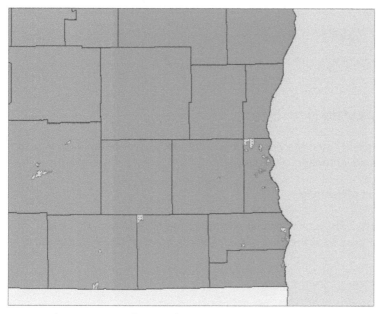

To get a better sense of where food deserts are in relation to large cities, you will next add a cities layer that includes cities with a population of at least 10,000.

12. **Add Wisconsin Cities.lyr to the map (located in the Data folder for chapter 17).**

The cities are symbolized by graduated symbols to show population ranges.

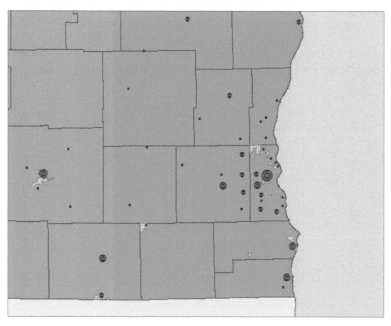

ON YOUR OWN

Turn on MapTips for Wisconsin Cities.

What is the name of the largest city in the area?

You have done some good visual analysis. Now, you are ready to relate tables so you can easily get all the detailed information you need.

13. Zoom to the full extent of the map.

14. In the table of contents, right-click Wisconsin Counties and click Joins and Relates > Relate. In the upper portion of the dialog box, read the description of the Relate command.

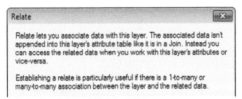

Relate lets you associate data with this layer. The associated data isn't appended into this layer's attribute table like it is in a Join. Instead you can access the related data when you work with this layer's attributes or vice-versa.

Establishing a relate is particularly useful if there is a 1-to-many or many-to-many association between the layer and the related data.

If you were to examine the attribute tables of Wisconsin Counties and Wisconsin Food Deserts, you would see a common field: CNTY_FIPS. In this case, the field in each table has the same name, but this is not always necessary, as long as both fields contain values that match. The common attribute is the county code, in accordance with FIPS.

15. On the Relate dialog box, for option 1, click the arrow and click CNTY_FIPS. For option 2, click Wisconsin Food Deserts. For option 3, click CNTY_FIPS. For option 4, maintain the default name. Click OK.

16. Open the Wisconsin Counties attribute table. Be sure the Show all records button at the bottom of the table is activated.

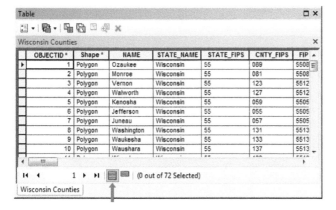

Notice there are 72 records—one for each county. No records or fields are added in a relate operation.

17. In the table, select the third record, for Vernon County, by clicking the gray tab to the left of the OBJECTID column.

18. At the top of the attribute table, click the Related Tables arrow and click the only related table, Relate1: FoodDeserts.

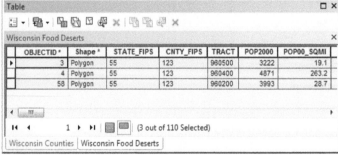

The related table appears, showing the three food desert tracts that are located within the boundaries of Vernon County. This allows county and state officials to quickly view the details of the food deserts under their jurisdiction.

19. Again, click the Related Tables arrow and click Relate1: Wisconsin Counties. Only the selected record is shown. In this way, you can switch back and forth between related tables, showing either all the records or only those that are currently selected.

TIP

You can also click the tabs at the bottom of the table to switch between the related tables.

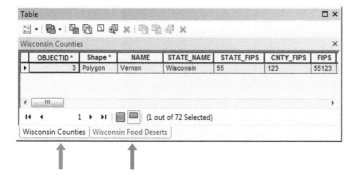

20. If necessary, move the table so you can see the ArcMap display. Click the Identify tool and click Vernon County, which is selected. Make sure you click a part of the county that is not a food desert—you want to select the county, not the food desert.

21. In the Identify window, notice the plus sign (+) symbol next to Vernon, in the upper panel. Expand the contents of Vernon. Now expand the contents of Wisconsin Food Deserts.

Conveniently, related features are visible in the Identify window.

22. Close the Identify window. At the bottom of the attribute table, click the Show all records button to return the table to its original view, and clear the selected features.

ON YOUR OWN

Find and select Milwaukee County and view the information pertaining to its food deserts.

How many food deserts does Milwaukee County have?

RELATIONSHIP CLASSES

If your data is stored in a geodatabase, you can permanently relate data in the database, not just in a map. To do this, you would create a relationship class between datasets. A relationship class is its own entity in the geodatabase, with user-defined rules, properties, and optionally, its own attribute table. You can set up a relationship class so that modifications trickle down to related features (for instance, if a utility pole is deleted, an attached transformer would be deleted as well). To learn more, consult ArcGIS Desktop Help.

23. Close the table.

24. To save your work, rename your map document and save to your MyData folder for chapter 17.

25. If you are continuing to the next exercise, leave ArcMap open. Otherwise, exit the application, and click No if prompted to save your changes.

ON YOUR OWN

Search ArcGIS.com for food desert maps.

Exercise 17c
Join data by location

In exercises 17a and 17b, you associated datasets according to common attributes; in this exercise, you will associate datasets according to common location. When you perform a spatial join operation, the output is a new feature class with new attributes based on the parameters selected on the tool dialog box.

For the final stage of your research project, you will create a map showing which counties have the highest number of food desert tracts. These counties should receive top priority for state resources dedicated to reducing food deserts.

1. In ArcMap, open the ex17c map document from your Chapter17 folder.

The map you have been working with shows Wisconsin counties and food deserts, and a background States layer.

1 2 3 4 5 6 7 8 9 10 11 12 13 14 15 16 17 18 19 20

You will join the Wisconsin Food Deserts layer to the Wisconsin Counties layer.

Instead of performing a table join between a spatial layer and a nonspatial table, as you did in exercise 17a, you will create a spatial join to combine two spatial layers. Specifically, you want to create a spatial join based on a containment relationship between two polygon layers (food desert tracts contained within counties). Although the Join dialog box can create spatial joins, there is another method that provides more options for handling data relationships—the Spatial Join geoprocessing tool.

The Spatial Join geoprocessing tool is in ArcToolbox. You will learn more about ArcToolbox and use many more of its tools in the upcoming chapters on analysis (chapters 18, 19, and 20).

2. On the Standard toolbar, click the ArcToolbox button 🗔. In ArcToolbox, expand the Analysis Tools > Overlay toolset.

3. Open the Spatial Join tool.

4. On the Spatial Join dialog box, for Target Features, choose Wisconsin Counties. For Join Features, choose Wisconsin Food Deserts. Set the Output Feature Class location to ..\Chapter17\MyData\MyWisconsin.gdb. Name the feature class **CountyStats**.

5. For Join Operation, maintain the default of JOIN_ONE_TO_ONE.

 You may wonder why you are using the one-to-one option here. This isn't about cardinality. Yes, you have many food deserts that you want to join to each county; however, you want only one row returned for each target feature. If you chose one-to-many, you would get a separate row for each county match.

6. Maintain the default for Keep All Target Features. This means even counties without food deserts will remain in the output feature class.

ON YOUR OWN
Click the Show Help button to read more about each tool parameter.

7. In the Field Map of Join Features panel, scroll down until you get the TOTALPOP (Double) field. Right-click it and click Merge Rule > Sum.

TIP

The TOTALPOP field is in the lower third of the list.

The output will now show the sum of all the populations that live within the food deserts in each county. Note that the default merge rule is "First." This means that when there are multiple join features that share space with a target feature, only the first attribute value will appear in the output table.

Next, you will change the merge rule for all the attributes related to the food desert study.

8. So you can see the average percentage of those living in food deserts who have low access to grocery stores, change the PERCENT_LOWA_Pop merge rule to Mean. Then, scrolling down as necessary, change the merge rule for the next nine fields, summing the numerical figures and averaging (mean) the percentages (those with field names that begin with PERCENT_).

Next, it is important to consider the correct match option to get the desired result. For instance, if you were to choose the INTERSECT option, the results would include all food deserts adjacent to, as well as within, county boundaries. You want to join just the food deserts that are contained within the boundaries of each county feature.

9. Change the Match Option to CONTAINS.

The Count field—which tallies all join features within each target feature—will be added automatically to the target table.

10. Click OK to run the Spatial Join tool. When the process is completed, the output layer is added to the map with a randomly chosen fill color.

11. Close ArcToolbox. Then move the Wisconsin Food Deserts layer to the top of the drawing order in the table of contents.

12. Open the attribute table of CountyStats. Notice the Join_Count field. Sort it by ascending order.

OBJECTID *	Shape *	Join_Count	TARGET_FID	NAME	STATE_NAME	S1
1	Polygon	0	1	Ozaukee	Wisconsin	55
6	Polygon	0	6	Jefferson	Wisconsin	55
8	Polygon	0	8	Washington	Wisconsin	55
10	Polygon	0	10	Waushara	Wisconsin	55
13	Polygon	0	13	Green	Wisconsin	55
14	Polygon	0	14	Green Lake	Wisconsin	55
20	Polygon	0	20	Marquette	Wisconsin	55
23	Polygon	0	23	Sheboygan	Wisconsin	55
24	Polygon	0	24	Sauk	Wisconsin	55
26	Polygon	0	26	Lafayette	Wisconsin	55

(0 out of 72 Selected)

CountyStats

13. Scroll through the rows and note how many food deserts are in each county.

There are a lot of attributes now. Remember, the food deserts layer was the result of a join to census tracts, so you now have attributes from counties, census tracts data, and the food desert research all within one table.

14. Scroll to the right. Notice that the counties that have a Join_Count value of 0 have null values for census tracts data and food desert data. If you select one of these rows, you can see on the map that the county has no food deserts.

15. Scroll farther to the right. The food desert attributes you are interested in begin with TOTALPOP (total number of people who live in food deserts in each county). Sort the TOTALPOP field by descending order, and scroll to the top of the table.

16. Examine the other food desert attributes, sorting fields as you like.

 Which county's food desert population has the highest percentage of children?

 Next, you want to symbolize the map to show the number of people in each county who have low access to grocery stores.

17. Close the table and clear any selections. Open the layer properties of CountyStats and click the Symbology tab.

18. Under Show, click Quantities > Graduated colors. In the Fields frame, for Value, scroll down and click LOWA_POP.

19. Choose any color ramp you like, and, in the Classification frame, click Classify.

20. On the Classification dialog box, in the Data Exclusion frame, click Exclusion. On the Data Exclusion Properties dialog box, on the Query tab, build the following query: LOWA_POP IS NULL.

21. Next, on the Data Exclusion Properties dialog box, click the Legend tab. Check the Show Symbol For Excluded Data check box, and set the symbol color to white or another color that is easily distinguishable from the color ramp you chose in step 19. For Label, type **No food deserts**.

22. Click OK to close the Data Exclusion Properties dialog box, and click OK again to close the Classification dialog box.

23. Back on the Symbology tab, above the center panel, click the Label tab and click Format Labels.

24. On the Number Format dialog box, in the Rounding frame, change the number of decimal places to **1** and click OK.

25. Click OK on the Layer Properties dialog box to apply the changes.

26. Turn off Wisconsin Food Deserts. Rename CountyStats Low Access Population and
rename the legend **Total by County**.

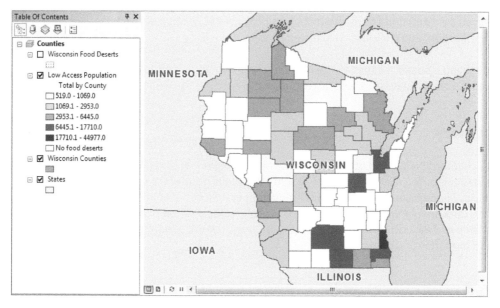

TIP

To map other attributes, such as the number of seniors in food deserts or the
percentage of the food desert population that doesn't have a vehicle, just open the
layer properties, and on the Symbology tab, change the field value. You don't have to
re-create the exclusion query.

ON YOUR OWN

Save Low Access Population as a layer file, and create additional graduated-color layer
files (with unique legend and layer file names) that capture other food desert statistics.
(**Hint:** Copy and paste, and then make adjustments.)

ON YOUR OWN

As you learned how to do in exercise 5c, share the map package to your ArcGIS Online
account.

27. To save your work, rename your map document and save to your MyData folder for chapter 17.

28. If you are continuing to the next chapter, leave ArcMap open. Otherwise, exit the application, and click No if prompted to save your changes.

Creating a spatial join has allowed you to create a map that effectively tells an important story. County, state, and federal resources can be better prioritized using this tool.

Summary

In this chapter, you learned about how and why you might join or relate tabular data to spatial data in ArcMap. You also learned how to join two spatial datasets to calculate the feature attributes of one layer that are spatially coincident to features in another layer. Joining and relating data are common tasks for many GIS professionals. Additional software functions that were reviewed include MapTips, identifying features, selecting features, layer symbology, and legend formatting.

Terms

attribute join
target table
join table
relate
spatial join

CHAPTER 18
Preparing data for analysis

Dissolve features
Clip layers
Create a data subset
Run tools in a model

Learning objectives

- Understand the basic steps of GIS analysis
- Perform common geoprocessing functions
- Select and export features
- Use ModelBuilder to create a geoprocessing model

GIS analysis allows you to investigate and understand spatial relationships between and among real-world features. Beginning with multiple datasets, you can perform various types of geoprocessing and mathematical operations to get new information. GIS analysis is widely used in many industries, from conservation to emergency response, to answer questions and make educated decisions. To use GIS properly, it is important to understand the problem at hand and then follow a defined process to obtain the solution. This chapter and chapter 19 will introduce some basic geoprocessing tools and forms of spatial analysis.

When you perform an operation on spatial data that creates a dataset (often a slightly modified version of the original), you are **geoprocessing**. Geoprocessing tasks are accomplished with geoprocessing tools. In ArcGIS, these tools are stored in ArcToolbox, which is accessible from ArcMap and ArcCatalog.

Typical analysis workflow includes the following steps:

1. *Frame the problem as a question.* Begin a GIS analysis by determining what type of information you need. This is most often accomplished by forming a relevant question: How many schools are within 500 feet of a busy intersection? How much land is covered by forest within a particular boundary? Which neighborhoods have the highest incidence of auto theft? Be as specific as possible about the question you want to answer. This will help you decide how to approach the analysis, which geoprocessing tools to use, which data you need, and how to present the results.

2. *Choose the appropriate data.* Choose good, quality data to use in your analysis. For example, if you want to know how many schools are within 500 feet of a busy intersection, you will need local schools and street data. Data can come from any number of sources—databases within your organization, commercial data providers, government organizations, the internet, ArcGIS Online, and more. Datasets are not always in exactly the condition you need for a project. For example, features may be either too detailed or too generalized for the map scale with which you are working. The data may not have all the attributes you want. You may have multiple datasets in different coordinate systems. Or you may have too much data—thousands of features that clutter your area of interest or extend well beyond it. Rarely is there an analysis project that doesn't require some amount of data preprocessing, such as assigning a projected coordinate system, clipping one dataset to the extent of another, deleting attribute fields, or joining tabular data to spatial data, as you did in chapter 17.

3. *Choose an appropriate analysis method.* Decide on an analysis method based on your proposed questions and how the results of the analysis will be used. For example, if you are doing a study of schools close to busy intersections in a city, you may map the individual schools and road network, create a buffer around intersections, and select schools that fall within the buffer zones. If you want to know which neighborhoods have the highest incidence of auto theft, you may select crime incidents based on type (auto theft), then perform a density analysis of the selected set, and then overlay the output with neighborhood boundaries. ArcToolbox contains hundreds of tools for processing and analyzing geospatial data. You can also build and use custom tools created with Python scripts.

4. *Execute the analysis.* You can choose to run your analysis one step at a time, or you can connect the steps in ModelBuilder and run them with a single click. ModelBuilder not only automates complex workflows, it also provides a means of diagramming and visualizing analysis steps. You can easily adjust tool parameters, and then run the tool again and compare output results.

5. *Examine the results.* Once you have run your analysis, you should examine the results with a critical eye. Did your original questions get answered? Were the results close to what you expected? Did you discover something new? Is the analysis sound? You can then determine what steps, if any, should be taken as a result of the analysis.

6. *Present your work.* At this point, you may want to take some time to display your analysis results. Although a lot of emphasis in GIS is on making maps, the software is flexible enough to allow you to display your results in a format that best suits your needs–displayed as a digital map, printed as a paper map, or combined with tables or charts. You can share your work in a presentation or by distributing it through ArcGIS Online or a network in your organization.

Data

The exercises in this chapter use the following datasets, stored in a geodatabase named OregonForest.gdb:

- curlook–point feature class of fire lookout towers in the state of Oregon *(Source: Oregon Department of Forestry)*
- forestland–polygon feature class of forested land in the state of Oregon *(Source: Oregon Department of Forestry)*
- fwz100k–polygon feature class of Fire Weather Zones in the state of Oregon *(Source: Oregon Department of Forestry)*
- populated_jurisdictions–polygon feature class of inhabited areas in the state of Oregon *(Source: Oregon Department of Forestry)*

Exercise 18a
Dissolve features

In ArcMap, a dissolve creates a layer in which all features in an input layer that have the same value for a specified attribute are dissolved into a single feature. In the example shown here on the maps, Wisconsin census tracts are dissolved into counties to form a new layer of counties made up of census tracts.

Wisconsin census tracts. Data from Data and Maps for ArcGIS 2010, courtesy of Tele Atlas, US Census, Esri (Pop2010 field).

Census tracts dissolved into counties. Data from Data and Maps for ArcGIS 2010, courtesy of ArcUSA, US Census, Esri (Pop2010 field).

Some 1,323 census tracts are dissolved into counties to form a new layer of 72 counties. The new layer's attribute table has the standard geometry (Shape) and feature identifier (OBJECTID) attributes, along with the attribute used in the dissolve (CNTY_FIPS). You can include other attributes as well. For each census tract, the input table contains attributes that store population counts and other demographic data. In the output table, these values can be summed for each tract.

Input census tracts table

Output counties table

The attributes of the input census tracts table (*left*) are summed in the output counties table (*right*). Data from Data and Maps for ArcGIS 2010, courtesy of Tele Atlas, US Census, Esri (Pop2010 field), and ArcUSA.

For this exercise, you work for a forestry consulting company that plans to conduct research on forest wildfires and their proximity to urban areas in Oregon. As the population grows, towns and cities expand outward–residential homes, infrastructure, and buildings are built in or near forests. This has caused an increased risk of human-caused fires and a greater risk of local and widespread forest fires, damage to property, and injuries, or even death.

Fire weather zones (FWZs) are used to forecast fire weather in the state of Oregon. There are 32 FWZs in the state. Each FWZ belongs to one of four county warning areas (CWAs), each of which has a weather forecast office that has the responsibility of forecasting fires and providing warnings to the communities it covers. Each FWZ carries its own fire risk based on the number of communities and the different types of forest and terrain.

1

2

3

4

5

6

7

8

9

10

11

12

13

14

15

16

17

18

19

20

The forests of Oregon, shown in green, cover approximately 30,472,000 acres (about 123,315 square kilometers). Populated communities and their fire risk assessments (low, medium, or high) are shown in yellow, orange, and red. Data courtesy of Oregon Department of Forestry.

As your consulting company's GIS analyst, your job in this chapter and chapter 19 is to calculate the risk area of a number of communities in low-, medium-, and high-risk situations in one CWA. You will also use active fire lookouts used by fire agencies in Oregon to create visibility buffers. Your analysis will help your company provide a proper risk assessment to the Oregon state government and supporting agencies.

You have a polygon layer of FWZs provided by the forest service. Its attributes include the names of each CWA and a number that uniquely identifies each zone. In this exercise, you will dissolve the FWZs into the four CWAs.

1. **Start ArcMap and open the ex18a map document from your C:\EsriPress\GTKArcGIS\ Chapter18 folder.**

The map shows the Forest polygon layer and the Fire Weather Zones polygon layer symbolized by a dark-red outline and hollow fill–the hollow fill allows the zones to be overlaid on top of the forest layer without obscuring it.

2. Open the attribute table of the Fire Weather Zones layer.

OBJECTID *	Shape *	AREA	PERIMETER	FWZ100K_	FWZ100K_ID	CTYWARNARE	FIREWXZONE
1	Polygon	37766257997.2	1680574.30951	2	394	PQR	601
2	Polygon	56140588582.3	1518881.22349	3	151	PQR	602
3	Polygon	73046316781.7	1310096.12297	4	125	PDT	635
4	Polygon	98586461697.7	1974894.30888	5	127	PDT	633
5	Polygon	171986611376	2325791.04396	6	132	PDT	631
6	Polygon	29145395679.2	964356.864051	7	132	PQR	609
7	Polygon	88381221756.8	3343966.54675	8	149	PQR	604
8	Polygon	36299443649.2	1285236.13579	9	153	PQR	607
9	Polygon	28535980707.1	1287446.39496	10	393	PQR	605
10	Polygon	15614262027.8	767286.013133	11	126	PDT	634

1 ▶ ▶I (0 out of 32 Selected)

Fire Weather Zones

The Fire Weather Zones layer has 32 zones, each represented by one record in the table. The FIREWXZONE attribute (where WX stands for "weather") contains a unique identifier for each FWZ (ranging from 601 to 638).

The CTYWARNARE attribute contains an abbreviation indicating the four CWAs: Boise (BOI), Medford (MFR), Pendleton (PDT), and Portland (PQR). Each FWZ belongs to one of these four CWAs. (Each CWA covers the extent of several counties.) The AREA attribute shows how large each zone is (in square feet). You want to dissolve the boundaries between FWZs so you end up with four polygons representing the four CWAs—each one will include the representative FWZs.

ON YOUR OWN

Click several records in the table. Determine which CWA the records belong to, and notice their locations on the map. When you are finished, clear your selection.

3. Close the table. On the menu bar, click Geoprocessing > Search For Tools.

The Search window appears. It may be tabbed to the side of the main ArcMap window, or docked or floating. This is the same Search window you used in chapter 4; however, because you started it from the Geoprocessing menu, it is automatically set to search for tools rather than maps or data.

The Search window displays many system toolboxes. Each toolbox has tools for related geoprocessing tasks. You can access tools directly from the ArcToolbox window, but if

you don't know exactly where a tool is, you should use the Search window instead. In this case, you know you want to use the Dissolve tool, but you don't know which toolbox it's in. So next, you will search for it with the Search window.

4. In the Search window, type **dissolve** and press Enter (or click the Search button, which looks like a magnifying glass).

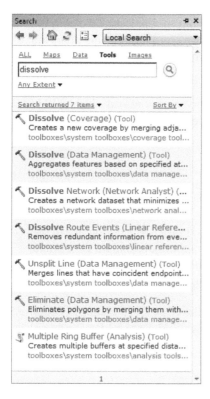

5. In the Search results, click Dissolve (Data Management). Be sure to click the title of the tool to open the tool dialog box. (The links below it open the description and location in ArcCatalog.) Then close or hide the Search window.

Although each tool has its own functionality and settings, they all use a similar dialog box with some common elements. For example, each tool has a Show Help button in the lower-right corner that describes the tool and its settings.

Tools that have lots of settings, such as the Dissolve tool, have a scroll bar on the right. As an alternative to scrolling, you can resize the dialog box by dragging on a side or a corner. Once you resize a dialog box, ArcMap remembers the new size and applies it to the next tool dialog box you open. (So if the dimensions of your Dissolve dialog box are different from the graphic, you may have previously resized this or some other tool. That's fine—the dialog box can be whatever size you like.)

The Dissolve tool requires that you select an input layer of features to dissolve, a location where the newly created layer will be saved on disk, and an attribute on which to base the dissolve of the input features. You will set these selections next.

6. **In the Input Features list, click Fire Weather Zones.**

The list of attributes on which dissolves can be based appears in the Dissolve_Field(s) box.

You can save the output as a shapefile. However, you will save it as a geodatabase feature class, because the rest of the Oregon forest data is in this format. To leave the original OregonForest geodatabase intact, you will save the output to a duplicate geodatabase named MyOregonForest that has been created for you.

7. **Click the Browse button next to the Output Feature Class box. On the Output Feature Class dialog box, browse to your ..\GTKArcGIS\Chapter18\MyData folder. Double-click MyOregonForest.gdb.**

8. **In the Name box, type CountyWarningAreas, and click Save.**

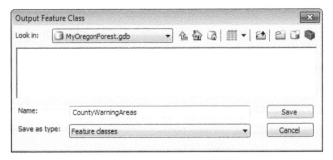

The output feature class information is updated.

9. **In the Dissolve_Field(s) box, check the CTYWARNARE check box.**

Each set of polygons with the same CWA name (meaning they are in the same CWA) will be dissolved into a single feature. The four possible areas created from the dissolve are BOI for Boise, MFR for Medford, PDT for Pendleton, and PQR for Portland.

ArcMap can also summarize the attribute values of dissolved features by a variety of statistics. To find out the area of each CWA, you will summarize the AREA attribute by the Sum statistic type. This will total the area (in square feet) of all FWZs dissolved into a CWA. (You can also get the mean, the range, the standard deviation, and other measures of numeric attributes.)

10. **On the Dissolve dialog box, scroll down to the Statistics Field(s) area. Click the Statistics Field(s) arrow and click AREA to add it to the list.**

The field name is displayed in the Field column. The red circle with a white X indicates that you still need to specify the type of statistic you want.

11. Click the first cell in the Statistic Type column. Then click the arrow and click SUM.

12. Click OK to run the tool.

ArcToolbox tools are executed in the background, which means you can continue working in ArcMap as a tool runs. You will see a progress bar at the bottom of the main window that displays the name of the currently running tool. When the tool process is completed, a pop-up notification appears in the notification area, at the far right of the taskbar. (It will disappear in a few seconds.)

TIP

Before the pop-up notification disappears, you can click the tool name (a link) to open the Results window. This window displays all the details of running an ArcToolbox tool, including what layers were involved, which fields were calculated, and the start, finish, and elapsed times. You can also access this information from the menu bar (Geoprocessing > Results).

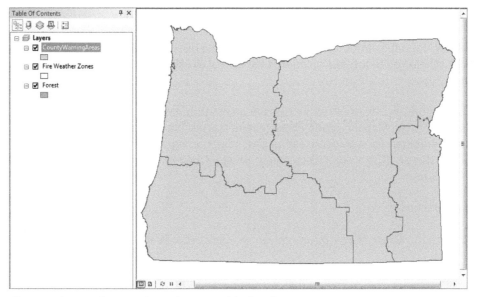

The new CountyWarningAreas layer is added to the map.

TIP

If you don't see the new layer in the table of contents, on the Standard toolbar, click the Add Data button, browse to your MyOregonForest geodatabase, and add the layer. Then on the menu bar, click Geoprocessing > Geoprocessing Options. On the Geoprocessing Options dialog box, in the lower portion, check the Add Results Of Geoprocessing Operations To The Display check box. Click OK. Any new geoprocessing layers will automatically be added to ArcMap.

When a new geoprocessing layer is created and added to ArcMap, it is displayed in a random color. (The colors of the layers you create in the exercises may not match the graphics in the book.)

13. Open the attribute table of the CountyWarningAreas layer. If necessary, resize the table to show all fields.

OBJECTID *	Shape *	CTYWARNARE	SUM_AREA	Shape_Length	Shape_Area
1	Polygon	BOI	279878427941	3102491.987324	279878427921.677
2	Polygon	MFR	736221086179.01	4944792.093709	736221086177.203
3	Polygon	PDT	1045625881886.1	6154950.839309	1045625881891.75
4	Polygon	PQR	641846600985.2	5536239.311169	641846600966.71

I◄ ◄ 1 ► ►I (0 out of 4 Selected)

CountyWarningAreas

The original 32 FWZs have been aggregated into four CWA polygons.

CTWARNARE is the attribute on which you based the dissolve. SUM_AREA is the statistical field you requested. It contains the sum of all FWZ values for area (in square feet).

Shape_Length and Shape_Area are measurement attributes that are automatically maintained by ArcMap for geodatabase feature classes. (Shape_Length measures feature perimeters for a polygon feature class.) Depending on the accuracy of the input attribute, there can be a slight discrepancy between the total calculated area and the area statistics.

14. Close the table.

15. Right-click the CountyWarningAreas layer and click Label Features.

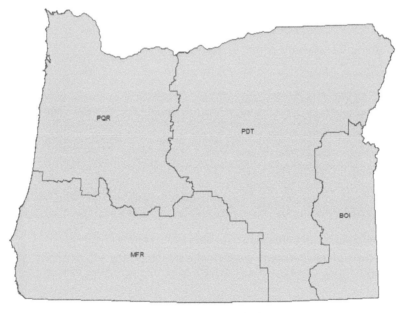

Each CWA is labeled with the appropriate label.

ON YOUR OWN

In the table of contents, drag the Fire Weather Zones layer to the top. Then symbolize CountyWarningAreas based on the CTYWARNARE field. Turn the Fire Weather Zones layer off and on. Doing so will help you visualize how the FWZs were dissolved into CWAs.

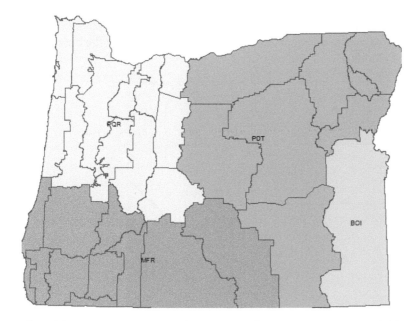

16. If you like, rename your map document and save to your MyData folder for chapter 18.

17. If you are continuing to the next exercise, leave ArcMap open. Otherwise, exit the application, and click No if prompted to save your changes.

Exercise 18b
Clip layers

A clip operation is a type of data extraction function. It trims features in one layer using the boundaries of polygon features from another layer, called the clip layer. It's like having a pair of scissors, or a cookie cutter, to trim off data you don't need for your project, leaving you with a clip of what you want. For example, you might have a feature class of all the streets in your county, yet your study area encompasses just one zip code within that county. If you have a polygon feature that represents that zip code, you can use it to clip just the streets you need. Clipping can be a matter of convenience—because smaller datasets are easier to process—but it can also be important for analysis. To find out the total length of roadway within a zip code, for instance, you must exclude streets and street segments that lie outside it.

1
2
3
4
5
6
7
8
9
10
11
12
13
14
15
16
17
18
19
20

The clip layer must be a polygon, although the dataset being clipped can be points, lines, or polygons. In the following example, a layer of interstate highways is clipped at the boundaries of a river basin in the state of Florida.

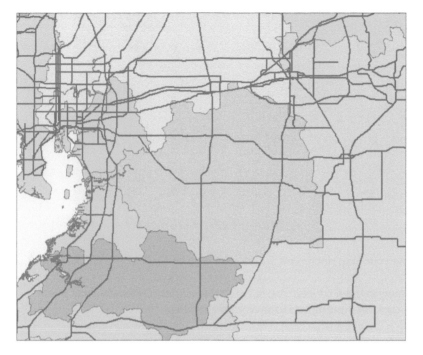

Input layer of US interstate highways and a clip layer of the Alafia River Basin boundary in Florida (center polygon). Data from the Florida Geographic Data Library, courtesy of the University of Florida GeoPlan Center.

1

2

3

4

5

6

7

8

9

10

11

12

13

14

15

16

17

18

19

20

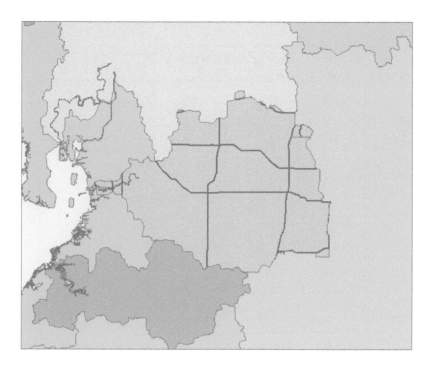

Highways clipped at the polygon boundary of the Alafia River Basin boundary. Data from the Florida Geographic Data Library, courtesy of the University of Florida GeoPlan Center.

For this exercise, your forestry consulting company has just determined that of the four CWAs, Portland (PQR) is the area of interest for this study. Your company will direct its attention to analyzing this area in depth.

Not every square foot of the CWAs is covered by forest or has a degree of fire danger. However, certain communities are built close to forestland, and these are the areas of concern. To aid in early smoke and fire detection, the state of Oregon strategically placed fire lookout towers throughout the forests. You have layers of forests, communities, and lookout towers that cover all four CWAs, but now you would like to work with datasets that cover the area of Portland only.

In this exercise, you will clip several layers to the Portland CWA, or PQR, boundary.

1. **In ArcMap, open the ex18b map document from your Chapter18 folder.**

1 2 3 4 5 6 7 8 9 10 11 12 13 14 15 16 17 18 19 20

The map displays the four CWAs. Below it is the Forest layer, which contains 3,771 forest polygons, and the Fire Weather Zones layer, which is turned off.

To clip the forest to the PQR boundary, you must first select the area boundary, then turn the selection into a new layer so you can look at the Portland area apart from the others.

2. In the table of contents, click the List By Selection button. Click the toggle button next to Fire Weather Zones and Forest to make them not selectable.

3. On the Tools toolbar, click the Select Features tool 🔤.

4. On the map, click PQR to select it. It is outlined in cyan.

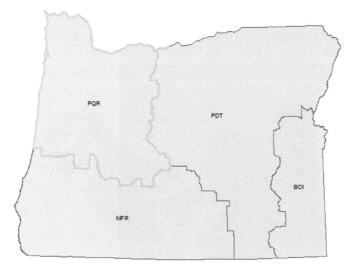

5. **In the table of contents, right-click the CountyWarningAreas layer and click Create Layer From Selected Features.**

 A layer called CountyWarningAreas selection is added to the table of contents. This layer references the same geodatabase feature class as the CountyWarningAreas layer. Remember, creating a selection layer doesn't create a new dataset. You can, however, export a selection layer, or a set of selected features, as a new dataset if you want. You will do this in exercise 18c.

6. **Click any white area of the map to clear the feature selection.**

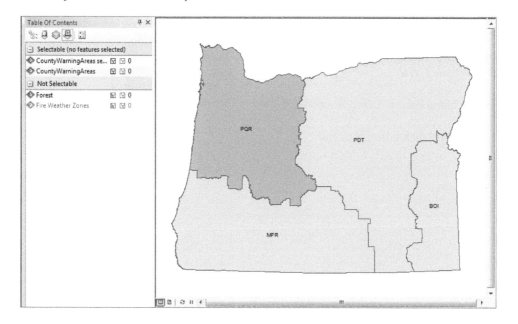

7. In the table of contents, click the List By Drawing Order button. Turn off the
CountyWarningAreas layer and drag the CountyWarningAreas selection layer to the
bottom of the table of contents.

The forest cover for the state is now visible again, and it covers the CountyWarningAreas
selection layer (shown here in a light purple color, although it may be a different color on
your map).

Commonly used geoprocessing tools are listed on the Geoprocessing menu (such as the
Dissolve tool you searched for previously). Next, you will use the Clip tool—accessed from
the Geoprocessing menu rather than ArcToolbox—to clip the forest to the boundary of
the Portland CWA.

8. On the menu bar, click Geoprocessing > Clip.

The Clip tool dialog box appears.

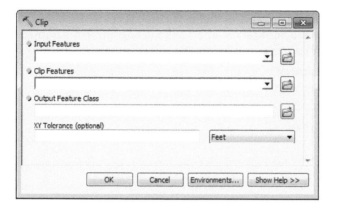

Using this dialog box, you will select the layer with the input features to be clipped, the layer to clip with, and an output layer. You will store the output in your MyOregonForest geodatabase.

9. In the Input Features list, click Forest. In the Clip Features list, click CountyWarningAreas selection.

10. Click the Browse button next to the Output Feature Class list. On the Output Feature Class dialog box, browse to the contents of MyOregonForest.gdb in your MyData folder for chapter 18. In the Name box, type **PortlandForest**, and click Save.

The output feature class information is updated on the Clip dialog box.

TIP

Geoprocessing tools are also available when editing data. You can access several geoprocessing tools such as the Clip and Buffer tools from the Editor toolbar menu.

11. Click OK to run the Clip operation. When the Clip tool processing is complete, a pop-up notification will appear in the notification area. Click the Clip link in the pop-up notification to view the Clip tool's results report.

12. Review the information in the Results window, then close it.

13. In the table of contents, turn off the Forest layer.

The PortlandForest layer (assigned a random color), which consists of forest polygon features clipped to the boundary of the Portland CWA, is added to the map.

14. Open the attribute table of the PortlandForest layer.

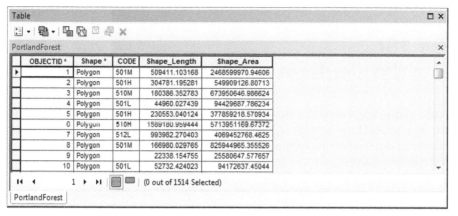

The clip operation has reduced the number of Forest layer features from 3,771 to 1,514. The table has attributes that include forest codes for some of the features. You will use this layer again in exercises 18c and 18d.

15. Close the table.

ON YOUR OWN

In the table of contents, turn on the Fire Weather Zones layer and drag it to the top. You can see how you have gone from a complete Fire Weather Zones layer to a dissolve to create CWAs and now to a clip of the Portland CWA.

16. If you want, rename your map document and save to your MyData folder for chapter 18.

17. If you are continuing to the next exercise, leave ArcMap open. Otherwise, exit the application, and click No if prompted to save your changes.

1
2
3
4
5
6
7
8
9
10
11
12
13
14
15
16
17
18
19
20

Exercise 18c
Create a data subset

Operations such as dissolving and clipping automatically create new datasets. Recall that another way to make a new dataset from an existing one is to make a selection on a layer and export the selected features, as you learned in previous chapters. This is a common task when creating maps and preparing data for analysis.

In this exercise, in addition to the Forest layer, you have a layer of fire lookouts covering all four CWAs. You could clip this layer just as you did the forest, but this time you will use another method to change the dataset's boundaries. Because lookout towers are points contained within polygons, you can use Select By Location to select the points inside the Portland CWA. Then you will export the selected set of points to create a dataset that includes Portland CWA lookout towers only. You will also learn how to import symbology from another layer.

1. In ArcMap, open the ex18c map document from your Chapter18 folder.

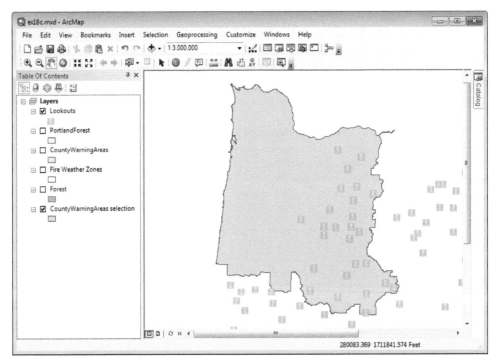

The map displays a layer of lookout tower locations and the Portland CWA created in exercise 18b (CountyWarningAreas selection). The Lookouts layer is symbolized by blue

and white lookout tower symbols—the colors in the book graphic may be different from your map if you experimented with changing colors in previous exercises.

2. On the menu bar, click Selection > Select By Location. If necessary, click the Selection method arrow and click Select Features From.

3. For Target layer(s), check the Lookouts check box.

4. For Source layer, click CountyWarningAreas selection.

5. For the Spatial selection method, click Are Completely Within The Source Layer Feature, and click Apply.

How many lookout towers in the Portland CWA are selected?
*(**Hint:** The count is displayed on the far left of the status bar after the Select operation is completed.)*

6. Close the Select By Location dialog box. In the table of contents, right-click the
 Lookouts layer and click Data > Export Data.

 The Export list is correctly set to Selected features. The option to use the same
 coordinate system as the layer's source data is also correct.

7. Click the Browse button next to the Output feature class box. On the Saving Data
 dialog box, make sure the Save As Type list is set to File And Personal Geodatabase
 Feature Classes. Browse to the contents of MyOregonForest.gdb in your MyData folder
 for chapter 18. In the Name box, replace Export_Output with **PortlandLookouts**, and
 click Save.

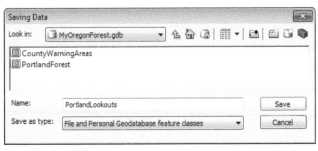

8. On the Export Data dialog box, click OK.

ArcMap exports the selected lookout towers to a new feature class and prompts you to add the data to the map.

9. Click Yes.

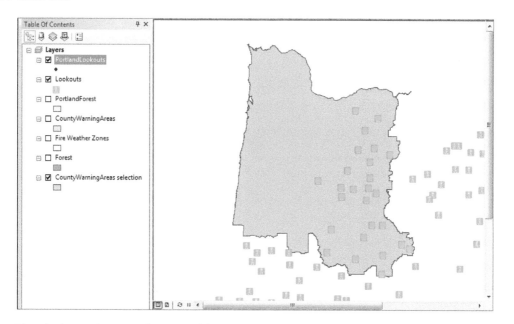

You don't see the point features of the new PortlandLookouts layer because they are covered by the highlighted selection on the Lookouts layer.

10. In the table of contents, turn off the Lookouts layer.

The new layer contains lookout towers within the Portland CWA only. The default symbol, however, is not as informative as the one in the Lookouts layer.

11. In the table of contents, open the PortlandLookouts Layer Properties dialog box. On the Symbology tab, click Import (upper-right corner of the Draw All Features Using The Same Symbol section).

The Import Symbology dialog box appears.

You want to import symbology from another layer on the map, so the first option is set correctly. The Layer list is also set correctly to Lookouts (to the only other point layer on the map).

12. Click OK.

Notice that the symbol is updated in the layer properties.

13. On the Layer Properties dialog box, click OK to apply the change.

ON YOUR OWN

Turn on the PortlandForest layer. Use the Import method to change the symbology to match the original Forest layer.

14. If you want, rename your map document and save to your MyData folder for chapter 18.

15. If you are continuing to the next exercise, leave ArcMap open. Otherwise, exit the application, and click No if prompted to save your changes.

1 2 3 4 5 6 7 8 9 10 11 12 13 14 15 16 17 18 19 20

You are almost done preparing the data for your analysis. You still have to prepare one more layer in exercise 18d. You will do some more clipping to create a layer that shows the populated jurisdictions (or communities) that are in the Portland CWA. However, this time you will learn how to run tools in ModelBuilder.

Exercise 18d
Run tools in a model

ModelBuilder is a graphic user interface for diagramming solutions to spatial analysis problems. The diagram you make is called a model. Models have many advantages. The ModelBuilder interface allows you to organize the inputs, tools, and outputs used to perform a GIS analysis. ModelBuilder is not a free-standing software application like ArcMap or ArcCatalog, but it is more than just a geoprocessing tool–it is a design environment for creating workflow diagrams. The ModelBuilder window has its own menu and toolbar. Many of the controls have the same icons and functions as ArcMap controls.

Individual geoprocessing tools in ArcToolbox, such as Dissolve, Clip, Buffer, and Union, are useful for completing particular tasks, but you can also string them together to create a complex tool that performs an entire workflow with just one click. ArcGIS will carry out the individual procedures in order, according to your plan, or model. A complete model is an interconnected set of processes, with each process consisting of input data, a tool, and output data, which can be fed into another process.

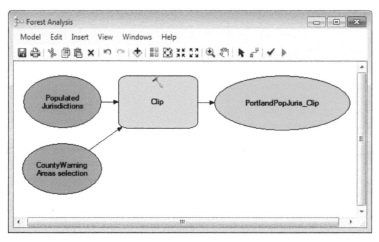

A model using the Clip tool in ModelBuilder.

The elements of a process are distinguishable by their symbology. By default, input data is represented by a blue oval. Tools are represented by yellow rectangles. Output data is represented by a green oval. The representative shapes, colors, and other properties can be changed, if you want.

TIP

Right-click a node (i.e., a data or tool) or edge (i.e., an arrow) element in your model to access the Display Properties dialog box. This allows you to change properties such as background and border color, change the shape and font of selected elements, and change the arrow style and size of selected connectors. You can also change the look of ToolTips that appear when the mouse is paused over an element.

Once you have diagrammed a model, you can run it with a single click. ModelBuilder runs each of the processes in turn, passing the output of one process to the next and from that process to the next after that, until a final dataset is produced.

Various combinations of input parameters can be set to output different kinds of results. This is the real power of modeling. For example, a model that calculates traffic noise in proximity to residential neighborhoods can have several main inputs, such as the decibel measure and the distance from noise origin to homes. The model can then be run quickly and repeatedly using various inputs and parameter settings to achieve different results. Without ModelBuilder, you would need to open each tool dialog box in ArcToolbox, set the tool parameters, and run each tool. Then you would have to reopen the tool dialog boxes each time to set different parameters. Each iteration would take you just as long (or almost as long) as the first time. ModelBuilder can save huge amounts of time if you perform repetitive spatial analyses. It also allows you to effectively evaluate the changing conditions and what-if scenarios that arise in decision-making scenarios.

Furthermore, models are a good way to evaluate the soundness of your methodology, to document it for others, and to develop and refine it over time. Sharing models with colleagues helps to standardize decision-making processes. If you worked for a government agency, for example, you could use models to exchange knowledge across the spectrum and establish uniform practices.

For more information about spatial modeling and ModelBuilder, consult ArcGIS Desktop Help.

Like tools, models are stored in a toolbox. In earlier chapters, you used some of the tools and toolboxes that come with ArcMap. In this exercise, you will learn how to create a toolbox and put a model inside it. A toolbox can be created within ArcToolbox, within an ordinary file folder, or within a geodatabase. Because models must be stored in toolboxes, these are also the three locations in which models can be created.

1
2
3
4
5
6
7
8
9
10
11
12
13
14
15
16
17
18
19
20

In this exercise, you will continue preparing your data for analysis and create a toolbox to store your model. Then you will create a model and add one process to it–clipping populated jurisdictions to the Portland CWA. Populated jurisdictions are areas that include residential and commercial areas. The data preparation you did in exercises 18a, 18b, and 18c is assumed to be complete, although you could include those operations in your model as well. Typically, ModelBuilder is used for multiple tasks, but here, you will create a single-operation model as an introduction to the application.

1. In ArcMap, open the ex18d map document from your Chapter18 folder.

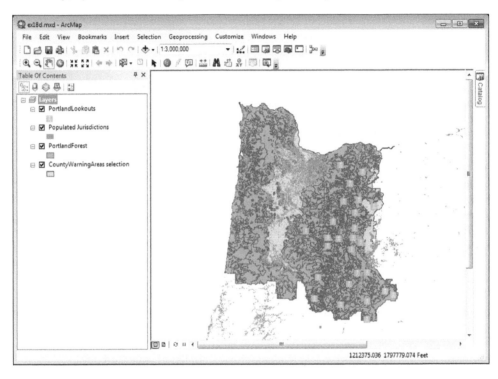

The map contains layers for lookout towers, forest, and the Portland CWA. It should look familiar, as it has the same data and symbology as the map document in exercise 18c. A new layer called Populated Jurisdictions has been added to the map for you. It contains communities and populated areas for the entire state.

2. Display the Catalog window. In the Catalog window, browse to your MyData folder for chapter 18. Right-click MyData and click New > Toolbox.

A new toolbox is added with its default name highlighted.

3. In the Catalog window, change the new toolbox name to **Oregon Forest Analysis** and press Enter. In the Catalog tree, expand your MyData folder if necessary.

1
2
3
4
5
6
7
8
9
10
11
12
13
14
15
16
17
18
19
20

You have created a toolbox in an ordinary file folder. If you have activated file extensions, such as in Windows Explorer, they appear with a .tbx file name extension.

TIP

Depending on your ArcGIS setup, you may have file extensions hidden. To change how file extensions are viewed in the Catalog window, open ArcCatalog, and on the menu bar, click Customize > ArcCatalog Options. On the ArcCatalog Options dialog box, on the General tab, uncheck the Hide File Extensions check box.

Next, you will create a model within the toolbox.

4. In the Catalog window, right-click the Oregon Forest Analysis toolbox and click New > Model.

This opens the Model window. Next, you will change the name in the title bar to something more meaningful.

5. On the ModelBuilder menu bar, click Model > Model Properties. On the General tab, in the Name box, replace Model with **ForestAnalysis**. (The model name cannot contain any spaces.) In the Label box, replace Model with **Forest Analysis**. (The model label can contain spaces.) Click OK.

The title bar is updated with the new name. Next, you will save and close the model so you can work with it in ArcMap.

6. On the ModelBuilder toolbar, click the Save button 💾.

7. Close the ModelBuilder window.

8. In the Catalog window, if necessary, click the plus sign (+) next to the Oregon Forest Analysis toolbox to open it and view the model inside.

You can access your model from the Catalog window, or you can add it to ArcToolbox. Next, you will add it to ArcToolbox, but it won't be a permanent change—it affects the appearance of ArcToolbox in the current session only.

9. **On the Standard toolbar, click the ArcToolbox button. (Depending on how you previously set up your ArcToolbox window, it may be floating, docked, or tabbed to the side of the main window.)**

TIP

You can still drag items between the Catalog and ArcToolbox windows, even when they are tabbed. With your mouse pointer, pause over the tab, and the window expands into view.

10. **In the Catalog window, click the Oregon Forest Analysis toolbox and drag it to the bottom of the ArcToolbox window. (Alternatively, you can right-click the white background in ArcToolbox and click Add Toolbox, navigate to the Oregon Forest Analysis toolbox, and click Open.)**

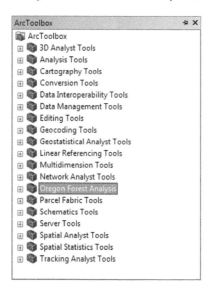

The Oregon Forest Analysis toolbox is automatically added and alphabetized with the other toolboxes.

To save a customization to ArcToolbox, right-click the background and click Save Settings. You can save to a file or change the default ArcToolbox settings.

11. Close or hide the Catalog window, since it will no longer be needed.

12. In the ArcToolbox window, click the plus sign (+) next to Oregon Forest Analysis to expand its contents. Right-click the Forest Analysis model and click Edit. When the ModelBuilder window appears, move it so it doesn't overlap either the ArcToolbox window or the ArcMap table of contents. (It may be easier to work with both windows if your ArcToolbox window is floating.)

Next, you will create a process to clip populated jurisdictions to the Portland CWA. You will drag the Clip tool from ArcToolbox and then the Populated Jurisdictions layer from the ArcMap table of contents to the model window.

13. In the ArcToolbox window, expand the Analysis Tools > Extract toolset.

14. Click the Clip tool and drag it to the center of the model window.

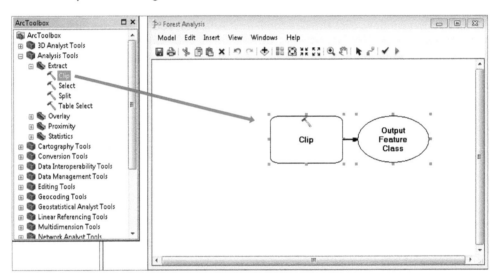

The Clip tool, represented by a rectangle, is added to the model. An arrow connects the tool to its output data, represented by an oval. Blue selection handles indicate that the elements are currently selected. The fact that they are not shaded means they are incomplete and not ready to run–they are waiting for input data to be attached, which you will provide next.

15. In the ArcMap table of contents, press and drag the Populated Jurisdictions layer to the model window and drop it directly on top of the Clip tool rectangle. A shortcut menu appears. Click Input Features.

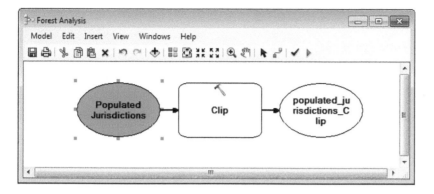

The Populated Jurisdictions layer, represented by a blue oval, is automatically connected to the tool it was dropped onto. Because clipping requires two layers, you will set the Portland CWA as a model parameter as well.

16. In the ArcMap table of contents, press and drag the CountyWarningAreas selection layer to the model window and drop it directly on top of the Clip tool rectangle. A shortcut menu appears. Click Clip Features.

The second input is placed directly on top of the first, so the first is no longer visible.

17. Drag the CountyWarningAreas selection element to a position below the Populated Jurisdictions element so you can see both inputs.

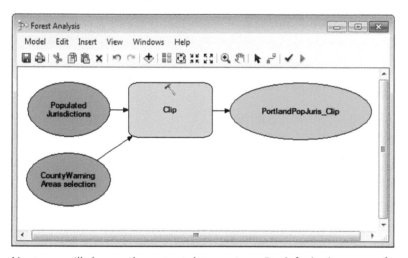

Next, you will change the output data settings. By default, the output feature class is set to your user Default.gdb. You will change the output location to the geodatabase in your MyData folder for chapter 18.

TIP

Each map document has a default geodatabase, which is the home location of the spatial content of your map. This location is used for adding datasets and for saving resultant datasets created by various editing and geoprocessing operations. To learn more, consult ArcGIS Desktop Help.

18. Double-click the output element in the model currently labeled as populated_jurisdictions_Clip. In the resulting window, click the Browse button and browse to the contents of MyOregonForest.gdb in your MyData folder for chapter 18. In the Name box, type **PortlandPopJurs**, and click Save.

The output feature class information is updated on the output feature class dialog box.

19. Click OK to close the dialog box.

All the elements are colored in, which means the process is ready to run. When you run the process, a feature class will be created and saved on disk. However, the output data is not automatically added as a layer to ArcMap. If you want the data added to the map, you will have to set a property on the output data element.

20. In the model, right-click the PortlandPopJurs output oval and click Add To Display so it will be turned on in the table of contents.

21. On the ModelBuilder toolbar, click the Run button ▶ to run the model (or on the ModelBuilder menu bar, click Model > Run).

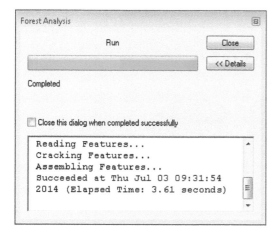

As a model runs, each of its tools turns red very briefly. Your model has only one tool, so if you weren't watching closely (or if the progress report covered the model), you might not have seen it. When the tool is finished running, it turns yellow again. Drop shadows behind the tool and output data elements indicate that the process has been run. Once a process has been run, you cannot run it again until you either change a parameter or choose Run Entire Model or Validate Entire Model from the Model menu.

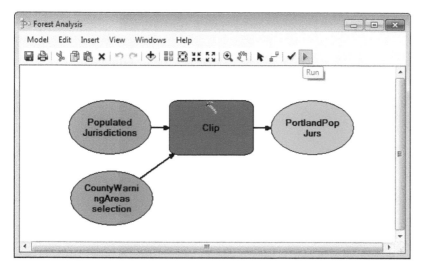

22. Close the progress report. Notice that after a process runs, drop shadows appear behind the model elements.

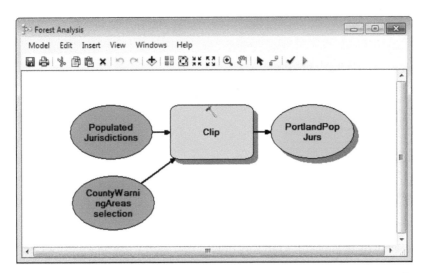

The PortlandPopJurs layer is added to ArcMap. (You may need to move the model window out of the way to see it.) It is symbolized in a random color.

23. Turn off the Populated Jurisdictions layer. Open the PortlandPopJurs layer properties and change the symbology to match the orange color of the Populated Jurisdictions layer.

24. On the ModelBuilder toolbar, click the Save button to save the model.

25. Close the ModelBuilder and ArcToolbox windows.

The model and all its settings are stored in the Oregon Forest Analysis.tbx file you created at the beginning of the exercise.

26. If you want, rename your map document and save to your MyData folder for chapter 18.

27. If you are continuing to the next chapter, leave ArcMap open. Otherwise, exit the application, and click No if prompted to save your changes.

In chapter 19, you will analyze spatial data. This involves comparing spatial relationships among features and trying to draw conclusions from analysis. You will learn how to use more geoprocessing tools and see how your analysis is coming together.

Summary

In this chapter, you learned some basic steps about how to approach GIS analysis. Many geoprocessing tools are commonly accessed from the ArcMap Geoprocessing menu, from a search using the Search window, or by using ArcToolbox directly. You learned how to use the Dissolve tool to aggregate fire weather zones into county warning areas. You also used the Clip tool twice—once on its own and once in a model. Learning how to use ModelBuilder efficiently will facilitate and automate many projects and workflows, from the simple to the complex. In chapter 20, you will build a larger model and connect more geoprocessing tools together.

Terms

geoprocessing
dissolve
clip
model

1
2
3
4
5
6
7
8
9
10
11
12
13
14
15
16
17
18
19
20

CHAPTER 19
Geoprocessing vector data

Buffer features
Overlay data
Calculate attribute values
Create graphs

Learning objectives

- Use geoprocessing functions such as buffering and overlaying
- Calculate attribute values
- Create graphs

Economic development issues, health-care reform, environmental impacts, industrial growth, and a large number of other concerns demand sophisticated problem-solving strategies. Most of the problems you solve with GIS involve comparing spatial relationships between and among features—in one layer or in different layers—and drawing conclusions. That's spatial analysis, in a nutshell. It can be as simple as visual comparison of map layers, or you can use a wide range of geoprocessing and analysis tools to modify, calculate, and generate new spatial data. Techniques of spatial analysis can include everything from measuring the distance between points to modeling the behavior of ecosystems to using a multivariable model to compute flow characteristics.

Spatial analysis today is highly visual, and so are many of the geoprocessing tools in ArcToolbox. Through the aid of ArcMap and ModelBuilder, you can create visual analysis models by piecing together inputs, processes, and outputs. Models facilitate data preparation and spatial analysis. In this chapter, you will continue your work with the Oregon forest fire analysis and create layers that are extremely useful in spatial analysis: buffers and overlays.

A buffer is an area drawn at a uniform distance around a feature—whether a point, a line, or a polygon. A buffer represents a critical zone, such as a floodplain, a protected species habitat, or a municipal services area, wherein features lying inside the buffer can have a different status from

features that lie outside of it. You can specify buffer distances manually or use attribute values to determine varying sizes of buffers, and even create multiple buffer rings. Buffers created on a single layer can often overlap. For example, cellular phone towers need to be built within a certain proximity of each other so that their service radii can cover enough of an urban area to keep calls from being dropped. Overlap between individual buffers is acceptable, although it may be more beneficial to dissolve the overlaps to form one large feature.

Buffers are created around point features, with some buffers overlapping.

Dissolved buffers create large polygons. Buffers that don't overlap remain the same.

A map **overlay** (e.g., union or intersect) identifies an overlap between features in two layers and creates a dataset in which the line of overlap defines a new feature. In a **union** overlay, nonoverlapping areas are included in the output dataset. A union dataset, then, has three types of features: those found only in the first input layer (with layer 1 attribute values), those found only in the second input layer (with layer 2 attribute values), and those created by areas of overlap between the two layers (with both layer 1 and layer 2 attribute values). In an **intersect** overlay, only the overlapping geometry is preserved, and features have attributes from both input layers.

Left: An area of soil and a spilled contaminant zone overlap each other. Overlay analysis–whether union (*center*) or intersect (*right*)–creates new features where input layers overlap (contaminated soil). In this case, both types of analysis identify areas where contaminated soil must be treated or cleaned up.

Merge combines multiple input datasets of the same data type (that is, several point feature classes can be merged, but a line feature class cannot be merged with a polygon feature class) into a single, new output dataset. All fields of the input datasets are included in the output dataset, even though some input fields may not match each other.

Left and *center*: West census tracts and east census tracts are two different input layers of the same data type. The merge process combines the layers into a new output layer that contains all the census tracts (*right*).

Layers overlaid on other layers can sometimes cause unexpected results, called slivers. Slivers are small polygons that are created because of uneven or unequal boundaries. In certain cases, boundaries may not match up perfectly, so attention must be paid to removing these extraneous features, if required. For example, a shared parcel lot boundary measured using two different methods (say, GPS versus field surveying) may not line up exactly. To combat inaccuracies, ArcGIS supports topology to enforce data integrity rules (such as no gaps allowed between polygons). In geoprocessing, and GIS in general, higher-quality data input results in higher-quality data output.

Data

The following datasets, stored in a geodatabase named OregonForest.gdb, are used in this chapter:

- CountyWarningAreas—polygon feature class of county warning areas composed of fire weather zones in the state of Oregon, derived from exercise 18a *(Source: Oregon Department of Forestry)*
- PortlandForest—polygon feature class of forested land in the state of Oregon, derived from exercise 18b *(Source: Oregon Department of Forestry)*
- PortlandLookouts—point feature class of fire lookout towers in the state of Oregon, derived from exercise 18c *(Source: Oregon Department of Forestry)*
- PortlandPopJurs—polygon feature class of inhabited areas in the state of Oregon, derived from exercise 18d *(Source: Oregon Department of Forestry)*
- World_Imagery—basemap *(Source: ArcGIS Online)*

Note: To use the ArcGIS Online basemap layer, you need a stable internet connection.

1 2 3 4 5 6 7 8 9 10 11 12 13 14 15 16 17 18 19 20

1
2
3
4
5
6
7
8
9
10
11
12
13
14
15
16
17
18
19
20

Exercise 19a
Buffer features

Buffers are based on the concept of proximity. Buffers are created as polygons in a new layer. Buffers can be drawn at a constant distance around every feature in a layer (e.g., 100 meters) or at a distance that varies according to attribute values. For example, buffers representing the size of earthquakes might vary according to an attribute that describes the strength of magnitude. Buffers can also be concentric rings representing multiple distances, such as the areas within 100, 500, and 1,000 meters of a well.

Buffers can be determined by distance in different ways.

Buffers can also be created for lines and areas. If features are close together, their buffers may overlap. You can preserve the overlaps or remove them by dissolving them.

Buffers for lines and areas can overlap or be dissolved.

You will continue your analysis of the fire safety of the Portland County Warning Area (PQR CWA). In chapter 18, you clipped forestland, populated jurisdictions, and lookout towers to the extent of the study area. Your goal is to determine which of the populated jurisdictions are in proximity to forestland and to lookout towers, because this will indicate which areas have been deemed to have a higher probability of risk due to a forest fire. In this exercise, you will buffer 1,000 feet around populated jurisdictions because you want to find the areas that are close to forestland, not just adjacent to it. You will also buffer the lookout towers to show the range of visibility of each tower. The visible-range value of each tower was precalculated and included as an attribute

(RADIUS2). Lookout towers were built in areas with a high probability of fire and with the least resources to handle a sudden fire outbreak (that is, with no fire stations nearby).

1. **Start ArcMap. Open the ex19a map document from your C:\EsriPress\GTKArcGIS\ Chapter19 folder.**

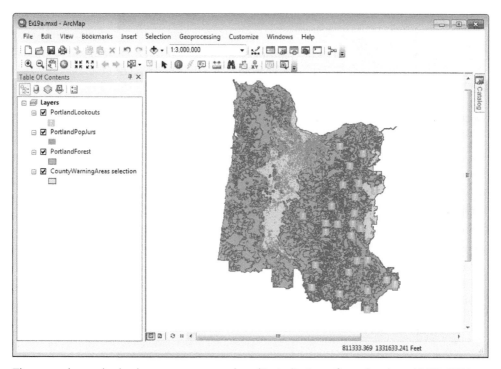

The map shows the lookout towers, populated jurisdictions, forestland, and PQR CWA (CountyWarningAreas selection). You will begin by buffering the populated jurisdictions using the Buffer geoprocessing tool.

2. **On the menu bar, click Geoprocessing > Buffer.**

 This menu provides access to commonly used geoprocessing tools. You can also access the tools through ArcToolbox. (Perform a search to locate them quickly.) The Buffer tool dialog box appears.

 Next, you will select the layer with the input features to be buffered, designate an output feature class, enter a buffer distance, and select a dissolve option.

3. **On the Buffer tool dialog box, click the Input Features arrow and click PortlandPopJurs. (Alternatively, drag the PortlandPopJurs layer from the ArcMap table of contents to this location.)**

1

2

3

4

5

6

7

8

9

10

11

12

13

14

15

16

17

18

19

20

TIP

> Only layers from an active data frame can be dragged to an ArcToolbox tool. If you drag a layer from an inactive data frame to a geoprocessing tool, warnings or errors are produced inside the tool, and you will not be able to proceed.

4. **Click the Browse button next to the Output Feature Class box. On the Output Feature Class dialog box, browse to ..\GTKArcGIS\Chapter19\MyData and double-click MyOregonForest.gdb. In the Name box, type PopJursBuffer, and click Save.**

The output feature class information is updated with the new location and name.

5. **For the buffer distance, make sure the Linear unit option is selected. In the Linear unit box, type 1000. Maintain Feet as the measurement unit. Scroll down and make sure Dissolve Type is set to NONE.**

6. **Click OK. (This process may take a few minutes because of the large number of communities.)**

 The new PopJursBuffer layer is added to the map. It looks similar to the original populated jurisdictions layer, except that it has a 1,000-foot buffer surrounding the features. (If you want, zoom in to view the buffer. Just be sure to return to your previous extent to continue the exercise.)

7. **Change the color of both the buffer fill and outline to Anemone Violet.**

8. **Open the attribute table of the PopJursBuffer layer.**

	OBJECTID *	Shape *	COMMUNITY	COMID	RISK	HAZARD	PROTECTION	VAL
▶	1	Polygon	Adair Village (City)	11	L	L	L	L
	2	Polygon	Amity (City)	15	L	L	L	L
	3	Polygon	Astoria (City)	19	H	L	L	H
	4	Polygon	Aumsville (City)	21	L	L	L	H
	5	Polygon	Aurora (City)	22	L	L	L	H
	6	Polygon	Banks (City)	40	L	L	L	H
	7	Polygon	Barlow (City)	41	L	L	L	H
	8	Polygon	Blue River WD (RFPD)	45	H	H	M	H
	9	Polygon	COBURG RFPD (RFPD)	70	L	L	L	L

I◄ ◄ 1 ► ►I ▤ ▥ | (0 out of 305 Selected)

PopJursBuffer

 In addition to its standard attributes, the output table has attributes from the input table. (Because you did not dissolve the buffers, the output and input tables have

the same number of records—there is one buffer for each populated jurisdiction. This correspondence of record to record makes it possible to copy attributes from one table to the other.) An attribute called BUFF_DIST is added to the table, which shows the buffer distance you specified.

9. Close the PopJursBuffer attribute table.

Next, you will buffer the lookout towers layer. The buffer sizes for this layer will vary according to the visibility range of each tower.

10. Open the attribute table of the PortlandLookouts layer.

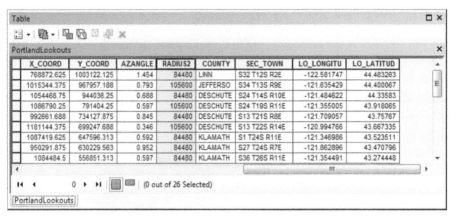

The attribute table contains an attribute called RADIUS2, which is the visibility range value (in feet) of each tower. Next, you will specify this attribute as the buffer distance value.

11. Close the PortlandLookouts attribute table.

12. On the menu bar, click Geoprocessing > Buffer.

13. On the Buffer tool dialog box, click the Input Features arrow and click PortlandLookouts. (Alternatively, drag the PortlandLookouts layer from the ArcMap table of contents to this location.)

14. Save the output feature class as **LookoutsBuffer** to MyOregonForest.gdb.

15. For the buffer distance, click the Field option. In the Field list, click RADIUS2.

16. Click the Dissolve Type arrow and click ALL. Then click OK.

Wherever buffer polygons overlap each other, the overlapping boundaries will be dissolved to make a single feature.

When the operation is completed, the new LookoutsBuffer layer is added to the map.

17. Change the color of the buffers so they are visible against the background. In the table of contents, drag the LookoutsBuffer layer below the PopJursBuffer layer.

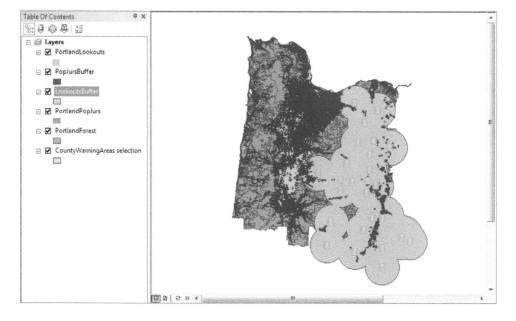

18. Open the attribute table of the LookoutsBuffer layer.

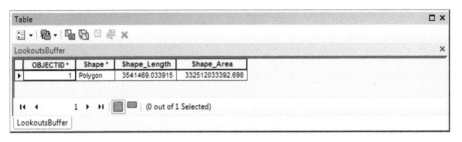

The table contains just a single record. Buffers created with the dissolve ALL option form a single feature, called a multipart polygon (a polygon with discontinuous boundaries). The attributes are the standard four for a geodatabase polygon feature class: OBJECTID, Shape, Shape_Length, and Shape_Area. Because of the dissolve, none of the PopJursBuffer layer attributes are passed to the output table.

> *What would happen if the LookoutsBuffer layer was not dissolved? How would that affect how it is used when including it in other forms of geoprocessing?*

19. Close the attribute table.

20. To save your work, rename your map document and save to your MyData folder for chapter 19.

21. If you are continuing to the next exercise, leave ArcMap open. Otherwise, exit the application, and click No if prompted to save your changes.

The two buffers, LookoutsBuffer and PopJursBuffer, define the areas where a probability of fire near urban areas exists. In exercise 19b, you will use overlays to define the areas in which the communities need to be isolated from forest areas.

Exercise 19b
Overlay data

A union overlay combines the features of two input layers to create a dataset that includes all combinations of polygons. In the following example, one polygon layer represents a land parcel and the other an oil spill. Part of the parcel lies outside the spill, and part of the spill lies outside the parcel; in the middle, the two layers overlap. When a union function is applied to the two layers, the two original polygons become three. The area of overlap becomes a new feature, and the nonoverlapping areas become a multipart polygon (individual polygons that are separated but still belong to the same group).

The parcel and oil spill layers are unioned to create three new polygons.

What kind of attributes does a union layer have? In addition to its own standard attributes, it includes all the attributes of both input layers. This doesn't mean, however, that every record has a value for every attribute. In this example, feature 1 (OBJECTID 1) in the output table has no Spill_ Type value because it is outside the oil spill. Feature 2 has no Owner or Landuse values because it is outside the parcel. Only feature 3, which spatially coincides with both input layers, has a value for every attribute.

The output table contains the attributes of both input layers. Output features get the attribute values of input features with which they are spatially coincident.

1

2

3

4

5

6

7

8

9

10

11

12

13

14

15

16

17

18

19

20

What happens to the unique identifier attribute of each layer? One of the standard attributes of the union layer, assuming it is a geodatabase feature class, is the OBJECTID field, which assigns a unique identifier (ID) number to each feature. In other data formats, the identifier attribute is called FID (known as feature ID) or OID (known as object ID). In this case, both the input layers also have OBJECTID fields. ArcMap will not delete these fields–they may be useful to you–but at the same time, there shouldn't be a table with three OBJECTID fields.

To resolve the conflict, the OBJECTID fields from the input layers are renamed in the union layer. The new name consists of the prefix "FID_" followed by the input layer name. Thus, the parcel layer's identifier attribute is renamed "FID_Parcel," and that of the oil spill layer is renamed "FID_OilSpill."

In an intersect overlay, the process is a little simpler, because only the area of overlap is preserved. An intersect of the oil spill and parcel layers would consist of a single feature, as shown in the graphic.

An intersect layer consists of a single feature.

As with a union layer, an intersect layer includes its own standard attributes, plus all the attributes of both input layers. In an intersect, each attribute is populated for every record. A union overlay requires that both input layers be polygon layers. In an intersect, the input layers can be either two polygon layers or a polygon layer and a line layer. In the latter case, shown in the graphic, the output is a line layer.

In this exercise, you will union the populated jurisdiction buffer layer and the forest layer to create a single output dataset that contains all the attributes of both inputs. Then you will union this layer with the lookout towers buffer layer. Because new features will be created wherever

forest, populated jurisdictions buffer, and lookouts buffer polygons overlap, each output feature will lie either entirely inside or entirely outside a buffer. Once you have performed these union operations, you will see how the output provides the correct attributes to determine which of the populated jurisdiction buffers are in proximity to both forestland and lookout towers. This provides an output that shows which areas have a high probability of a forest fire within the visibility of lookout towers.

1. **In ArcMap, open the ex19b map document from your Chapter19 folder.**

The map shows the buffer for populated jurisdictions and the Portland forest layer. The other layers are turned off.

2. **On the menu bar, click Geoprocessing > Union.**

The Union tool and other analysis tools can also be accessed from ArcToolbox. On the Standard toolbar, click the ArcToolbox button. In the ArcToolbox window, expand Analysis Tools and then the Overlay toolset. Then double-click the Union tool.

Next, you will specify the layers to overlay and designate an output feature class.

3. On the Union tool dialog box, click the Input Features arrow and click PopJursBuffer. (Alternatively, drag the layer from ArcMap to the Union dialog box.)

4. Click the Input Features arrow again and click PortlandForest.

The selected layers are added to the list of layers that will be processed in a union.

5. Save the output feature class as **AreaAtRisk** to **MyOregonForest.gdb.**

By default, the output layer attribute table contains its standard attributes, plus the attributes of both input layers. Optionally, you can omit the input layer identifier attributes. Alternatively, you can omit all input attributes, except the identifiers. This creates a smaller output table and is convenient if you don't need to work with attributes. (If you later decide that you need the input attributes, you can join or relate back to the input layer tables using the common identifier attribute.)

For now, you will include all the layer attributes.

6. For the JoinAttributes parameter, make sure ALL is selected. Click OK to run the tool.

When the operation is completed, the AreaAtRisk layer is added to the map. The new layer, which consists of all the areas from the two input layers, defines the zone where communities are in proximity to forestland and have a probability of either causing a fire or being at risk of damage from a fire. The layer is symbolized uniformly with a single color.

7. Open the attribute table of the AreaAtRisk layer.

FED_TRIBAL	BUFF_DIST	FID_PortlandForest	AREA	PERIMETER	FORSTLND_	FORSTLND_I	CODE	SYM
Y	1000.002	-1	0	0	0	0		
Y	1000.002	-1	0	0	0	0		
Y	1000.002	-1	0	0	0	0		
Y	1000.002	-1	0	0	0	0		
	0	1	25167	497990.375	2	12	501M	
	0	2	55360	303881.969	3	13	501H	
	0	3	67535	178216.953	4	15	510M	
	0	4	94601	45036.75	5	18	501L	
	0	5	37850	230620.766	6	27	501H	

(0 out of 8845 Selected)

AreaAtRisk

The table contains all the attributes included in both unioned layers. Along with the four standard polygon feature class attributes (OBJECTID, Shape, Shape_Length, and Shape_Area), the table contains the renamed identifier attributes from the input layers: FID_PopJursBuffer and FID_PortlandForest.

In the exercise introduction, you saw that not every attribute may be populated for every record in a union attribute table. Identifier attributes, however, are completely populated for every record. Every record in the AreaAtRisk attribute table has not only an OBJECTID value, but also an FID_PortlandForest value and an FID_PopJursBuffer value, regardless of which input feature is spatially coincident with the output feature.

8. Adjust or move the columns to examine or compare the attributes in the table.

For example, the 10th record in the AreaAtRisk table has an OBJECTID value of 10. This value is the feature's new identifier. The record's FID_PopJursBuffer value is 13. This means that the feature coincides spatially with the feature that has the OBJECTID of 13 in the PopJursBuffer table. The record's FID_PortlandForest value is −1. The −1 value means the feature does not coincide spatially with any feature in the PortlandForest layer. If you scroll down the attribute table to OBJECTID 246, you will notice that the record has an FID_PopJursBuffer value of −1. This means the feature does not coincide spatially with any feature in the PopJursBuffer layer. So, at this point in the analysis, the areas at risk are all features that have an FID_PopJursBuffer value other than −1. Later in this exercise, you will see how to use this information to your advantage.

9. Close the table. Turn on the LookoutsBuffer layer, and drag it above the AreaAtRisk layer. This gives you a better view of how this layer will look after the union overlay.

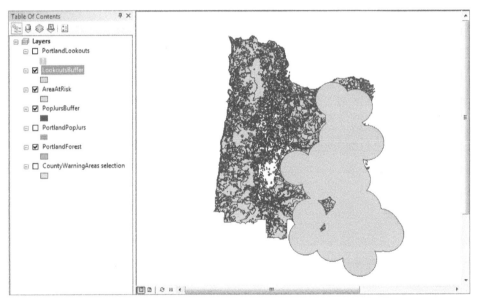

To find the areas at risk near lookout towers, you will union the AreaAtRisk layer with LookoutsBuffer. After the overlay, you will be able to select the polygons that represent communities at risk in the PQR CWA that are within visible range of a lookout tower.

10. On the menu bar, click Geoprocessing > Union.

11. On the Union tool dialog box, click the Input Features arrow and click AreaAtRisk to add it to the list of input layers. Do the same for LookoutsBuffer.

12. Save the output feature class as **Final** to MyOregonForest.gdb.

Leave the Join Attributes (optional) list set to ALL. You will include all input attributes in the Union layer. (You will need several of the attributes to calculate values and to use for symbology in exercise 19c.)

13. Click OK to run the tool.

When the operation is complete, the Final layer, which has more than 9,000 features, is added to the map.

At this scale, it's hard to tell if you are looking at the result of a spatial analysis or just a jumble of lines. Before zooming in, you will look at the attribute table of the Final layer.

14. Open the attribute table of the Final layer.

FID_AreaAtRisk	FID_PopJursBuffer	COMMUNITY	COMID	RISK	HAZARD	P
1	1	ADAIR RFPD (RFPD)	1	L	M	L
2	2	ALBANY FIRE DEPT (RFPD)	4	L	L	L
3	4	AMITY FIRE DIST (RFPD)	6	L	L	L
4	5	AUMSVILLE RFPD (RFPD)	8	L	L	L
5	6	AURORA RFPD (RFPD)	9	L	L	L
6	8	Albany (City)	14	L	L	L
7	9	Amity (City)	15	L	L	L
8	10	Astoria (City)	19	H	L	L
9	11	Aumsville (City)	21	L	L	L

(0 out of 9018 Selected)

Final

The FID_AreaAtRisk attribute is the renamed OBJECTID from the AreaAtRisk layer. If a record has a value other than −1 in this field, it means the output feature coincides spatially with a buffer feature–in other words, it is an area at risk of fire. If a record has the value of −1, it means the output feature does not coincide with a buffer and is therefore not at risk. Similarly, if a record in the FID_PortlandForest attribute has a value other than −1, it means the output feature coincides spatially with forested areas. If not, it means the output feature does not coincide with the forest area and so is not considered at risk.

Because you are also attempting to correlate areas at risk to areas that are visible from lookout towers, you will have to consider the FID_LookoutsBuffer attribute. If a record has a value of 1 in this field, it means it coincides with areas at risk. If it has a value other than 1, it does not coincide.

The area of communities at risk in the Portland forest within visibility of a lookout tower, therefore, is the area composed of all polygons in the Final layer that have a value other than −1 in the FID_AreaAtRisk field, a value other than −1 in the FID_PortlandForest layer, and a value of 1 in the FID_LookoutsBuffer field. However, at this point, it may still be difficult to see exactly which areas are at risk. So next, you will create a definition query to isolate these areas specifically.

15. Close the Final attribute table. In the table of contents, turn off all layers except Final.

16. Double-click the Final layer. Click the Definition Query tab, and click Query Builder. In the Query Builder window, use the inputs to build the query: **FID_AreaAtRisk <> -1 AND FID_PortlandForest <> -1 AND FID_LookoutsBuffer = 1.** Then click OK.

17. Move the Layer Properties window aside to see the main ArcMap window and click Apply.

The applied query isolates the forested areas and populated communities within visibility of a lookout tower. However, you need to symbolize the layer to see the locations of communities at risk.

18. On the Layer Properties dialog box, click the Symbology tab. Under Show, click Categories, and click Unique values. In the Value Field list, click FID_PopJursBuffer, and click Add Values.

19. On the Add Values dialog box, in the Select the value(s) to add list, click −1, and click OK.

20. Double-click the −1 value color chip (which is assigned a random color from the color ramp), and change the fill color to Macaw Green and the outline color to Tarragon Green (under Macaw Green).

21. Similarly, double-click the <all other values> color chip, and change the fill color to Fire Red and the outline color to No Color.

22. Finally, click the Label value for <all other values> and replace it with **Communities at risk**. Click the <Heading> label and type **Not at risk**. For the second item in the list, change the Label value −1 to **Forested area**. Make sure your settings match the graphic, and click OK.

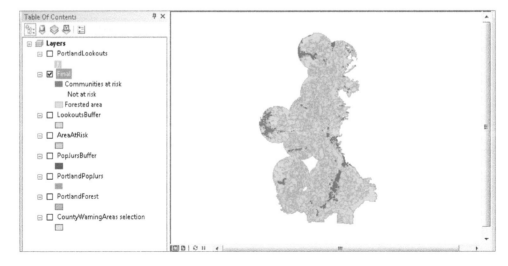

23. On the menu bar, click Bookmarks > Close-up.

To see which areas are at risk in visible proximity of lookout towers, you will turn on labels.

24. In the table of contents, double-click the Final layer. On the Layer Properties dialog box, click the Labels tab. Check the Label Features In This Layer check box. In the Text String frame, click the Label Field arrow and click FID_PopJursBuffer. Then click OK.

On the map, red polygons labeled with values not equal to –1 represent community areas at risk (exactly how you symbolized them).

25. To save your work, rename your map document and save to your MyData folder for chapter 19.

26. If you are continuing to the next exercise, leave ArcMap open. Otherwise, exit the application, and click No if prompted to save your changes.

In exercise 19c, you will recalculate the Final layer's area attribute to get an accurate total of how much area could potentially be affected by forest fire.

Exercise 19c
Calculate attribute values

You can write an expression to calculate attribute values for all records in a table or for just selected ones. For numeric attributes, the expression can include constants, functions, or values from other fields in the table. For text attributes, the expression can include character strings that you type or text values from other fields.

In this exercise, you will adjust the definition query to display communities at risk only, and then view the risk ratings. This isolates these areas for easier calculation. Then, you will calculate values to determine how much area is actually affected.

1. **In ArcMap, open the ex19c document in your Chapter19 folder.**

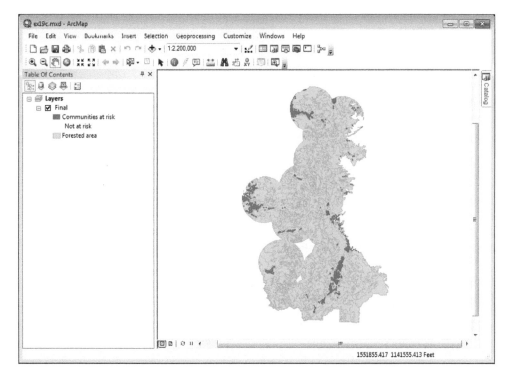

The map contains the Final layer you created in exercise 19b. The other layers were removed from the document so you can concentrate on just the Final layer.

2. **Go to File > ArcGIS Online. Search for world imagery and add it as a basemap by clicking Add. If you get a coordinate system warning, click OK.**

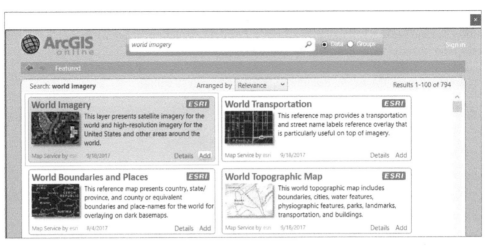

In exercise 19b, you unioned the populated jurisdictions buffer and the Portland forest to create a layer containing communities in proximity to forestland. You then unioned this layer to the lookout towers buffer to show which areas are within the visibility of lookout towers and used this definition query to represent it:

"FID_AreaAtRisk" <> –1 AND "FID_PortlandForest" <> –1 AND "FID_LookoutsBuffer" = 1

Next, you will adjust the existing definition query to display only the communities at risk that are within the visibility of a lookout tower (temporarily removing the forested area).

3. **In the table of contents, double-click the Final layer. On the Layer Properties dialog box, click the Definition Query tab.**

4. **Click Query Builder to open the Query Builder dialog box.**

5. **In the expression box, click the white space so that the pointer appears at the end of the expression. Click the And button, and then in the Fields box, double-click FID_ PopJursBuffer to add it to the expression box. Click the not equal to (< >) button, and click Get Unique Values. In the Unique Values list, double-click –1. Click OK.**

The query is displayed in the Definition Query box.

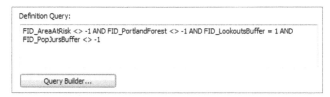

6. On the Layer Properties dialog box, click OK.

On the map, only features that satisfy the query are displayed. These are the communities at risk in the Portland forest within visibility of a lookout tower. The layer attribute table shows only the records that correspond to these features.

7. Open the attribute table of the Final layer.

FID_AreaAtRisk	FID_PopJursBuffer	COMMUNITY	COMID	RISK	HAZARD	PROTECTI
1877	14	BEND FD (RFPD)	27	H	H	H
1878	14	BEND FD (RFPD)	27	H	H	H
1879	14	BEND FD (RFPD)	27	H	H	H
1906	16	BLACK BUTTE RANCH RFPD (RFPD)	29	H	H	M
1907	16	BLACK BUTTE RANCH RFPD (RFPD)	29	H	H	M
1908	16	BLACK BUTTE RANCH RFPD (RFPD)	29	H	H	M
1909	16	BLACK BUTTE RANCH RFPD (RFPD)	29	H	H	M
1934	19	BROWNSVILLE RFD (RFPD)	36	H	L	L
1935	19	BROWNSVILLE RFD (RFPD)	36	H	L	L

(0 out of 649 Selected)

[Final]

The table shows 649 records, rather than the original 1,598. However, take a look at the COMMUNITY field of the first three records in the attribute table. These three records are for a single community. And it is not the only case of duplicate communities. How did this happen?

During the unions, the populated jurisdictions buffer layer (PopJursBuffer) was overlapped by the forest layer (PortlandForest), which produced small area slivers. To recombine the communities, you will perform a dissolve.

8. **Close the Final attribute table.**

9. **On the menu bar, click Geoprocessing > Dissolve.**

10. **On the Dissolve tool dialog box, in the Input Features list, click Final. Save the output as FinalDissolve to MyOregonForest.gdb. In the Dissolve Field(s) box, check the check boxes for COMMUNITY, RISK, HAZARD, PROTECTION, VALUE, and OVERALL, and click OK.**

The FinalDissolve layer is added to the map in a random color.

Each set of polygons with the same community name was dissolved into a single feature. Next, you will quickly symbolize the layer to see the overall risk rating.

11. Open the FinalDissolve Layer Properties dialog box. On the Symbology tab, click Import.

12. On the Import Symbology dialog box, click the Layer Browse button and browse to ..\GTKArcGIS\Chapter19\Data, click FinalDissolve.lyr, and click Add.

13. On the Import Symbology dialog box, click OK. On the Import Symbology Matching Dialog box, make sure the Value Field list is set to OVERALL and click OK.

The layer is categorized based on the OVERALL risk rating attribute.

14. Click OK.

15. Open the attribute table of the FinalDissolve layer.

COMMUNITY	RISK	HAZARD	PROTECTION	VALUE	OVERALL	Shape_Length
▶ Bend (City)	H	H	L	H	H	153635.101138
BEND FD (RFPD)	H	H	H	H	H	355161.363752
BLACK BUTTE RANCH RFPD (RFPD)	H	H	M	H	H	63178.752541
Blue River WD (RFPD)	H	H	M	H	H	30185.742407
BROWNSVILLE RFD (RFPD)	H	L	L	L	M	117784.060367
Camp Sherman (RFPD)	H	H	L	L	H	60553.278745
CLACKAMAS (County)	H	M	L	H	M	177514.083936
CLOVERDALE RFPD (RFPD)	H	H	H	H	H	225817.003084
CRESCENT RFPD (RFPD)	H	H	M	H	H	68354.281376

FinalDissolve

I◀ ◀ 1 ▶ ▶I (0 out of 49 Selected)

There are 49 communities at risk on the map. In addition, the other attributes chosen as dissolve fields–RISK, HAZARD, PROTECTION, VALUE, and OVERALL–have risk rating values of high (H), medium (M), or low (L). These were included by the agency to show each populated jurisdiction's risk rating, hazard rating, protection capability risk rating, value at risk rating, and overall risk rating. Throughout the geoprocessing tasks, ArcMap has automatically updated the Shape_Area attribute, which shows the area of each community. The Shape_Area attribute value is shown in square feet, but next you will create a new field and recalculate it in acres. (**Hint:** One acre is equal to 43,560 square feet.)

16. Above the table title bar, click the Table Options button and click Add Field.

17. On the Add Field dialog box, in the Name box, type **ACRES**. In the Type list, click Double and click OK.

TIP

The Double data type provides double-precision floating-point numbers, which means they are precise to many decimal places. For more about data types stored in ArcGIS, consult ArcGIS Desktop Help.

18. Scroll to the far right of the attribute table. Right-click the field name ACRES and click Calculate Geometry. Click Yes if there are messages warning that you cannot undo the calculation.

The Calculate Geometry dialog box appears.

19. Make sure Area is selected in the Property list. In the Units list, select Acres US [ac] and click OK.

This new field now contains values in acres.

20. Right-click the ACRES field name and click Statistics.

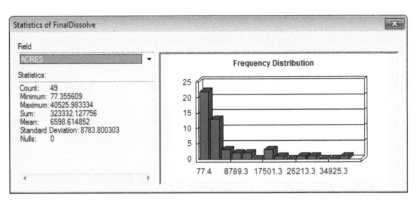

Thus, the total area of communities potentially at risk of a forest fire within visibility of a lookout tower is just over 323,332 acres. Remember that this also includes a 1,000-foot buffer that puts these communities in contact with the Portland forest.

21. **Close the Statistics window and the FinalDissolve attribute table.**

22. **To save your work, rename your map document and save to your MyData folder for chapter 19.**

23. **If you are continuing to the next exercise, leave ArcMap open. Otherwise, exit the application, and click No if prompted to save your changes.**

 Your agency will use this information to establish future fire safety procedures. A more detailed risk analysis would consider additional factors, such as the location of existing roads, the slope of the land, and the density of surrounding vegetation.

Exercise 19d
Create graphs

The ArcGIS Graph Wizard allows you to create many kinds of graphs, including column, pie, area, and scatterplot graphs. Using the wizard, you can set properties for such elements as titles, axes, and graph markers (for instance, the bars in a bar graph or the symbols in a scatterplot). Graphs can be saved with a map document or as files with a .grf extension that can be added to any map document.

In exercise 19c, you dissolved the Final layer to show all 49 communities within visibility of a lookout tower. You also calculated a new ACRES field in the attribute table. In this exercise, you will present the ACRES values in a graph and add the graph to a map layout.

1. In ArcMap, open the ex19d map document from your Chapter19 folder.

The map shows the at-risk areas you created in exercises 19b and 19c, overlaid on an imagery basemap.

TIP

The imagery basemap is a web map being served from ArcGIS Online. If you don't have a stable internet connection, or if you find your map takes too long to load or navigate, you can turn it off.

2. On the View menu, click Graphs > Create Graph.

The Create Graph Wizard appears. The panel on the left has all the settings to create the graph. A preview of the graph will appear on the right.

3. Click the Graph type arrow and click Horizontal Bar. Make sure Layer/Table is set to FinalDissolve. Click the Value field arrow and click ACRES.

Next, you will label the y-axis. As you make changes, the graph preview will be updated.

4. Click the Y label field arrow and click COMMUNITY.

The community values appear on the y-axis, although it looks a bit messy.

5. Uncheck the Add To Legend check box. Click Next.

On the second panel of the wizard, you will change the default graph title, add a footer, and remove the axis titles.

6. In the General graph properties frame, in the Title box, replace Graph of FinalDissolve with **Overall Fire Risk Area**. In the Footer box, type **Includes 1000-ft buffer.**

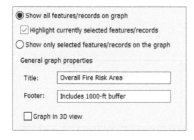

7. In the Axis properties frame (in the lower portion of the wizard), make sure the Left tab is selected. In the Title box, replace COMMUNITY with **Community**. Click the Bottom tab, and in the Title box, replace ACRES with **Acres**. Make sure the Visible check box is checked.

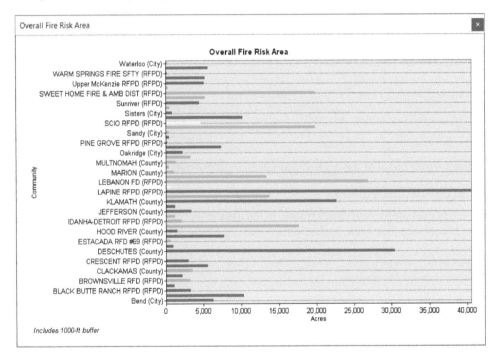

8. Click Finish.

The graph appears in a window that floats on the application window. The graph makes it easy to compare the area values. If you resize the graph window, values that appear along the left axis may change.

La Pine Rural Fire Protection District (RFPD) has the greatest area at risk. You will add the graph to the map layout next.

9. Right-click the graph title bar. On the shortcut menu, click Add to Layout. Close the graph window.

ArcMap switches to layout view.

The graph appears in the middle of the layout page, marked with cyan selection handles. (Other layout elements, such as the map title and axis labels, have been added for you.)

10. On the Tools toolbar, make sure the Select Elements tool is selected. Drag and resize the graph to fill the lower part of the layout, as shown in the graphic. Click outside the layout page to clear the graph selection.

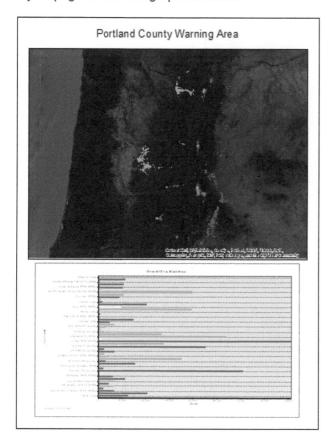

Next, you will add a few layers to complete the look of the map.

11. Add the CountyWarningAreas layer from ..\Data\OregonForest.gdb. Remove the fill, give the outline stroke an appropriate color, and label it.

1
2
3
4
5
6
7
8
9
10
11
12
13
14
15
16
17
18
19
20

12. Add the PortlandLookouts layer and give it an appropriate symbol.

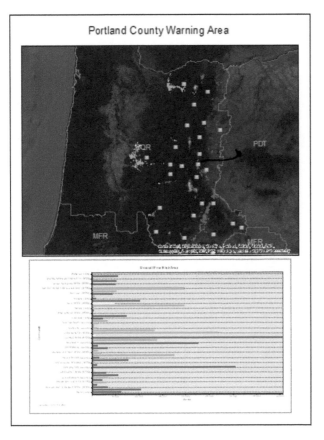

13. To save your work, rename your map document and save to your MyData folder for chapter 19.

14. If you are continuing to the next chapter, leave ArcMap open. Otherwise, exit the application, and click No if prompted to save your changes.

Summary

In this chapter, you built on the foundation gained in chapters 17 and 18. You became familiar with some common geoprocessing tools and got practice running tools. You used common analysis tools including Dissolve, Clip, Buffer, and Union. You now have the skills to go back on your own and try re-creating this chapter's workflow using ModelBuilder to automate some of the geoprocessing tasks. It keeps track of the processes you run, as well as their results and their interdependencies. ModelBuilder gives you a convenient way to build a large project from its component parts and run its processes separately or together. Your model can be changed at any time to incorporate new data, add new conditions, or try different assumptions (best case, worst case, what if). You will use more of ModelBuilder in chapter 20 for performing raster analysis.

Terms

buffer
overlay
union
intersect
merge
multipart polygon

1
2
3
4
5
6
7
8
9
10
11
12
13
14
15
16
17
18
19
20

CHAPTER 20
Using ArcGIS Spatial Analyst

Create raster surfaces
Combine raster surfaces

Learning objectives

- Understand the fundamentals of raster models
- Explore tools available with the ArcGIS Spatial Analyst extension
- Continue using ModelBuilder to run a series of geoprocessing tools
- Create density and distance raster surfaces using ArcGIS Spatial Analyst
- Modify geoprocessing environment settings
- Convert polygons to a raster layer
- Reclassify rasters so they can be logically combined
- Overlay rasters to calculate a multiple-criteria suitability surface

In chapters 18 and 19, you learned how to use geoprocessing tools such as Dissolve, Buffer, and Clip. As you have seen so far, the exercises in this book work mainly with vector data such as shapefiles and geodatabase feature classes. However, the raster data model is just as important to learn. As you learn GIS, it is useful to know that both vector and raster data models have inherent advantages and disadvantages. This chapter provides an introduction to fundamental concepts of the raster data model and to raster analysis using the ArcGIS Spatial Analyst extension to ArcMap.

As discussed in chapter 1, the first thing you must understand about a raster is that it is composed of pixels, instead of discrete x,y coordinates, that define geographic entities. The pixels form a mesh of grid cells, which are used to record and define geographic phenomena on the surface of the earth. Each raster cell represents a portion of the earth, such as a square meter or square mile, and usually has an attribute value associated with it, such as soil type or vegetation class. A surface can be discrete data, meaning it shows distinct and discernible regions on a map, such as soil types, or it can be continuous data, meaning there are smooth transitions between variations in the range of data. Elevation data is one of the most popular uses of a continuous raster. Typical

raster datasets are very detailed and store a large amount of information, because information about every cell must be recorded.

This raster of continuous data represents elevation. Data courtesy of Office of Geographic Information (MassGIS), and the Commonwealth of Massachusetts Information Technology Division.

This raster of discrete data represents land cover. Data courtesy of Lake Tahoe Data Clearinghouse, US Geological Survey.

Map algebra—a language that combines GIS layers—is fundamental to raster analysis. You can evaluate raster cell values using functions and mathematical, logical, or Boolean operators. An output raster is the result of a cell-by-cell function performed on two input rasters.

A 3 × 3 grid added to another 3 × 3 grid (*left*) produces a result grid (*right*), in which the cell values represent the total of the inputs.

ArcGIS Spatial Analyst contains close to 200 geoprocessing tools to perform comprehensive, raster-based spatial analysis. These tools allow you to perform basic mathematical and logical operations, as well as raster dataset creation and processing. The most common way to access ArcGIS Spatial Analyst functionality is to run individual tools from ArcToolbox and use the graphic user interface of ModelBuilder to create models and automate workflows or, for advanced GIS users, to work in the Python window.

ArcGIS Spatial Analyst performs raster analysis using Esri grids. In this context, a grid is a raster data storage format native to Esri and consists of two types: integer and floating point. Integer grids typically represent discrete data, and floating-point grids typically represent continuous data. Rasters can exist in a variety of formats such as TIFF, MrSID (Multi-Resolution Seamless Image Database), and DEM (digital elevation model) but you must convert them to the grid format before you can use this data in analysis. When calculations are performed, the input grid type is usually carried through to the output. Although every cell in a grid has a value assigned to it, you can designate cells without an actual value as NoData cells. However, NoData and zero (0) values are not the same thing. Zero is still a valid value, but you cannot use NoData cells in calculations. So, raster calculations ignore NoData cells.

Examples of the types of spatial analysis you can perform with ArcGIS Spatial Analyst include terrain analysis, surface and density modeling, suitability modeling, and image classification.

The exercises in this chapter require ArcGIS Spatial Analyst, an extension of ArcGIS software. If you are using trial ArcGIS software, you have access to this extension. Make sure your ArcGIS Spatial Analyst extension has been authorized before proceeding.

1 2 3 4 5 6 7 8 9 10 11 12 13 14 15 16 17 18 19 20

HOW TO AUTHORIZE EXTENSIONS

If you did not authorize ArcGIS Spatial Analyst when you installed the software, follow these instructions:

1. On the Start menu, go to ArcGIS > ArcGIS Administrator.
2. In the ArcGIS Administrator window, click Desktop in the left panel and Authorize Now in the right panel.
3. In the Software Authorization Wizard, click the second option ("I have already authorized the software and need to authorize additional extensions"), and click Next.
4. Click the appropriate authorization method. (The first option, using the internet, is preferable.)
5. On the next two panels, enter the authorization information requested.
6. On the Authorize Software Extensions panel, click the Spatial Analyst text box and enter the authorization code from the back of the book. Then click Next.
7. Select any additional extensions you want to evaluate, and click Next.
8. When authorization is complete, click Finish to close the wizard.

As this is the final chapter in the analysis section, it is assumed you have already completed chapters 18 and 19 and have gained a general familiarity with ModelBuilder and geoprocessing tools. As such, the exercises in this chapter are written in a streamlined manner, designed to challenge your recall.

Data

The exercises in this chapter use the following datasets, stored in the New_Clinic geodatabase:

- Blockgroups–Selected US Census block group polygons for Santa Rosa, California. Other layers were clipped to the extent of this layer. *(Sources: Tele Atlas, US Census, Esri)*
- CityBusStops–Points representing bus stops in Santa Rosa, California. *(Source: County of Sonoma GIS Central)*
- ClientList–Point locations of current and past clients of a local clinic serving low-income families with young children. *(Source: County of Sonoma GIS Central)*
- streets–Street lines symbolized by type. *(Source: County of Sonoma GIS Central)*

Exercise 20a
Create raster surfaces

In this exercise, you will create three raster surfaces, each representing a suitability criterion for an analysis study. In exercise 20b, you will combine the raster surfaces to create a final suitability surface.

For this exercise, you are a GIS consultant for the State of California. A state-funded medical clinic serving low-income families with young children in the town of Santa Rosa, in the North Bay Area of California, needs to relocate to a larger location capable of serving more clients. You have been asked to create a suitability map based on location and accessibility by existing and potential clients. Once your study is complete, a real estate search will be conducted in the high-suitability areas with proper zoning.

The new location must meet the following criteria: it is located near a high concentration of existing clients, within easy walking distance of a bus stop, and where there is a high percentage of potential clients (specifically, families with children under age five).

1. In ArcMap, open the **ex20a** map document from your **C:\EsriPress\GTKArcGIS\ Chapter20** folder.

The map is centered on the town of Santa Rosa. The map includes the point locations of past and current local clients of the existing clinic, bus stops, streets, and census block groups for Santa Rosa.

2. **Turn on the ArcGIS Spatial Analyst extension (on the menu bar, Customize > Extensions). When you are done, close the Extensions dialog box.**

3. **Open the geoprocessing environment settings (on the menu bar, click Geoprocessing > Environments). In the Raster Analysis section, change Cell Size to As Specified Below and enter the value of 100.**

4. **Set the mask to Study Area Block Groups and click OK.**

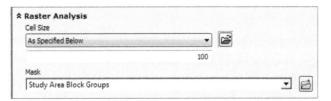

These environment settings apply to your current ArcMap session and all the tools in your current model. (However, the environment can be overridden for specific tools if needed—you will see how this works later in the exercise.) A mask is a means of identifying the areas that will be included in the analysis.

TIP

Some geoprocessing tools will give you different outputs depending on your environment settings. Sometimes the best way to determine the correct environment settings is through trial and error. If the geoprocessing output isn't what you expected, you can modify the environment settings. The raster cell size of 100 is your new default, but most raster tools allow you to choose a different size.

5. Open the Catalog window embedded in ArcMap and create a new, empty toolbox in your ..\Chapter20\MyData folder. (Right-click MyData and click New > Toolbox.) Maintain the default name or choose a unique name.

6. Open ModelBuilder and ArcToolbox.

7. In ArcToolbox, browse to Spatial Analyst Tools > Density toolset. Click the Point Density tool and drag it to the ModelBuilder window.

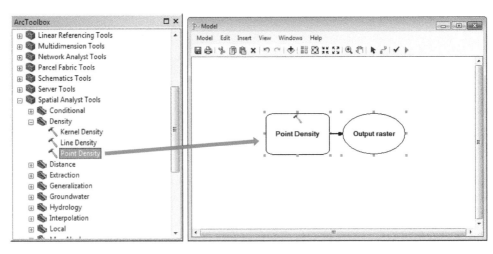

Next, you will create a raster showing existing client concentrations. The areas where the magnitude of points per neighborhood is highest will be displayed as hot spots–the areas that best fulfill the criteria of a high concentration of existing clients. A neighborhood is a user-defined area around output raster cells whose calculated values are dependent on the values of the input features.

8. In the ModelBuilder window, double-click the empty Point Density tool.

The tool dialog box appears.

9. On the Point Density dialog box, for Input point features, click the arrow and choose Clients. Keep the Population field set to NONE. For Output raster, browse to your ..\Chapter20\MyData folder and name the raster **density.** Set the output cell size to 100. In the Neighborhood Settings frame, change the neighborhood radius to **2000 map units,** and click OK.

You can also set the Point Density tool to calculate hot spots based on population attribute values, if there are any.

The Point Density tool is ready to run in ModelBuilder. Next, you will run it to see the results of the client density calculation.

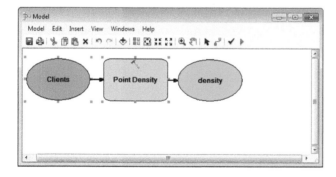

10. Run the model (on the model menu bar, click Model > Run). When the process is complete, right-click the density element in the model and click Add To Display. Keep your model open for the extent of this exercise.

11. In the table of contents, drag the density raster layer above the Study Area Block Groups layer. Turn off the bus stops layer. Turn the Clients layer on and off to get a better understanding of the analysis results.

The Point Density tool created a raster surface that shows where points are clustered. The darker areas have a higher concentration of current clients, and the lighter areas have a lower concentration. Each client location was counted once.

12. Turn the bus stops layer back on and return to the ModelBuilder window. From ArcToolbox, drag the Euclidean Distance tool (Spatial Analyst Tools > Distance toolset) to your model below the Point Density tool.

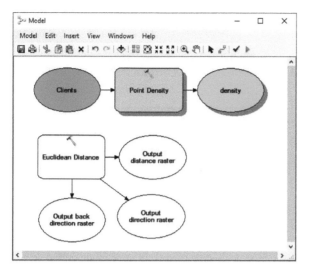

You want to calculate the distance from bus stops. You aren't interested in areas that are farther than a half-mile walk from a bus stop (and the most desirable areas will be much closer than that), so you will choose a maximum output distance of a half mile (2,640 feet).

13. In ModelBuilder, double-click the empty Euclidean Distance tool. On the Euclidean Distance dialog box, for the input, choose City Bus Stops. For the output, browse to your ..\Chapter20\MyData folder and name the raster **distance**. For Maximum distance, enter **2640**. (This distance is calculated in map units, which is feet.) Make sure Output cell size is 100, and leave Output direction raster (optional) and Output back direction raster (optional) blank.

1
2
3
4
5
6
7
8
9
10
11
12
13
14
15
16
17
18
19
20

TIP

Raster file names cannot exceed 13 characters and cannot include special characters.

ON YOUR OWN

Show the Help panel for the Output direction raster parameter.

Can you think of a scenario where it would be important to calculate the direction from the source layer?

14. Click OK to apply the parameters of the Euclidean Distance tool. In ModelBuilder, the Euclidean Distance tool is ready to run. (The optional outputs will remain empty.)

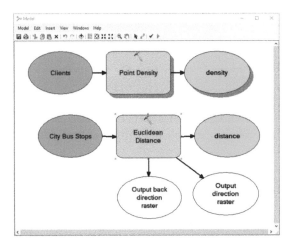

TIP

In ModelBuilder, click the Full Extent button to see the model in its entirety.

Trial and error has shown that the default processing extent of the Euclidean Distance tool results in an undesirable output with some edges cut off at the input points. Because you want to see a complete buffer around all bus stop points, even those on the outer edge of the input layer, you will change an environment setting for this tool only. (Again, trial and error has shown that changing the environment for the entire model would cause subsequent tools–ones you haven't added yet–to fail.) Changing the setting for only one tool will provide more desirable results.

15. **In the model, right-click Euclidean Distance and click Properties. Select the Processing Extent check box, and click Values. On the Environment Settings dialog box, expand Processing Extent. Change the extent to Same as layer Study Area Block Groups. Click OK, and OK again to close both windows.**

16. **Save your model. On the model menu bar, click Model > Save As. Save the model to the toolbox you created (Toolbox.tbx), with the default or a custom name, in your ..\Chapter20\MyData folder.**

TIP

Models are tools, so they must reside in toolboxes.

17. Run the model and add the distance output to the map display.

18. Turn off the Clients layer to get a better understanding of the results.

Areas closest to bus stops are yellow; the farthest away (within the half-mile limit) are purple.

The final criterion is a high percentage of children under age five. You have demographic data for census block groups. To balance total population differences in block groups, a percentage calculation is preferred over an absolute count. If you examine the attribute table of the block groups, you will see fields called AGE_UNDER5 and POP2010, but there is no percentage field. You will create one. This requires two separate processes: first, you will add the field, and then you will calculate its desired values.

19. From ArcToolbox, drag the Add Field and Calculate Field tools (Data Management Tools > Fields toolset) to the ModelBuilder window. Drop them below your existing processes.

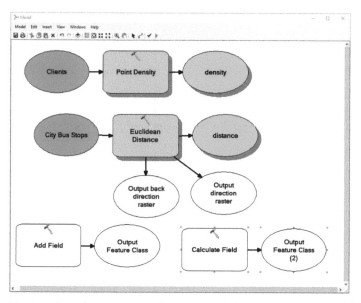

Notice when there are two elements with the same name, ModelBuilder appends a number to the element (2, 3, and so on) to differentiate between them.

TIP

You can resize or rearrange model elements to fit the model window. Click an element to select it. Drag it where you like or click one of the blue selection handles to resize it. Your model elements may differ in color and position from the exercise graphics somewhat. Just make sure the tool parameters (inputs and outputs) match.

20. Open the Add Field tool, set the following parameters, and click OK:

- Input Table: Study Area Block Groups
- Field Name: **PERC_UNDER5**
- Field Type: FLOAT
- Select the Field Is Nullable check box

21. Open the Calculate Field tool. For Input Table, click the arrow and click the intermediate layer Study Area Block Groups (2), preceded by a blue recycle icon ♻ indicating intermediate data.

Intermediate data means the layers that reside in the model, rather than those that reside in the map document. In other words, the new field can't be calculated until it has

been added, so the Calculate tool now works on the new Study Area Block Groups layer that is created as an output of the Add Field operation.

22. For Field Name, click the arrow, scroll to the bottom of the list, and click PERC_UNDER5. (If you don't see this field name, you chose the wrong input.)

23. Click the Calculator button next to the Expression parameter. In the Field Calculator, build the following expression, and click OK:

[AGE_UNDER5] / [Pop2010]

24. Click OK to close the Calculate Field tool.

The field tools are connected and ready to run as a single operation in ArcMap. You will add a conversion process to the model before you run it. You must turn the block group layer into a raster so you can combine it with the other two suitability rasters.

25. Drag the Polygon to Raster tool (Conversion Tools > To Raster toolset) to the model and drop it to the right of the Calculate Field process. Open the tool dialog box, set the following parameters, and click OK:

- Input features: Study Area Block Groups (3)
- Value field: PERC_UNDER5
- Output Raster Dataset: ..\Chapter20\MyData**children**
- Priority field: NONE
- Cellsize: **100**

26. Run the model to execute the conversion process. Add the output (children) to the ArcMap display.

27. Change the symbology of the output layer using a stretched color ramp, with darker colors representing higher percentages.

28. Turn off the City Bus Stops layer and turn on Clients.

Notice that high concentrations of clients are typically found where there are also high concentrations of children under the age of five.

ON YOUR OWN

Turn the three surfaces on and off. Can you visually determine areas that will satisfy all three criteria?

29. Save your model. You have saved all the suitability surfaces to your MyData folder, so there is no need to save your map. If you are continuing to exercise 20b and you were successful creating the three suitability surfaces, you may continue using this map document. Otherwise, you can use the map document provided for you.

You have created three suitability surfaces to site a new location for the Santa Rosa medical clinic. In exercise 20b, you will reclassify and combine the surfaces for a final suitability layer.

Exercise 20b
Combine raster surfaces

Before you combine raster surfaces, it's important to reclassify the cell values. Reclassification allows you to logically compare the inputs more easily. It's also helpful to standardize the extent of the inputs, although it's not always necessary. Raster calculations can be as simple as adding cell values together or they can be complex mathematical equations.

In this exercise, you will finish the analysis project you began in exercise 20a. You have already created three suitability surfaces for the clinic placement study. Next, you will complete the steps to create the final output, in which the three surfaces are combined.

1. In ArcMap, open the ex20b map document from your ..\Chapter20\Data folder. Or if you just completed exercise 20a, continue using the same map document and zoom to the full extent of the map layers.

Now you have the Santa Rosa map and three suitability surfaces.

2. Set the geoprocessing environment (Geoprocessing > Environments) for Raster Analysis: set the cell size to 100, and set the mask to Study Area Block Groups, if not already specified. Click OK.

3. Turn the suitability surfaces on and off. Notice the extent of the layers.

ON YOUR OWN

Turn on the Effects toolbar and experiment with the Swipe tool ⊞ as another way to view multiple layers.

The distance raster covers a smaller area than the density and children rasters. That's because you entered a maximum calculation distance of a half mile from bus stops. Areas beyond a half mile are unsuitable locations for the new clinic.

Before you combine the three surfaces, ensure that they cover the same area. In exercise 18b, you used the Clip tool to cut out features. With rasters, you can use a similar operation called extract.

4. If the model from exercise 20a is not already open, expand the Catalog window in ArcMap and expand ..\Home\MyData\Toolbox.tbx. Right-click the model you created in exercise 20a and click Edit to open it in ModelBuilder. (If you didn't create a model, or if you aren't sure yours is correct, edit the model in ..\Chapter20\Results\Toolbox.tbx.)

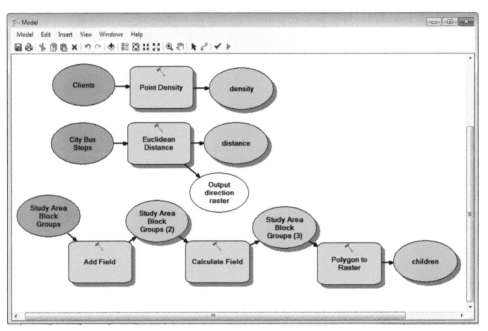

You will use the Extract by Mask tool twice—once to reduce the extent of the density raster and a second time to reduce the extent of the children raster—you want both to match the extent of the distance raster. Then you will use the Reclassify tool on all three surfaces to give them a common scale, assigning the highest value to the most desirable characteristics. Finally, you will use the Map Algebra tool to add the three surfaces together. The output of this operation will be your final suitability map.

5. Drag the Extract by Mask tool from ArcToolbox (Spatial Analyst Tools > Extraction toolset) to the model and open the tool dialog box. Set the following parameters, and click OK:

 - Input raster: density (the intermediate dataset preceded by the blue icon)
 - Input raster or feature mask data: distance (intermediate layer)
 - Output raster: ..\Chapter20\MyData\ext_density

6. Again, drag the Extract by Mask tool to the model, set the following parameters, and click OK:

 - Input raster: children (intermediate)
 - Input raster or feature mask data: distance (intermediate layer)
 - Output raster: \Chapter20\MyData\ext_children

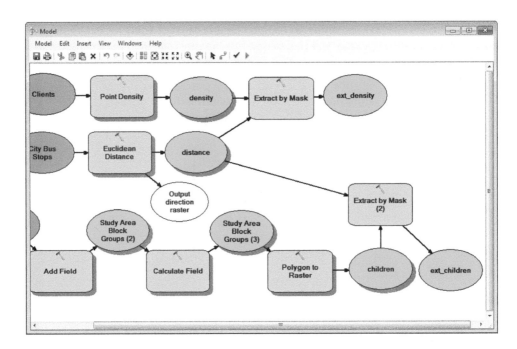

TIP

Your model will look somewhat different from the graphic, depending on how you have arranged it.

7. **Run the model, and add the two new outputs to the map.**

 Your three suitability surfaces are now called distance, ext_density, and ext_children.

8. **Turn off the layers named children, density, and Study Area Block Groups. Zoom in slightly to see the new outputs.**

Now you must reclassify the datasets so they can all be used together. Next, you will assign a common scale to each.

9. From ArcToolbox, drag the Reclassify tool (Spatial Analyst Tools > Reclass toolset) to the model three times.

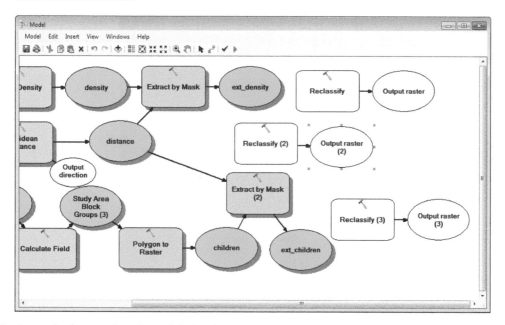

10. Open the first Reclassify tool dialog box. For Input raster, click the ext_density intermediate dataset.

11. **Click the Classify button. Maintain the default classification method (Natural Breaks), but change the number of classes to 7 and click OK.**

The output will have seven (7) classes, with the highest number (7) assigned to the areas that have the highest density of existing clients.

12. **For Output raster, browse to your ..\Chapter20\MyData folder and name the output recl_density. Click OK to close the tool dialog box.**

Next, you will reclassify the distance raster. The process is similar, but this time you flip the output values so the highest number (7) is assigned to the lowest distance (because a shorter distance from bus stops is more desirable).

13. Open the Reclassify (2) tool dialog box. Set the following parameters, and click OK:

- Input raster: distance (intermediate layer)
- Classify: Natural Breaks, **7** classes
- **Reverse the values (click Reverse New Values)**
- Output raster: \Chapter20\MyData**recl_distance**

14. Open the Reclassify (3) tool dialog box. Set the following parameters, and click OK:

- Input raster: ext_children (intermediate layer)
- Classify: Natural Breaks, **7** classes
- Output raster: \Chapter20\MyData**recl_children**

You will use the Auto Layout tool to automatically organize your elements.

15. Click the Auto Layout tool 📊.

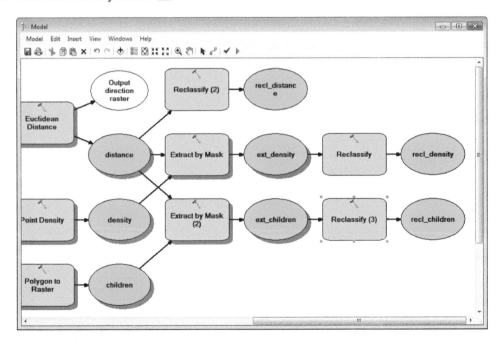

16. Run the model and add the three new output layers to the map.

Now you will add together the values of the three outputs to get a final suitability layer that indicates areas in which all three criteria are strongly represented.

17. From ArcToolbox, drag the Raster Calculator tool (Spatial Analyst Tools > Map Algebra toolset) to the model. Open the tool dialog box and use the list and operator buttons to add the values of recl_density, recl_distance, and recl_children (be sure to choose the intermediate data), as shown in the graphic. The tool automatically creates a map algebra expression when you select the variables and operators. Save the output as ..\Chapter20\MyData\final, and click OK.

The model is complete.

18. Click the Auto Layout button again if you like. To see the model in its entirety, expand the window and click the Full Extent button.

TIP

For a weighted overlay, you can use a multiplier to apply a weight, or importance value, to the inputs. Or, you can use a tool from the Overlay toolset under Spatial Analyst Tools, such as Weighted Overlay or Weighted Sum. For more information, consult ArcGIS Desktop Help.

19. Save your model.

20. Run the final process and add the output to the map. Resymbolize the suitability surface using a suitable color ramp.

21. Change the color for the value 21 to something that will stand out, such as purple.

22. Turn off all layers except Streets and final.

The most suitable areas for a new clinic have the highest values in this output. The highest is 21, because three (3) surfaces with seven (7) classes each were added together.

23. You have successfully completed a raster analysis project to determine site suitability for a new clinic that can serve more families. If you want to save your map, rename your map document and save to your MyData folder for chapter 20.

ON YOUR OWN

To run the entire analysis with a single click:

- Close the model and open 20a.mxd (the original map document without any processed layers).
- Locate and edit the model you just created. Click Model > Delete Intermediate Data. Then validate the model to clear all processes (Model > Validate Entire Model).
- Run the model.
- Give yourself a pat on the back for creating and executing a sophisticated raster analysis.

24. Exit ArcMap. Save the map if you like.

1
2
3
4
5
6
7
8
9
10
11
12
13
14
15
16
17
18
19
20

Summary

If you want to combine datasets to create a new, meaningful output, you may want to take advantage of the powerful analysis tools offered by ArcGIS Spatial Analyst. Although you can perform a lot of analysis on vector data, combining multiple layers into a new output layer requires you to convert your data to the raster data model. Now you have some idea of how rasters work and how to create common analysis surfaces such as point density and Euclidean distance. Most importantly, before you combine rasters, you must check the datasets to ensure they have the same cell size, extent, and classification scheme.

Terms

cells
discrete data
continuous data
map algebra
Esri grids
NoData
environment settings
mask
neighborhood
intermediate data
reclassification

Data source credits

Chapter 1
No data.

Chapter 2
No data.

Chapter 3
EsriPress\GTKArcGIS\Chapter03\Data\World.gdb\
Background from Data and Maps for ArcGIS 2010,
courtesy of Esri.
EsriPress\GTKArcGIS\Chapter03\Data\World.gdb\Cities
from Data and Maps for ArcGIS 2010, courtesy of
ArcWorld.
EsriPress\GTKArcGIS\Chapter03\Data\World.gdb\
Countries from The World Bank: Air Pollution in World
Cities (PM10 Concentrations): David R. Wheeler, Uwe
Deichmann, Kiran D. Pandey, Kirk E. Hamilton, Bart
Ostro, and Katie Bolt; from Data and Maps for ArcGIS
2010, courtesy of ArcWorld, US Census International
Division, *CIA World Factbook*.
EsriPress\GTKArcGIS\Chapter03\Data\World.gdb\Latlong
from Data and Maps for ArcGIS 2010, courtesy of Esri.

Chapter 4
EsriPress\GTKArcGIS\Chapter04\Data\World.gdb\
Background from Data and Maps for ArcGIS 2010,
courtesy of Esri.
EsriPress\GTKArcGIS\Chapter04\Data\World.gdb\Cities
from Data and Maps for ArcGIS 2010, courtesy of
ArcWorld.
EsriPress\GTKArcGIS\Chapter04\Data\World.gdb\
Countries from The World Bank: Air Pollution in World
Cities (PM10 Concentrations): David R. Wheeler, Uwe
Deichmann, Kiran D. Pandey, Kirk E. Hamilton, Bart
Ostro, and Katie Bolt; from Data and Maps for ArcGIS
2010, courtesy of ArcWorld, US Census International
Division, *CIA World Factbook*.
EsriPress\GTKArcGIS\Chapter04\Data\World.gdb\Latlong
from Data and Maps for ArcGIS 2010, courtesy of Esri.
EsriPress\GTKArcGIS\Chapter04\DataUS_Cities_PM.shp
from The World Bank: Air Pollution in World Cities
(PM10 Concentrations): David R. Wheeler, Uwe
Deichmann, Kiran D. Pandey, Kirk E. Hamilton, Bart

Ostro, and Katie Bolt; from Data and Maps for ArcGIS
2010, courtesy of ArcWorld.

Chapter 5
EsriPress\GTKArcGIS\Chapter05\Data\Chicago.gdb\bike_
racks courtesy of City of Chicago.
EsriPress\GTKArcGIS\Chapter05\Data\Chicago.gdb\bike_
routes courtesy of City of Chicago.

Chapter 6
EsriPress\GTKArcGIS\Chapter06\Data\Canada_Mexico.
shp from Data and Maps for ArcGIS 2010, courtesy of
ArcWorld Supplement.
EsriPress\GTKArcGIS\Chapter06\Data\Cities.shp from
Data and Maps for ArcGIS 2010, courtesy of US Census.
EsriPress\GTKArcGIS\Chapter06\Data\ColoradoCounties.
shp from Data and Maps for ArcGIS 2010, courtesy of
Esri, derived from Tele Atlas, US Census, Esri (Pop2010).
EsriPress\GTKArcGIS\Chapter06\Data\Lakes.shp from
Data and Maps for ArcGIS 2010, courtesy of ArcWorld.
EsriPress\GTKArcGIS\Chapter06\Data\Latlong.shp from
Data and Maps for ArcGIS 2010, courtesy of Esri.
EsriPress\GTKArcGIS\Chapter06\Data\ritidian_point.jpg
courtesy of US Geological Survey.
EsriPress\GTKArcGIS\Chapter06\Data\USstates.shp from
Data and Maps for ArcGIS 2010, courtesy of ArcUSA, US
Census, Esri (Pop2010).

Chapter 7
EsriPress\GTKArcGIS\Chapter07\Data\Florida.gdb\
Bathymetry from the Florida Geographic Data Library,
courtesy of Florida Fish and Wildlife Conservation
Commission's Fish and Wildlife Research Institute
(FWC-FWRI) and the National Oceanic and Atmospheric
Administration (NOAA).
EsriPress\GTKArcGIS\Chapter07\Data\Florida.gdb\Cities
from National Atlas of the United States, courtesy of US
Geological Survey.
EsriPress\GTKArcGIS\Chapter07\Data\Florida.gdb\Coral
from the Florida Geographic Data Library, courtesy of
Florida Fish and Wildlife Conservation Commission's
Fish and Wildlife Research Institute (FWC-FWRI).

EsriPress\GTKArcGIS\Chapter07\Data\Florida.gdb\ Counties from the Florida Geographic Data Library, courtesy of US Census Bureau.

EsriPress\GTKArcGIS\Chapter07\Data\Florida.gdb\Hwy1 from the Florida Geographic Data Library, courtesy of Florida Department of Transportation.

EsriPress\GTKArcGIS\Chapter07\Data\Florida.gdb\Lakes from the Florida Geographic Data Library, courtesy of Florida Department of Environmental Protection.

EsriPress\GTKArcGIS\Chapter07\Data\Florida.gdb\Rivers from the Florida Geographic Data Library, courtesy of Florida Department of Environmental Protection.

EsriPress\GTKArcGIS\Chapter07\Data\Florida.gdb\ Shipwrecks from The Office of Coast Survey's Automated Wreck and Obstruction Information System, courtesy of NOAA.

EsriPress\GTKArcGIS\Chapter07\Data\Florida.gdb\ Southeast from Data and Maps for ArcGIS 2010, courtesy of ArcUSA, US Census, Esri (Pop2010 field).

EsriPress\GTKArcGIS\Chapter07\Data\Florida.gdb\ StParks from the Florida Geographic Data Library, courtesy of Florida Department of Environmental Protection.

EsriPress\GTKArcGIS\Chapter07\Data\florida_coast.tif Made with Natural Earth.

EsriPress\GTKArcGIS\Chapter07\Data\shdrl, from the Florida Geographic Data Library, courtesy of the University of Florida GeoPlan Center.

EsriPress\GTKArcGIS\Chapter07\Data\usgsdem, from the Florida Geographic Data Library, courtesy of US Geological Survey. Raster images based on the National Elevation Dataset courtesy of the US Geological Survey.

Chapter 8

EsriPress\GTKArcGIS\Chapter08\Data\TexasCounty. gdb\Anno_4_9 from Data and Maps for ArcGIS 2010, courtesy of Esri, derived from Tele Atlas, US Census, Esri (Pop2010 field). US Department of Justice, Federal Bureau of Investigation. Uniform Crime Reporting Program Data [United States]: County-level Detailed Arrest and Offense Data, 2002 [Computer file]. ICPSR ed. Ann Arbor, MI: Inter-university Consortium for Political and Social Research [producer and distributor], 2004.

EsriPress\GTKArcGIS\Chapter08\Data\TexasCounty. gdb\counties from Data and Maps for ArcGIS 2010, courtesy of Esri, derived from Tele Atlas, US Census, Esri (Pop2010 field). US Department of Justice, Federal Bureau of Investigation. Uniform Crime Reporting Program Data [United States]: County-level Detailed Arrest and Offense Data, 2002 [Computer file]. ICPSR ed. Ann Arbor, MI: Inter-university Consortium for Political and Social Research [producer and distributor], 2004.

EsriPress\GTKArcGIS\Chapter08\Data\TexasCounty.gdb\ countiesAnno from Data and Maps for ArcGIS 2010, courtesy of Esri, derived from Tele Atlas, US Census, Esri (Pop2010 field). US Department of Justice, Federal Bureau of Investigation. Uniform Crime Reporting Program Data [United States]: County-level Detailed Arrest and Offense Data, 2002 [Computer file]. ICPSR ed. Ann Arbor, MI: Inter-university Consortium for Political and Social Research [producer and distributor], 2004.

EsriPress\GTKArcGIS\Chapter08\Data\TexasCounty. gdb\HarrisAdjacentCounties from Data and Maps for ArcGIS 2010, courtesy of Esri, derived from Tele Atlas, US Census, Esri (Pop2010 field). US Department of Justice, Federal Bureau of Investigation. Uniform Crime Reporting Program Data [United States]: County-level Detailed Arrest and Offense Data, 2002 [Computer file]. ICPSR ed. Ann Arbor, MI: Inter-university Consortium for Political and Social Research [producer and distributor], 2004.

Chapter 9

Esri Press\GTKArcGIS\Chapter09\Data\Florida.gdb\ Bathymetry from the Florida Geographic Data Library, courtesy of Florida Fish and Wildlife Conservation Commission, Fish and Wildlife Research Institute (FWC-FWRI) and NOAA.

Esri Press\GTKArcGIS\Chapter09\Data\Florida.gdb\Cities from National Atlas of the United States, courtesy of US Geological Survey.

Esri Press\GTKArcGIS\Chapter09\Data\Florida.gdb\Coral from the Florida Geographic Data Library, courtesy of Florida Fish and Wildlife Conservation Commission, Fish and Wildlife Research Institute (FWC-FWRI).

Esri Press\GTKArcGIS\Chapter09\Data\Florida.gdb\ Counties from the Florida Geographic Data Library, courtesy of US Census Bureau.

EsriPress\GTKArcGIS\Chapter09\Data\Florida.gdb\Hwy1 from the Florida Geographic Data Library, courtesy of Florida Department of Transportation.

EsriPress\GTKArcGIS\Chapter09\Data\Florida.gdb\Lakes from the Florida Geographic Data Library, courtesy of Florida Department of Environmental Protection.

EsriPress\GTKArcGIS\Chapter09\Data\Florida.gdb\Rivers from the Florida Geographic Data Library, courtesy of Florida Department of Environmental Protection.

EsriPress\GTKArcGIS\Chapter09\Data\Florida.gdb\ Shipwrecks from the Office of Coast Survey's Automated Wreck and Obstruction Information System (AWOIS) courtesy of NOAA.

EsriPress\GTKArcGIS\Chapter09\Data\Florida.gdb\ Southeast from Data and Maps for ArcGIS 2010, courtesy of ArcUSA, US Census, Esri (Pop2010 field).

EsriPress\GTKArcGIS\Chapter09\Data\Florida.gdb\ StParks from the Florida Geographic Data Library, courtesy of Florida Department of Environmental Protection.

EsriPress\GTKArcGIS\Chapter09\Data\florida_coast.tif courtesy of Natural Earth.

Chapter 10

EsriPress\GTKArcGIS\Chapter10\Data\agency-logo.gif courtesy of Esri.

EsriPress\GTKArcGIS\Chapter10\Data\australia.shp from Data and Maps for ArcGIS 2010, courtesy of ArcWorld Supplement.

EsriPress\GTKArcGIS\Chapter10\Data\bioregions.shp courtesy of State of Queensland (Department of Science, Information Technology, Innovation and the Arts) 2010, 3.0 Australia (CC BY 3.0).

EsriPress\GTKArcGIS\Chapter10\Data\biosubregions. shp courtesy of State of Queensland (Department of Science, Information Technology, Innovation and the Arts) 2010, 3.0 Australia (CC BY 3.0).

EsriPress\GTKArcGIS\Chapter10\Data\country_ background.shp from Data and Maps 2010, courtesy of ArcWorld Supplement.

EsriPress\GTKArcGIS\Chapter10\Data\Metadata_ biosubregions.xml courtesy of State of Queensland (Department of Science, Information Technology, Innovation and the Arts) 2010, 3.0 Australia (CC BY 3.0).

EsriPress\GTKArcGIS\Chapter10\Data\OB_LR.tif courtesy of Natural Earth.

Chapter 11

EsriPress\GTKArcGIS\Chapter11\Ex11aData\health_areas. shp from NYC Open Data Portal courtesy of New York City Department of City Planning.

EsriPress\GTKArcGIS\Chapter11\Ex11aData\MN_ neighborhoods.shp from NYC Open Data Portal courtesy of New York City Department of City Planning.

EsriPress\GTKArcGIS\Chapter11\Ex11aData\MN_streets. shp from NYC Open Data Portal courtesy of New York City Department of City Planning.

EsriPress\GTKArcGIS\Chapter11\Ex11aData\subwayroutes. shp based on tabular data from MTA, courtesy of Steven Romalewski, Director, CUNY Mapping Service at the Center for Urban Research.

EsriPress\GTKArcGIS\Chapter11\Ex11aData\subwaystops. dbf based on tabular data from MTA, courtesy of Steven Romalewski, Director, CUNY Mapping Service at the Center for Urban Research.

EsriPress\GTKArcGIS\Chapter11\Ex11bData\CityData.gdb\ health_areas from NYC Open Data Portal courtesy of New York City Department of City Planning.

EsriPress\GTKArcGIS\Chapter11\Ex11bData\CityData. gdb\MN_neighborhoods from NYC Open Data Portal courtesy of New York City Department of City Planning.

EsriPress\GTKArcGIS\Chapter11\Ex11bData\CityData.gdb\ MN_streets from NYC Open Data Portal courtesy of New York City Department of City Planning.

EsriPress\GTKArcGIS\Chapter11\Ex11bData\CityData.gdb\ subwayroutes courtesy of Steven Romalewski, Director, CUNY Mapping Service at the Center for Urban Research.

EsriPress\GTKArcGIS\Chapter11\Ex11bData\CityData.gdb\ subwaystops based on tabular data from MTA, courtesy of Steven Romalewski, Director, CUNY Mapping Service at the Center for Urban Research.

Chapter 12

EsriPress\GTKArcGIS\Chapter12\Data\CityData.gdb\ health_areas from NYC Open Data Portal courtesy of New York City Department of City Planning.

EsriPress\GTKArcGIS\Chapter12\Data\CityData.gdb\MN_ neighborhoods from NYC Open Data Portal courtesy of New York City Department of City Planning.

EsriPress\GTKArcGIS\Chapter12\Data\CityData.gdb\MN_ streets from NYC Open Data Portal courtesy of New York City Department of City Planning.

EsriPress\GTKArcGIS\Chapter12\Data\CityData.gdb\ subwayroutes based on tabular data from MTA, courtesy of Steven Romalewski, Director, CUNY Mapping Service at the Center for Urban Research.

EsriPress\GTKArcGIS\Chapter12\Data\CityData.gdb\ subwaystops based on tabular data from MTA, courtesy of Steven Romalewski, Director, CUNY Mapping Service at the Center for Urban Research.

Chapter 13

EsriPress\GTKArcGIS\Chapter13\Data\CityDataA.gdb\ health_areas from NYC Open Data Portal courtesy of New York City Department of City Planning.

EsriPress\GTKArcGIS\Chapter13\Data\CityDataA.gdb\ MN_neighborhoods from NYC Open Data Portal courtesy of New York City Department of City Planning.

EsriPress\GTKArcGIS\Chapter13\Data\CityDataA.gdb\ MN_streets from NYC Open Data Portal courtesy of New York City Department of City Planning.

EsriPress\GTKArcGIS\Chapter13\Data\CityDataA.gdb\ subwayroutes based on tabular data from MTA, courtesy of Steven Romalewski, CUNY Mapping Service at the Center for Urban Research.

EsriPress\GTKArcGIS\Chapter13\Data\CityDataA.gdb\ subwaystops based on tabular data from MTA, courtesy of Steven Romalewski, CUNY Mapping Service at the Center for Urban Research.

EsriPress\GTKArcGIS\Chapter13\Data\CityDataC.gdb\ health_areas from NYC Open Data Portal courtesy of New York City Department of City Planning.

Chapter 14

EsriPress\GTKArcGIS\Chapter14\Ex14aData\High School Catchment.shp courtesy of The School District of Philadelphia, Office of Capital Programs.

EsriPress\GTKArcGIS\Chapter14\Ex14aData\Philadelphia Streets.shp from GIS Manager Andy Mehos, Streets Department/Office of Innovation and Technology, courtesy of the City of Philadelphia.

EsriPress\GTKArcGIS\Chapter14\Ex14aData\Schools.csv Philadelphia City Planning Commission in cooperation with The School District of Philadelphia, the Archdiocese of Philadelphia, and the Mayor's Office.

EsriPress\GTKArcGIS\Chapter14\Ex14bData\High School Catchment.shp courtesy of The School District of Philadelphia, Office of Capital Programs.

EsriPress\GTKArcGIS\Chapter14\Ex14bData\Philadelphia Streets.shp from GIS Manager Andy Mehos, Streets Department/Office of Innovation and Technology, courtesy of the City of Philadelphia.

EsriPress\GTKArcGIS\Chapter14\Ex14bData\Schools.csv courtesy of the Philadelphia City Planning Commission in cooperation with The School District of Philadelphia, the Archdiocese of Philadelphia, and the Mayor's Office.

EsriPress\GTKArcGIS\Chapter14\Ex14cData\High School Catchment.shp courtesy of The School District of Philadelphia, Office of Capital Programs.

EsriPress\GTKArcGIS\Chapter14\Ex14cData\Philadelphia Streets.shp from GIS Manager Andy Mehos, Streets Department/Office of Innovation and Technology, courtesy of the City of Philadelphia.

EsriPress\GTKArcGIS\Chapter14\Ex14cData\Schools.csv courtesy of the Philadelphia City Planning Commission in cooperation with The School District of Philadelphia, the Archdiocese of Philadelphia, and the Mayor's Office.

Chapter 15

EsriPress\GTKArcGIS\Chapter15\Data\cabbagetown.shp courtesy of Generation5, a Motricity Business © 2011 Motricity Canada Inc.

EsriPress\GTKArcGIS\Chapter15\Data\cabbagetown1.jpg photo by Michael Law.

EsriPress\GTKArcGIS\Chapter15\Data\cabbagetown2.jpg photo by Michael Law.

EsriPress\GTKArcGIS\Chapter15\Data\cabbagetown3.jpg photo by Michael Law.

EsriPress\GTKArcGIS\Chapter15\Data\cabbagetown4.jpg photo by Michael Law.

EsriPress\GTKArcGIS\Chapter15\Data\cabbagetown5.jpg photo by Michael Law.

EsriPress\GTKArcGIS\Chapter15\Data\cabbagetown6.jpg photo by Michael Law.

EsriPress\GTKArcGIS\Chapter15\Data\cabbagetown7.jpg photo by Michael Law.

EsriPress\GTKArcGIS\Chapter15\Data\cabbagetown8.jpg photo by Michael Law.

EsriPress\GTKArcGIS\Chapter15\Data\toronto_ background.shp courtesy of Generation5, a Motricity Business © 2011 Motricity Canada Inc.

EsriPress\GTKArcGIS\Chapter15\Data\toronto_streets.shp courtesy of Open Data, Geospatial Competency Centre, City of Toronto.

Chapter 16

EsriPress\GTKArcGIS\Chapter16\Data\BayArea.shp courtesy of County of Sonoma, Information Systems Department (ISD)/GIS.

EsriPress\GTKArcGIS\Chapter16\Data\Cities.shp from Data and Maps for ArcGIS 2010, courtesy of National Atlas of the United States.

EsriPress\GTKArcGIS\Chapter16\Data\Guerneville.jpg courtesy of County of Sonoma, Information Systems Department (ISD)/GIS.

EsriPress\GTKArcGIS\Chapter16\Data\GuernevilleParcels. shp courtesy of County of Sonoma, Information Systems Department (ISD)/GIS.

EsriPress\GTKArcGIS\Chapter16\Data\RussianRiver.shp courtesy of County of Sonoma, Information Systems Department (ISD)/GIS.

EsriPress\GTKArcGIS\Chapter16\Data\ RussianRiverFloodZone.shp courtesy of County of Sonoma, Information Systems Department (ISD)/GIS.

Chapter 17

EsriPress\GTKArcGIS\Chapter17\Data\Wisconsin.gdb\ FoodDeserts from Data and Maps for ArcGIS 2010, courtesy of ArcUSA, US Census, Esri (Pop2010 field); US Department of Agriculture, Economic Research Service.

EsriPress\GTKArcGIS\Chapter17\Data\Wisconsin.gdb\ USstates from Data and Maps for ArcGIS 2010, courtesy of Esri, derived from Tele Atlas, US Census, Esri (Pop2010 field).

EsriPress\GTKArcGIS\Chapter17\Data\Wisconsin.gdb\ WisconsinCities from Data and Maps for ArcGIS 2010, courtesy of US Census.

EsriPress\GTKArcGIS\Chapter17\Data\Wisconsin.gdb\ WisconsinCounties from Data and Maps for ArcGIS 2010, courtesy of Esri, derived from Tele Atlas, US Census, Esri (Pop2010 field).

EsriPress\GTKArcGIS\Chapter17\Data\Wisconsin.gdb\ WisconsinTracts from Data and Maps for ArcGIS 2010, courtesy of Tele Atlas, US Census, Esri (Pop2010 field).

Chapter 18

EsriPress\GTKArcGIS\Chapter18\Data\OregonForest.gdb\ curlook courtesy of Oregon Spatial Data Library and Oregon Department of Forestry.

EsriPress\GTKArcGIS\Chapter18\Data\OregonForest.gdb\ forestland courtesy of Oregon Spatial Data Library and Oregon Department of Forestry.

EsriPress\GTKArcGIS\Chapter18\Data\OregonForest.gdb\ fwz100k courtesy of Oregon Spatial Data Library and Oregon Department of Forestry.

EsriPress\GTKArcGIS\Chapter18\Data\OregonForest.gdb\ populated_jurisdictions courtesy of Oregon Spatial Data Library and Oregon Department of Forestry.

EsriPress\GTKArcGIS\Chapter18\Results\MyOregonForest. gdb\CountyWarningAreas courtesy of Oregon Spatial Data Library and Oregon Department of Forestry.

EsriPress\GTKArcGIS\Chapter18\Results\MyOregonForest. gdb\PortlandForest courtesy of Oregon Spatial Data Library and Oregon Department of Forestry.

EsriPress\GTKArcGIS\Chapter18\Results\MyOregonForest. gdb\PortlandLookouts courtesy of Oregon Spatial Data Library and Oregon Department of Forestry.

EsriPress\GTKArcGIS\Chapter18\Results\MyOregonForest. gdb\PortlandPopJurs courtesy of Oregon Spatial Data Library and Oregon Department of Forestry.

Chapter 19

EsriPress\GTKArcGIS\Chapter19\Data\OregonForest.gdb\ CountyWarningAreas courtesy of Oregon Spatial Data Library and Oregon Department of Forestry.

EsriPress\GTKArcGIS\Chapter19\Data\OregonForest.gdb\ PortlandForest courtesy of Oregon Spatial Data Library and Oregon Department of Forestry.

EsriPress\GTKArcGIS\Chapter19\Data\OregonForest.gdb\ PortlandLookouts courtesy of Oregon Spatial Data Library and Oregon Department of Forestry.

EsriPress\GTKArcGIS\Chapter19\Data\OregonForest. gdb\PortlandPopJurs courtesy of Oregon Spatial Data Library and Oregon Department of Forestry.

EsriPress\GTKArcGIS\Chapter19\Results\OregonForest. gdb\AreaAtRisk courtesy of Oregon Spatial Data Library and Oregon Department of Forestry.

EsriPress\GTKArcGIS\Chapter19\Results\OregonForest. gdb\CountyWarningAreas courtesy of Oregon Spatial Data Library and Oregon Department of Forestry.

EsriPress\GTKArcGIS\Chapter19\Results\OregonForest. gdb\Final courtesy of Oregon Spatial Data Library and Oregon Department of Forestry.

EsriPress\GTKArcGIS\Chapter19\Results\OregonForest. gdb\FinalDissolve courtesy of Oregon Spatial Data Library and Oregon Department of Forestry.

EsriPress\GTKArcGIS\Chapter19\Results\OregonForest. gdb\LookoutsBuffer courtesy of Oregon Spatial Data Library and Oregon Department of Forestry.

EsriPress\GTKArcGIS\Chapter19\Results\OregonForest. gdb\PopjursBuffer courtesy of Oregon Spatial Data Library and Oregon Department of Forestry.

EsriPress\GTKArcGIS\Chapter19\Results\OregonForest. gdb\PortlandForest courtesy of Oregon Spatial Data Library and Oregon Department of Forestry.

EsriPress\GTKArcGIS\Chapter19\Results\OregonForest. gdb\PortlandLookouts courtesy of Oregon Spatial Data Library and Oregon Department of Forestry.

EsriPress\GTKArcGIS\Chapter19\Results\OregonForest. gdb\PortlandPopjurs courtesy of Oregon Spatial Data Library and Oregon Department of Forestry.

Chapter 20

EsriPress\GTKArcGIS\Chapter20\Data\New_Clinic.gdb\ Blockgroups from Data and Maps for ArcGIS 2010, courtesy of Tele Atlas, US Census, Esri (Pop2010 field).

EsriPress\GTKArcGIS\Chapter20\Data\New_Clinic. gdb\CityBusStops courtesy of County of Sonoma, Department of Health Services.

EsriPress\GTKArcGIS\Chapter20\Data\New_Clinic.gdb\ ClientList courtesy of County of Sonoma, Department of Health Services.

EsriPress\GTKArcGIS\Chapter20\Data\New_Clinic.gdb\ streets courtesy of County of Sonoma, Department of Health Services.

Data license agreement

Downloadable data that accompanies this book is covered by a license agreement that stipulates the terms of use.

675

Glossary

This glossary contains definitions for the ArcGIS terms used throughout the book and listed at the end of each chapter. Most of these definitions can also be found in Esri's online GIS Dictionary, https://support.esri .com/en/other-resources/gis-dictionary.

address locator A dataset in ArcGIS that stores the address attributes, associated indexes, and rules that define the process for translating nonspatial descriptions of places, such as street addresses, into spatial data that can be displayed as features on a map. An address locator contains a snapshot of the reference data used for geocoding and includes the parameters for standardizing addresses, searching for match locations, and creating output.

annotation In ArcGIS, text or graphics that can be individually selected, positioned, and modified. Annotation may be manually entered or generated from labels. Annotation can be stored as features in a geodatabase or as map annotation in a data frame.

ArcCatalog The ArcCatalog application provides a Catalog window that is used to organize and manage various types of geographic information—such as geodatabases, raster files, map documents, and geoprocessing toolboxes—for ArcGIS Desktop.

ArcGIS Pro An application for creating and working with spatial data on your desktop. It provides tools to visualize, analyze, compile, and share your data, in both 2D and 3D environments.

ArcMap A central application used in ArcGIS, ArcMap is used to display and explore GIS datasets, assign symbology, and create map layouts for printing or publication.

attribute domain In a geodatabase, a mechanism for enforcing data integrity. An attribute domain defines what values are allowed in a field in a feature class or a nonspatial attribute table. If the features or nonspatial objects have been grouped into subtypes, different attribute domains can be assigned to each of the subtypes.

attribute join Appending the fields of one table to those of another through an attribute common to both tables. *See also* spatial join and relate.

attribute query A request for records of features in a table based on their attribute values.

attributes Nonspatial information about a geographic feature in a GIS, usually stored in a table and linked to the feature by a unique identifier. For example, attributes of a river might include its name, length, and sediment load at a gauging station. Attribute tables also store information about feature geometry, such as length and area.

basemaps Maps depicting background reference information, such as landforms, roads, landmarks, political boundaries, or satellite imagery, onto which other thematic information is placed.

buffer A zone around a map feature measured in units of distance or time. A buffer is useful for proximity analysis.

cache In ArcGIS Online, a temporary storage of map tiles in the desktop computer's RAM, which may result in improved display performance.

cells The smallest units of information in raster data, usually square in shape. In a map or GIS dataset, each cell represents a portion of the earth, such as a square meter or square mile, and usually has an attribute value associated with it, such as soil type or vegetation class.

clip A command that extracts features from one feature class that reside entirely within a boundary defined by features in another feature class.

cloud computing A scalable technology that furnishes technological capabilities—commonly maintained off premises—that are delivered on demand as applications or services via the internet.

continuous data Data such as elevation or temperature that varies without discrete steps. Because computers store data discretely, continuous data is usually represented by triangulated irregular networks (TINs), rasters, or contour lines, so that any location has either a specified value or one that can be derived.

coordinate systems Reference frameworks consisting of a set of points, lines, and/or surfaces, and a set of rules used to define the positions of points in space in either two or three dimensions. The Cartesian coordinate system and the geographic coordinate system used on the earth's surface are common examples of coordinate systems.

data dictionary A catalog or table containing information about the datasets stored in a database. In a GIS, a data dictionary might contain the full names of attributes, meanings of codes, scale of source data, accuracy of locations, and map projections used.

data frames Map elements that define a geographic extent, a page extent, a coordinate system, and other display properties for one or more layers in ArcMap. A dataset can be represented in one or more data frames. In data view, only one data frame is displayed at a time; in layout view, all of a map's data frames are displayed at the same time.

data view An all-purpose view in ArcMap and ArcReader for exploring, displaying, and querying geographic data. This view hides all map elements, such as titles, north arrow, and scale bars.

datum The reference specifications of a measurement system, usually a system of coordinate positions on a surface (a horizontal datum) or heights above or below a surface (a vertical datum).

datum transformation A transformation definition that is used to transform the coordinates of a point defined in one datum to coordinates in a different datum.

decimal degrees Values of latitude and longitude expressed in decimal format rather than in degrees-minutes-seconds.

defined interval Also known as class interval. A set of categories for classification that divide the range of all values so that each piece of data is contained within a nonoverlapping category.

degrees-minutes-seconds The unit of measure for describing latitude and longitude. A degree is 1/360th of a circle. A degree is further divided into 60 minutes, and a minute is divided into 60 seconds.

digitizing The process of converting the geographic features on an analog map into digital format using a digitizing tablet, or digitizer, which is connected to a computer. Features on a paper map are traced with a digitizer puck, a device similar to a mouse. The x,y coordinates of these features are automatically recorded and stored as spatial data. Digitizing also refers to creating features with sketch tools in ArcMap.

discrete data Data that represents phenomena with distinct boundaries. Property lines and land-use areas are examples of discrete data.

display scale The scale at which data is rendered on a computer screen or on a printed map.

dissolve A geoprocessing command that removes boundaries between adjacent polygons that have the same value for a specified attribute.

dot density map A quantitative, thematic map on which dots of the same size are randomly placed in proportion to a numeric attribute associated with an area. Dot density maps convey the intensity of an attribute.

edge A line between two points that forms a boundary.

edit session In ArcMap, the environment in which spatial and attribute editing take place. In an edit session, a user can modify feature locations, geometry, or attributes. Modifications are not saved unless the user explicitly chooses to save them.

edit sketch In ArcMap, a temporary sketch that is used to perform a variety of tasks. Creating an edit sketch is a standard way to edit feature geometry.

endpoints The points that define the length of a line segment.

environment settings Settings that can apply to all tools within the application, all tools within a model or script, or to a particular process within a model or script. Environment settings include current workspace, output spatial reference, output spatial grids, cell size, and tile size. They are generally set before running tools.

equal interval A data classification method that divides a set of attribute values into groups that contain an equal range of values.

Esri grids An Esri data format for storing raster data that defines geographic space as an array of equally sized square cells arranged in rows and columns. Each cell stores a numeric value that represents a geographic attribute (such as elevation) for that unit of space. When the grid is drawn as a map, cells are assigned colors according to their numeric values. Each grid cell is referenced by its x,y coordinate location.

false easting The linear value added to all x-coordinates of a map projection so that none of the values in the geographic region being mapped are negative.

false northing The linear value added to all y-coordinates of a map projection so that none of the values in the geographic region being mapped are negative.

feature class In ArcGIS, a collection of geographic features with the same geometry type (such as point, line, or polygon), the same attributes, and the same spatial reference. Feature classes can be stored in a geodatabase, shapefile, or coverage, among other data formats. Feature classes allow homogeneous features to be grouped into a single unit for data storage purposes. For example, highways, primary roads, and secondary roads can be grouped into a line feature class named "roads." In a geodatabase, feature classes can also store annotation and dimensions.

feature dataset A collection of feature classes stored together which share the same spatial reference; that is, they share a coordinate system, and their features fall within a common geographic area. Feature classes with different geometry types may be stored in a feature dataset.

geocoding A GIS operation for converting street addresses into spatial data that can be displayed as features on a map, usually by referencing address information from a street segment data layer.

geodatabase A database or file structure used primarily to store, query, and manipulate spatial data. Geodatabases store geometry, a spatial reference system, attributes, and behavioral rules for data. Various types of geographic datasets can be collected within a geodatabase, including feature classes, attribute tables, raster datasets, network datasets, topologies, and many others. Geodatabases can be stored in IBM

DB2, IBM Informix, Oracle, Microsoft Access, Microsoft SQL Server, and PostgreSQL relational database management systems or in a system of files, such as a file geodatabase.

geodatabase topology The arrangement that constrains how point, line, and polygon features share geometry. For example, street centerlines and census blocks share geometry, and adjacent soil polygons share geometry. Topology defines and enforces data integrity rules (e.g., there should be no gaps between polygons). It supports topological relationship queries and navigation (e.g., navigating feature adjacency or connectivity), works with sophisticated editing tools, and allows feature construction from unstructured geometry (e.g., constructing polygons from lines).

geographic coordinate system A reference system that uses latitude and longitude to define the locations of points on the surface of a sphere or spheroid. A geographic coordinate system definition includes a datum, prime meridian, and angular unit.

geoid A hypothetical surface representing the form the earth's oceans would take if there were no land and the water were free to respond to the earth's gravitational and centrifugal forces. The resulting geoid is irregular and varies from a perfect sphere by as much as 75 meters above and 100 meters below its surface.

geometrical interval A classification scheme in which the class breaks are based on class intervals that are in a geometric series. The geometric coefficient in this classifier can change once (to its inverse) to optimize the class ranges.

geoprocessing A GIS operation used to manipulate GIS data. A typical geoprocessing operation takes an input dataset, performs an operation on that dataset, and returns the result of the operation as an output dataset. Common geoprocessing operations include geographic feature overlay, feature selection and analysis, topology processing, raster processing, and data conversion. Geoprocessing helps define, manage, and analyze the information used to form decisions.

georeferencing Aligning geographic data to a known coordinate system so it can be viewed, queried, and analyzed with other geographic data. Georeferencing may involve shifting, rotating, scaling, skewing, and in some cases, warping, rubber sheeting, or orthorectifying the data.

GIS Acronym for geographic information system. An integrated collection of computer software and data used to view and manage information about geographic places, analyze spatial relationships, and model spatial processes. A GIS provides a framework for gathering and organizing spatial data and related information so that it can be displayed and analyzed.

GPS Acronym for Global Positioning System. A system of radio-emitting and -receiving satellites used for determining positions on the earth. The orbiting satellites transmit signals that allow a GPS receiver anywhere on earth to calculate its own location through trilateration. Developed and operated by the US Department of Defense, the system is used in navigation, mapping, surveying, and other applications in which precise positioning is necessary.

graduated-color map A map on which a range of colors indicates a progression of numeric values. For example, increases in population density might be represented by the increased saturation of a single color, or temperature differences might be shown by a sequence of colors from blue to red.

graduated-symbol map A map with symbols that change in size according to the value of the attribute they represent. For example, denser populations might be represented by larger dots or larger rivers by thicker lines.

graticule A network of longitude and latitude lines on a map or chart which relates points on a map to their true locations on the earth.

groups In ArcGIS Online, a way to collaborate with other ArcGIS users and to exchange content related to a specific project or common activity.

hillshade Shadows drawn on a map to simulate the effect of the sun's rays over the varied terrain of the land.

histogram A graph showing the distribution of values in a set of data. Individual values are displayed along a horizontal axis, and the frequency of their occurrence is displayed along a vertical axis.

hyperlink A reference (link) from one point in an electronic document to another document or another location in the same document (the target). Activating the link, usually by clicking it with the mouse, causes the browser to display the target of the link.

intermediate data Any data in a process that did not exist before the process existed and that will not be maintained after the process is executed.

intersect A geometric integration of spatial datasets that preserves features or portions of features that fall within areas common to all input datasets.

join table The table in a join operation, often a nonspatial table, that is appended to the target layer's attribute table.

label In ArcGIS, descriptive text, usually based on one or more feature attributes. Labels are placed dynamically on or near features based on user-defined rules and in response to changes in the map display. Labels cannot be individually selected and modified by the user. Label placement rules and display properties (such as font size and color) are defined for an entire layer.

latitude and longitude A reference system used to locate positions on the earth's surface. Distances east to west are measured with lines of longitude (also called meridians), which run north to south and converge at the North and South Poles. Distance measurements begin at the prime meridian and are measured positively 180 degrees to the east and negatively 180 degrees to the west. Distances north to south are

measured with lines of latitude (also called parallels), which run east to west. Distance measurements begin at the equator and are measured positively 90 degrees to the north and negatively 90 degrees to the south.

layer In ArcGIS, a reference to a data source, such as a shapefile, coverage, geodatabase feature class, or raster, that defines how the data should be symbolized on a map. Layers can also define additional properties, such as which features from the data source are included. Layers can be stored in a map document (.mxd) or saved individually as a layer file (.lyr).

layer file A file with a .lyr extension that stores the path to a source dataset and other layer properties, including symbology.

layer package A file that includes both the layer properties and the source data referenced by the layer. Layer packages contain symbolization, labeling, table properties, and the data. They can be created in ArcMap, ArcGlobe, and ArcScene and can be viewed in any of these applications, as well as in ArcGIS Explorer.

layout The arrangement of elements on a map, including a title, legend, north arrow, scale bar, and geographic data.

layout view In ArcMap and ArcReader, a view that shows the virtual page on which geographic data and map elements, such as titles, legends, and scale bars, are placed and arranged for printing.

legend The description of features included in a map, usually displayed in the map layout. Legends often use graphics of symbols or examples of features from the map with a written description of what each symbol or graphic represents.

location query Also called a spatial query, it is a statement or logical expression that selects geographic features based on location or spatial relationship. For example, a spatial query might find which points are contained within a polygon or set of polygons, find features within a specified distance of a feature, or find features that are adjacent to each other.

manual classification The process of sorting or arranging entities into groups or categories; on a map, the process of representing members of a group by the same symbol, usually defined in a legend.

map algebra A language that defines a syntax for combining map themes by applying mathematical operations and analytical functions to create new map themes. In a map algebra expression, the operators are a combination of mathematical, logical, or Boolean operators (+, >, AND, tan, and so on) and spatial analysis functions (slope, shortest path, spline, and so on), and the operands are spatial data and numbers.

map extent The limit of the geographic area shown on a map, usually defined by a rectangle. In a dynamic map display, the map extent can be changed by zooming and panning.

map package A file that contains a map document (.mxd) and the data referenced by the layers it contains; the file can be uploaded to a web portal and shared.

map projection A method by which the curved surface of the earth is portrayed on a flat surface. This generally requires a systematic mathematical transformation of the earth's graticule of lines of longitude and latitude onto a plane. Some projections can be visualized as a transparent globe with a light bulb at its center (although not all projections emanate from the globe's center) casting lines of latitude and longitude onto a sheet of paper. Generally, the paper is either flat and placed tangent to the globe (a planar or azimuthal projection) or formed into a cone or cylinder and placed over the globe (a cylindrical or conical projection). Every map projection distorts distance, area, shape, or direction, or some combination thereof.

map topology A temporary set of topological relationships between coincident parts of simple features on a map, used to edit shared parts of multiple features.

map units Ground units of measurement (e.g., feet, miles, meters, or kilometers) in which coordinates of spatial data are stored.

mashup In internet mapping, the combination of content from more than one data source into one dynamic map service.

mask A means of identifying areas to be included in analysis. Such a mask is often referred to as an analysis mask and may be either a raster layer or a feature layer.

merge Combines multiple input datasets of the same data type into a single, new output dataset. This tool can combine point, line, or polygon feature classes or tables.

metadata Information that describes the content, quality, condition, origin, and other characteristics of data or other pieces of information. Metadata for spatial data may describe and document its subject matter; how, when, where, and by whom the data was collected; availability and distribution information; its projection, scale, resolution, and accuracy; and its reliability with regard to some standard. Metadata consists of properties and documentation. Properties are derived from the data source (e.g., the coordinate system and projection of the data), and documentation is entered by a person (e.g., keywords used to describe the data).

model In geoprocessing in ArcGIS, one process or a sequence of connected processes in ModelBuilder.

multipart polygon In ArcGIS, a digital representation of a place or thing that has more than one part but is defined as one feature because it references one set of attributes. In a layer of states, for example, the state of Hawaii could be considered a multipart feature because its separate geometric parts are classified as a single state. A multipart feature can be a point, line, or polygon.

natural breaks A method of manual data classification that seeks to partition data into classes based on natural groups in the data distribution. Natural breaks occur in the histogram at the low points of valleys. Breaks are assigned in the order of the size of the valleys, with the largest valley assigned the first natural break.

neatline The border delineating and defining the extent of geographic data on a map. It demarcates map units so that depending on the map projection, the neatline does not always have 90-degree corners. In a properly made map, it is the most accurate element of the data; other map features may be moved slightly or exaggerated for generalization or readability, but the neatline is never adjusted.

neighborhood In GIS, the user-defined area used to establish new output values. Neighborhood functions establish new values for locations using the values of other locations within a given distance or direction.

NoData In raster data, the absence of a recorded value. NoData does not equate to a zero (0) value. Although the measure of a particular attribute in a cell may be zero, a NoData value indicates that no measurements have been taken for that cell at all.

normalization The process of dividing one numeric attribute value by another to minimize differences in values based on the size of areas or the number of features in each area. For example, normalizing (dividing) total population by total area yields population per unit area, or density.

north arrow A map symbol that shows the direction of north on the map, thereby showing how the map is oriented.

on-the-fly projection ArcMap can display data stored in one projection as if it were in another projection. The new projection is used for display and query purposes only. The actual data is not altered.

operational layers Web mapping layers that users view and work with to perform their tasks.

operators The symbolic representation of a process or operation performed against one or more operands in an expression, such as + (plus, or addition) and > (greater than). When evaluated, operators return a value as their result. If multiple operators appear in an expression, they are evaluated in order of their operator precedence.

overlay In geoprocessing, the geometric intersection of multiple datasets to combine, erase, modify, or update features in a new output dataset.

projected coordinate system A reference system used to locate x, y, and z positions of point, line, and area features in two or three dimensions. A projected coordinate system is defined by a geographic coordinate system, a map projection, any parameters needed by the map projection, and a linear unit of measure.

projects In ArcGIS Pro, a collection of related geographic datasets, maps, layouts, tools, settings, and resources. A project is stored as an item in ArcGIS Online or on disk as an .aprx file.

proportional-symbol map A map that contains symbols whose sizes differ in relation to the phenomenon being mapped.

pyramid In raster datasets, a reduced-resolution layer that copies the original data in decreasing levels of resolution to enhance performance. The coarsest level of resolution is used to quickly draw the entire dataset. As the display zooms in, layers with finer resolution are drawn. Drawing speed is maintained because fewer pixels are needed to represent the successively smaller areas.

quantile A data classification method that distributes a set of values into groups that contain an equal number of values.

query expression A type of expression that evaluates to a Boolean (true or false) value and is typically used to select those rows in a table whose values cause the expression to evaluate to true. Query expressions are generally part of an SQL statement.

raster A spatial data model that defines space as an array of equally sized cells arranged in rows and columns and composed of single or multiple bands. Each cell contains an attribute value and location coordinates. Unlike a vector structure, which stores coordinates explicitly, raster coordinates are contained in the ordering of the matrix. Groups of cells that share the same value represent the same type of geographic feature.

reclassification The process of taking input cell values and replacing them with new output cell values. Reclassification is often used to simplify or change the interpretation of raster data by changing a single value to a new value or grouping ranges of values into single values (e.g., assigning a value of one [1] to cells that have values of 1 to 50, two [2] to cells that range from 51 to 100, and so on).

relate Establishing a connection between two tables using an attribute common to both.

scale bar A map element used to graphically represent the scale of a map. A scale bar is typically a line marked like a ruler in units proportional to the map's scale.

shapefiles A vector data storage format for storing the location, shape, and attributes of geographic features. A shapefile is stored in a set of related files and contains one feature class.

shortcut (context) menu The menu of commands that is accessed by right-clicking an object.

snapping An automatic editing operation in which points or features within a specified distance (tolerance) of other points or features are moved to match or coincide exactly with each other's coordinates.

spatial join A type of table join operation in which fields from one layer's attribute table are appended to another layer's attribute table based on the relative locations of the features in the two layers.

spheroid When used to represent the earth, a 3D shape obtained by rotating an ellipse about its minor axis, either with dimensions that approximate the earth as a whole or with a part that approximates the corresponding portion of the geoid.

standard deviation A data classification method that finds the mean value and then places class breaks above and below the mean at intervals of either 0.25, 0.5, or 1 standard deviation until all the data values are contained within the classes. Values that are beyond three standard deviations from the mean are aggregated into two classes—greater than three standard deviations above the mean and less than three standard deviations below the mean.

standard meridians Great circles on the earth that pass through the poles, often used synonymously with longitude. Meridians run north to south between the poles. By convention, meridians are labeled with positive numbers that ascend as one moves eastward from the prime meridian and negative numbers as one moves westward from the prime meridian until the Eastern and Western Hemispheres meet at the 180-degree line. Meridians can also, however, be labeled with all positive or negative numbers, including positive numbers increasing westward from the prime meridian.

standard parallels Lines of latitude in a conic or cylindrical projection in normal aspect where the projection surface touches the globe. A tangent conic or cylindrical projection has one standard parallel, while a secant conic or cylindrical projection has two. At the standard parallel, the projection shows no distortion.

style An organized collection of predefined colors, symbols, properties of symbols, and map elements. A style promotes standardization and consistency in mapping products.

symbology The set of conventions, rules, or encoding systems that define how geographic features are represented with symbols on a map. A characteristic of a map feature may influence the size, color, and shape of the symbol used.

target table The table to which new attributes are appended. Records in a target table are often referred to as target features.

tiles A subset of a spatial dataset (commonly a raster) composed of a manageable rectangular set, or rows and columns of pixels, typically used to process or analyze a large raster dataset without consuming vast quantities of computer memory.

topography The study and mapping of land surfaces, including relief (relative positions and elevations) and the position of natural and constructed features.

union A topological overlay of two or more polygon spatial datasets which preserves the features that fall within the spatial extent of either input dataset–that is, all features from both datasets are retained and extracted into a new polygon dataset.

vector A coordinate-based data model that represents geographic features as points, lines, and polygons. Each point feature is represented as a single coordinate pair, and line and polygon features are represented as ordered lists of vertices. Attributes are associated with each vector feature, as opposed to a raster data model, which associates attributes with grid cells.

vertex (vertices) One of a set of ordered x,y coordinate pairs that defines the shape of a line or polygon feature.